Sponsored by the
European Association of Neurosurgical Societies

Advances and Technical Standards in Neurosurgery

Vol. 29

Edited by
J. D. Pickard, Cambridge (Editor-in-Chief),
C. Di Rocco, Roma, V. V. Dolenc, Ljubljana,
R. Fahlbusch, Erlangen, J. Lobo Antunes, Lisbon,
M. Sindou, Lyon, N. de Tribolet, Lausanne,
C. A. F. Tulleken, Utrecht, M. Vapalahti, Kuopio

SpringerWienNewYork

With 101 partly coloured Figures

Printed in Austria

Springer-Verlag Wien New York is a part of Springer Science Business Media
springeronline.com

Library of Congress Catalogue Card Number 74-10499

Typesetting: Asco Typesetters, Hong Kong
Printing: Druckerei Theiss GmbH, A-9431 St. Stefan

Printed on acid-free and chlorine-free bleached paper

SPIN: 10929469

ISSN 0095-4829

ISBN 3-211-14027-1 Springer-Verlag Wien New York

Preface

As an addition to the European postgraduate training system for young neurosurgeons, we began to publish in 1974 this series of *Advances and Technical Standards in Neurosurgery* which was later sponsored by the European Association of Neurosurgical Societies.

This series was first discussed in 1972 at a combined meeting of the Italian and German Neurosurgical Societies in Taormina, the founding fathers of the series being Jean Brihaye, Bernard Pertuiset, Fritz Loew and Hugo Krayenbuhl. Thus were established the principles of European cooperation which have been born from the European spirit, flourished in the European Association, and have been associated throughout with this series.

The fact that the English language is now the international medium for communication at European scientific conferences is a great asset in terms of mutual understanding. Therefore we have decided to publish all contributions in English, regardless of the native language of the authors.

All contributions are submitted to the entire editorial board before publication of any volume for scrutiny and suggestions for revision.

Our series is not intended to compete with the publications of original scientific papers in other neurosurgical journals. Our intention is, rather, to present fields of neurosurgery and related areas in which important recent advances have been made. The contributions are written by specialists in the given fields and constitute the first part of each volume.

In the second part of each volume, we publish detailed descriptions of standard operative procedures and in depth reviews of established knowledge in all aspects of neurosurgery, furnished by experienced clinicians. This part is intended primarily to assist young neurosurgeons in their postgraduate training. However, we are convinced that it will also be useful to experienced, fully trained neurosurgeons.

We hope therefore that surgeons not only in Europe, but also throughout the world, will profit by this series of *Advances and Technical Standards in Neurosurgery*.

The Editors

Contents

A. Advances

B. Technical Standards

Preoperative Clinical Evaluation, Outline of Surgical Technique and Outcome in Temporal Lobe Epilepsy. A. IMMONEN[1], L. JUTILA[2], R. KÄLVIÄINEN[2], E. MERVAALA[3], K. PARTANEN[4], J. PARTANEN[3], R. VANNINEN[4], A. YLINEN[5], I. ALAFUZOFF[6], L. PALJÄRVI[6], H. HURSKAINEN[1], J. RINNE[1], M. PURANEN[4], and M. VAPALAHTI[1], [1]Department of Neurosurgery, Kuopio University Hospital, Kuopio, Finland, [2]Department of Neurology, Kuopio University Hospital, Kuopio, Finland, [3]Department of Clinical Neurophysiology, Kuopio University Hospital, Kuopio, Finland, [4]Department of Clinical Radiology, Kuopio University Hospital, Kuopio, Finland, [5]Department of Neuroscience and Neurology, University of Kuopio, Kuopio, Finland, [6]Department of Clinical Pathology, Kuopio University Hospital, Kuopio, Finland

Motor Evoked Potential Monitoring for the Surgery of Brain Tumours and Vascular Malformations. G. Neuloh and J. Schramm, Department of Neurosurgery, University of Bonn, Germany

List of Contributors

Alafuzoff, I., Department of Clinical Pathology, Kuopio University Hospital, PO Box 1777, FIN – 70211 Kuopio, Finland

Bricolo, A., Department of Neurological Sciences and Vision, Section of Neurosurgery, University Hospital, Piazzale Stefani 1, 37100 Verona, Italy

Fahlbusch, R., Department of Neurosurgery, University Erlangen-Nurnberg, Schwabachanlage 6, 91054 Erlangen, Germany

Ganslandt, O., Department of Neurosurgery, University Erlangen-Nurnberg, Schwabachanlage 6, 91054 Erlangen, Germany

Guenot, M., Department of Functional Neurosurgery, Hopital Neurologique P. Wertheimer, 59, Bd Pinel, 69003 Lyon, France

Guimaraes-Ferreira, J., Instituto de Anatomia Normal, Faculdade de Medicina de Lisboa, Av. Prof. Egas Moniz, 1649-028 Lisboa, Portugal

Hurskainen, H., Department of Neurosurgery, Kuopio University Hospital, PO Box 1777, FIN – 70211 Kuopio, Finland

Immonen, A., Department of Neurosurgery, Kuopio University Hospital, PO Box 1777, FIN – 70211 Kuopio, Finland

Isnard, J., Department of Functional Neurology and Epileptology, Hoptial Neurologique P. Wertheimer, 59, Bd Pinel, 69003 Lyon, France

Jutila, L., Department of Neurology, Kuopio University Hospital, PO Box 1777, FIN – 70211 Kuopio, Finland

Kalviainen, R., Department of Neurology, Kuopio University Hospital, PO Box 1777, FIN – 70211 Kuopio, Finland

Lanteri, P., Department of Neurological Sciences and Vision, Section of Neurosurgery, University Hospital, Piazzale Stefani 1, 37100 Verona, Italy

Lauritzen, C., Craniofacial Unit, Department of Plastic Surgery, Sahlgrenska University Hospital, 41345 Goteborg, Sweden

Mervaala, E., Department of Clinical Neurophysiology, Kuopio University Hospital, PO Box 1777, FIN – 70211 Kuopio, Finland

Miguéns, J., Department of Neurosurgery, Faculdade de Medicina de Lisboa, Av. Prof. Egas Moniz, 1649-028 Lisboa, Portugal

Neuloh, G., Department of Neurosurgery, Sigmund-Freud Sr. 25, 53105 Bonn, Germany

Nimsky, C., Department of Neurosurgery, University Erlangen-Nurnberg, Schwabachanlage 6, 91054 Erlangen, Germany

Paljarvi, L., Department of Clinical Pathology, Kuopio University Hospital, PO Box 1777, FIN – 70211 Kuopio, Finland
Partanen, J., Department of Clinical Neurophysiology, Kuopio University Hospital, PO Box 1777, FIN – 70211 Kuopio, Finland
Partanen, K., Department of Clinical Radiology, Kuopio University Hospital, PO Box 1777, FIN – 70211 Kuopio, Finland
Puranen, M., Department of Clinical Radiology, Kuopio University Hospital, PO Box 1777, FIN – 70211 Kuopio, Finland
Rinne, J., Department of Neurosurgery, Kuopio University Hospital, PO Box 1777, FIN – 70211 Kuopio, Finland
Sala, F., Department of Neurological Sciences and Vision, Section of Neurosurgery, University Hospital, Piazzale Stefani 1, 37100 Verona, Italy
Schramm, J., Department of Neurosurgery, Sigmund-Freud Sr. 25, 53105 Bonn, Germany
Sindou, M., Department of Functional Neurosurgery, Hopital Neurologique P. Wertheimer, 59, Bd Pinel, 69003 Lyon, France
Valatx, J. L., INSERM U480, Claude Bernard University, 8, avenue Rockefeller, 69373 Lyon Cedex 08, France
Vanninen, R., Department of Clinical Radiology, Kuopio University Hospital, PO Box 1777, FIN – 70211 Kuopio, Finland
Vapalahti, M., Department of Neurosurgery, Kuopio University Hospital, PO Box 1777, FIN – 70211 Kuopio, Finland
Ylinen, A., University of Kuopio, PO Box 1627, FIN – 70211 Kuopio, Finland

Advances

Disorders of Consciousness: Anatomical and Physiological Mechanisms

J. L. Valatx

I.N.S.E.R.M. U480, Claude Bernard University, Lyon, France

With 3 Figures

Contents

Abstract

The anatomical and physiological mechanisms of consciousness are reviewed, focussing on the wakefulness mechanisms, which are one aspect, albeit an indispensable one, of consciousness. In trying to understand disorders of consciousness, it must be remembered that wake physiology is linked to sleep physiology and a brief summary of the phenomenology and regulation of the wake sleep cycle is therefore presented.

Wakefulness is produced and maintained by a complex neural network composed of at least ten groups of neurones, which spread out from the medulla oblongata to the telencephalic structures. Some of these elements (reticular formation and diffuse thalamic nuclei) contribute to the arousal and general activation of the brain, while others (aminergic and peptidergic neurones) contribute to the quality of wakefulness, as well as to general activation. Disorders of wakefulness may be due to dysfunction of one or several elements of the wake network or sleep disorders, as well as to sleep deprivation.

Keywords: Consciousness; sleep physiology; wakefulness; biogenic amines; orexine.

Introduction

Any discussion of the physiology of consciousness requires a precise definition of consciousness. Classically, «to be conscious» means to be awake, as opposed to unconscious or asleep. In addition, «to be conscious of» something means to be aware of it, and refers to mental activity or cognition (perception, computation, meditation, movement representation, etc) (Rees *et al.* 2002).

Recent research on cognition has focused attention on the neural mechanisms of mental activity as a result of the development of functional neuroimaging (Rees *et al.* 2002). However, for mental performance, the brain must be highly aroused. Wakefulness mechanisms are thought to be entirely distinct from those for mental processes, but sometimes it is difficult to separate them. Nevertheless, it is important to draw the attention of neurosurgeons to the wakefulness mechanisms, since they are confronted on a daily basis with disorders of consciousness and require a precise knowledge of the mechanisms involved.

The study of wakefulness cannot be separated from that of sleep, as the quality of consciousness depends on the quality and sufficiency of preceding sleep, and, conversely, sleep quality depends on what has happened during the day.

In this chapter, I will first briefly summarise the organisation of the wake-sleep cycle, knowledge of which is required to understand certain disorders of consciousness and sleep in hospitalised patients, then describe the different components of the wakefulness network, highlighting their specific effects on consciousness.

The Wake-Sleep Cycle

From a historical point of view, wakefulness and sleep have been thought to be produced either by a single centre, that of wake, with sleep being a passive stage due to arrested activity of the wakefulness centre (Bremer 1935, 1970), or by two centres, one for wakefulness and one for sleep. The latter theory, which was based on the pathophysiological observations of Von Economo (1929) during an epidemic of lethargic encephalitis, placed the waking centre in the posterior hypothalamus and the sleep centre in the anterior hypothalamus. This theory was confirmed by animal studies (Ranson *et al.* 1939, Nauta 1946). The discovery of the role of the reticular formation focused research on brainstem waking mechanisms, the so-called ascending reticular activating system (Moruzzi *et al.* 1949). The description of paradoxical sleep by Jouvet (1959) led to the three states concept of vigilance, i.e. wakefulness, slow sleep and paradoxical sleep (Jouvet 1962).

When fluorescence techniques demonstrated the presence of monoamine-containing neurons in the brain (Dalhström *et al.* 1964), many studies were devoted to the role of these new molecules (Jouvet 1972) and some were implicated in the wake mechanisms. The use of immunohistological and molecular methods then resulted in the discovery of numerous peptides and hormones which colocalised with the classical neurotransmitters. As a result of these observations, wake sleep mechanisms are now thought of in terms of networks or systems composed of several groups of neurons.

Phenomenology

Most human beings are diurnal and are therefore aroused and active during the light period of the day and inactive and sleepy at night. However, since the advent of electricity, many people now stay awake for longer, resulting in sleep reduction, which, in turn, causes disorders of wakefulness quality and cognitive performance (described as «Edison disease» by Jouvet).

At sunset, the decrease in light intensity is perceived by a very small set of retinal cells, the photic system (Moore 1982, 1992), which is different from the classic vision system, and this indicates to the biological clock (suprachiasmatic nuclei) that night is very close. In turn, the brain sends signals (yawning, blinking, inattention, etc), indicating it is time to sleep. If sleep is resisted, these signals disappear after about twenty minutes and appear again after an interval corresponding roughly to one sleep cycle (about 90 minutes).

When the person does not resist sleep, a ritual, often unconscious, behaviour precedes and prepares sleep onset, e.g. listening to pleasant music, drinking an infusion or a glass of milk, walking the dog or having a warm bath. Such a ritual allows the gentle transition from active to quiet wake which is necessary in order to fall asleep in less than 10 minutes. The person then goes to a familiar bedroom and takes up a sleep posture, which depends on the ambient temperature; if cool or cold, the body is curled up to preserve body temperature, while, if warm or hot, it is fully stretched out to facilitate evaporation from the skin.

At that moment the eyes are closed, respiration is regular and no body movements are observed. Polygraphic records (EEG, EMG, ECG, blood pressure, and body temperature) show a slowdown of all parameters. In human beings, this phase of sleep has been classified on the basis of the recorded frequency of cerebral activity into four stages, light sleep (stages 1 and 2), with very few slow waves, and deep sleep (stages 3 and 4), with numerous high amplitude slow waves. These four stages constitute the so-called «slow wave sleep» or SWS (Rechtschaffen *et al.* 1968).

After 65–75 minutes of SWS, the second phase of sleep occurs; this is characterised by irregular respiration, rapid eye movements, muscle atonia, an elevated arousal threshold and fast low amplitude cerebral activity similar to that seen in wake. Because of the association of the signs of deep sleep and signs of wake, this phase of sleep was named «paradoxical» sleep (PS) by Jouvet (1959). When aroused from this stage of sleep, in 80% of cases, the subject will say that he was dreaming. Throughout PS, erection of the penis or clitoris is observed, unlinked to dream content. The mean duration of PS is about 20 minutes. Most sleep researchers now call this «rapid eye movement» (REM) sleep, SWS also being called «non-REM» sleep. Thus, SWS and PS constitute a sleep cycle of about 90 minutes, and a regular night's sleep consists of several successive cycles, the number of which depends on the amount of sleep required to be in good form on awakening.

The mean duration of physiological sleep is 8.00 ± 2.00 hrs, with a wide range from 3 to 12 hours. Such a variability may be explained by several factors, such as environment (temperature, light, food, etc) or genetics (Valatx 1995a, 2001).

Regulation

Basic research has demonstrated that each state of vigilance is produced and regulated by a distinct neuronal network. Information is transferred within a given network by various neurotransmitters, which can be used by other networks. The network specificity (function) lies in the network configuration and not in a given transmitter.

It is possible to generate a model of the regulation of the wake-sleep cycle which integrates most of the results. A very simplistic diagram is shown in Fig. 1. It consists of five elements, the wake, SWS and PS networks, a sleep onset system and the biological clock.

All the components of the PS network have been localised in the brainstem, pons and medulla oblongata (Jouvet 1967, Sakai *et al.* 2001). It is comparable to a pacemaker (Valatx 1995b) which is controlled by inhibitory inputs from the wake and SWS systems, since PS makes the subject very vulnerable (muscle paralysis and increased sensory thresholds)

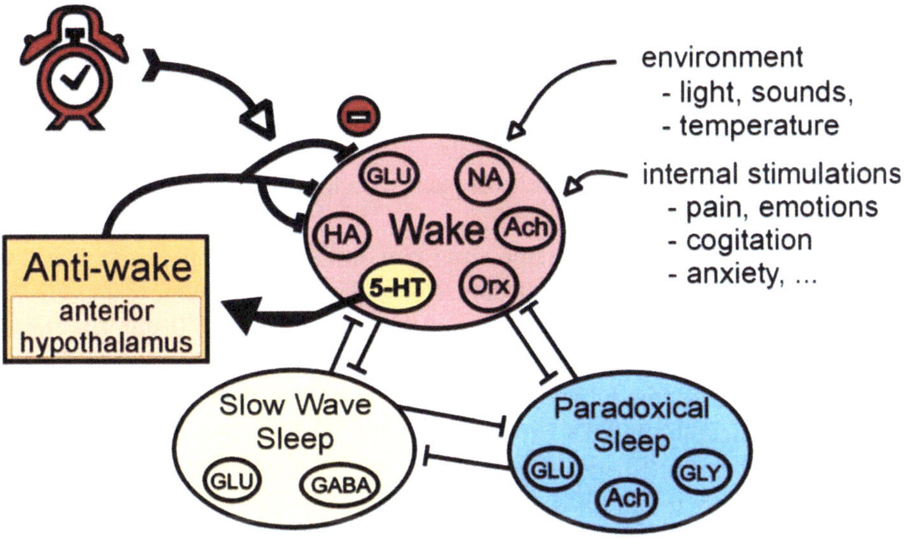

Fig. 1. Schematic representation of the sleep-wake regulation. This model consists of five neuronal networks, wake, slow wave sleep, paradoxical sleep, anti-wake, and biological clock. Remark the anti-wake network is triggered by one element of the wake system (5-HT). (See explanations in the text). Some of transmitters involved in each state of vigilance are indicated. *Ach* Acetylcholine, *GABA* gamma-aminobutyric acid, *Glu* glutamate, *HA* histamine, *NA* noradrenaline, *Orx* orexine, *5-HT* serotonine

and must therefore only occur under conditions of optimal security, i.e. during sleep, which occurs in a safe place.

The SWS network is mainly located in forebrain structures and also behaves as a pacemaker controlled by the wake system (Stériade 1991).

The wake system is a very complex network (see below). Schematically, waking is the consequence of two parallel mechanisms, sleep inhibition and general cerebral activation. Sleep inhibition seems to involve mechanisms specific to each sleep state, located in the brainstem and the hypothalamus (Buda *et al.* 1994). Brain activation is achieved by several structures (see below) and by stimulation from the environment (light, noise, odour, temperature and social interactions) and/or the internal milieu (pain, thirst, hanger, emotions, cognition, etc).

In order to sleep, the wake network must be stopped. This is performed by the sleep onset system, which we may call the «anti-wake system», located in the basal preoptic hypothalamus. These neurons are triggered during the waking state by serotoninergic inputs from the pontine dorsal raphe nucleus and send inhibitory projections to all components of the wake system. They also receive inputs from the medullar nucleus of the solitary tract, explaining vegetative influences (stimulation from nerves of the digestive tract, respiratory system, or blood pressure) on vigilance (Puizillout 1984). For example, influence of a meal on vigilance or massage of the neck at the level of the carotid bifurcation induce sleep or a karate chop at this same level causes unconsciousness. Carotid means, in old Greek, "what provokes deep sleep."

The last component is the biological clock, which triggers sleep onset and awakening (Born *et al.* 1999).

On the basis of this scheme, one can understand that insomnia and hypersomnia are not sleep disorders, but wake disorders, due, respectively, to excessive or defective stimulation of the wake network. In the treatment of insomnia, the first step is to help the patient to identify external, internal and/or psychic factors which delay sleep onset, then try to reduce or eliminate them. Sleeping tablets can be prescribed to help in getting through a difficult period, but for no longer than three weeks. For hypersomnia, the only efficient treatment now available is the prescription of drugs stimulating the wake network, for example, modafinil which has no side-effects, rather than amphetamine derivatives, which have well known side-effects.

Wakefulness Network

Several reviews have been published on this subject (Moruzzi 1972, Jouvet 1972, Stériade 1991, Lin 2000).

For decades, the study of brain mechanisms was based on the recording of gross cerebral activity (electroencephalogram, EEG) and lesions provoked by local ablation, electrocoagulation or vascular damage (occlusion or haemorrhage). The miniaturisation of biosensors then made it possible to record from single neurones (unit activity) and to measure levels of various transmitters *in vivo*. Lesioning became more precise, using new neurotoxic substances which caused specific degeneration of neurones without affecting glia cells and neural fibres of passage.

During waking, the cortical EEG varies depending on the level of mental activity. It is a low amplitude high frequency activity called desynchronisation, as opposed to sleep synchronisation, which consists of high amplitude low frequency waves. During quiet wake with closed eyes, the oscillation frequency is about 10 Hz (alpha rhythm), while, during attention and cognition, it fluctuates from 20–30 Hz (beta rhythm) to 40–60 Hz (gamma rhythm).

Recordings of electrical activity of single neurones show that there are several modalities of potential discharge (firing); (i) continuous and stable firing across all vigilance stages, (ii) firing specific for one particular stage, e.g., neurones active during wake and silent during sleep and (iii) neurones showing an abrupt change from an irregular to a bursting discharge pattern on a change in vigilance state or movement (Sakai 1998, Urbain *et al.* 2002).

Research in animals has shown that the cerebral cortex cannot by itself maintain such fast activity. Indeed, complete deafferentation of the neocortex results in continuous synchronisation of the EEG, similar to that seen during deep sleep (0.5–2.5 Hz, delta rhythm). This indicates that cortical neurones need stimulation from other parts of the brain to maintain wake activity.

Several subcortical and brainstem structures are involved in the maintenance of the cortical activation required for cognitive performances (Fig. 2). Wakefulness is therefore caused by two simultaneous mechanisms, the blockade of slow waves and the stimulation of neurones. A mismatch of these mechanisms might explain some lack of cognitive performance.

The wake structures can be schematically grouped into two subsystems, the dorsal reticulo-thalamo-cortical tract and the ventral reticulo-hypothalamo-cortical tract.

The dorsal reticulo-thalamo-cortical tract is composed of the thalamic diffuse system (intralaminar nuclei), which sends glutamatergic fibres to

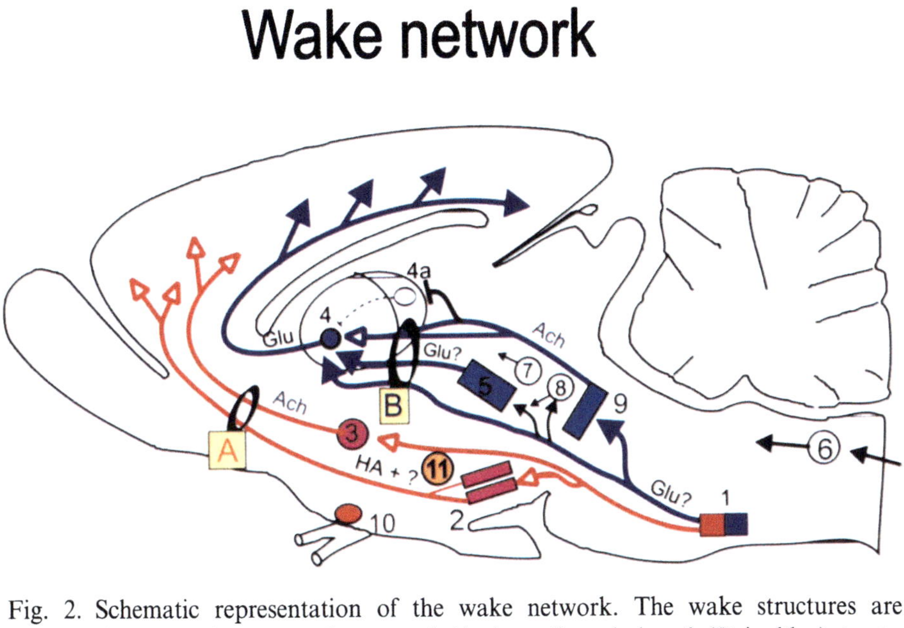

Fig. 2. Schematic representation of the wake network. The wake structures are grouped in two subsystems, the ventral (A, in red) and dorsal (B in blue) tracts. *1* Bulbar magnocellular nucleus, *2* posterior hypothalamus, *3* Meynert nucleus, *4* thalamic intralaminar nuclei, *4a* thalamic reticular nucleus, *5* mesencephalic reticular formation, *6* solitary tract nucleus, *7* dorsal raphe nucleus, *8* locus coeruleus nucleus, *9* mesopontine and laterodorsal tegmentum nuclei, *10* suprachiasmatic nuclei, *11* orexine nuclei

the entire cortex, and of the mesencephalic and bulbar reticular formation, which impinges upon the thalamic nuclei. Brainstem neuronal groups using various transmitters (acetylcholine, noradrenaline and serotonine) are also part of this dorsal tract (see below).

Since the pioneer work of Moruzzi and Magoun (1949), *the brainstem reticular formation* as a whole was considered the main source of stimulation of the cortex via the medial and intralaminar thalamic nuclei, the neurones of which synthesise an excitatory transmitter, glutamate (Stériade *et al.* 1991). However, recent studies have shown that the reticular formation can be divided into several nuclei involved in different functions (pain, temperature, sleep, wake and respiration). Two nuclei seem to be particularly involved in arousal, these being *the mesencephalic reticular nucleus* (MRn) in the upper part of this formation, directly connected to the thalamus through excitatory synapses (glutamate), and, in the lower part, *the magnocellular nucleus* (MCn) of the medulla oblongata which receives inputs from sensory and vegetative structures and thus stimulates not only the dorsal tract through the MRn, but also the ventral tract and brainstem cholinergic and monoaminergic nuclei. Lesions of the MCn seem not to be compatible with survival, mainly because it is difficult to lesion the wake neurons without also lesioning those involved in respiration control with which they are closely mixed.

Neurotoxic lesions of the MRn or non-specific thalamic nuclei cause a decrease in cortical activation lasting only a few days, indicating the existence of other arousal systems which compensate for the deficit (Denoyer *et al.* 1991). However, behavioural studies in these lesioned animals have yet to be done to check for possible changes in wakefulness quality.

The ventral reticulo-hypothalamo-cortical tract, beginning at the bulbar MCn, is involved in activation through the posterior hypothalamus and basal forebrain.

In the basal forebrain, the nucleus basalis of Meynert contains at least two sets of intermingled cholinergic and GABAergic neurones, which discharge tonically during wakefulness (Jones *et al.* 1999). The cholinergic neurones send fibres to the entire neocortex. Acetylcholine stimulates cortical neurones and blockade of Ach receptors by atropine induces a slowing (synchronisation) of the cortical EEG (Longo 1966). However, atropine-induced slowing must be differentiated from that seen during natural sleep. Although atropinised animals appear awake, it is not complete wakefulness, as synchronisation of cortical activity profoundly alters conditioned behaviour (Rougeul *et al.* 1969) and short term memory (Deutsch *et al.* 1967, Berger *et al.* 1969). The GABAergic neurones of the Meynert nucleus are involved in cortical activation through a mechanism called «desinhibition» and act by inhibiting cortical inhibitory interneurones (Freund *et al.* 1992). Moreover, through their inhibitory projections to the reticular thalamic nuclei, the Ach-containing and GABA-containing Meynert neurones can indirectly activate cortical neurones by blockade of sleep spindles (Stériade *et al.* 1991).

The Posterior Hypothalamus

The pioneering clinical observations of Von Economo (1929) and several animal studies (Ranson 1939, Nauta 1946) suggested that the posterior hypothalamus as a whole was a key region for wake control, as its surgical lesion caused a long lasting coma, while its stimulation caused hyperarousal. This region is under the control of the anterior hypothalamus, since lesion of the anterior hypothalamus or its inhibition by local application of muscimol, a GABA-A receptor agonist, causes a long lasting insomnia (Sallanon *et al.* 1989), which can be reversed by local injection of muscimol into the posterior hypothalamus; this can also provoke a transitory increase in SWS or PS, depending on the precise injection site (Lin *et al.* 1989). Recent animal studies have shown that this region contains several distinct elements involved in wake regulation, intermingled with others involved in other functions, such as regulation of blood pressure or food intake. Many transmitters are produced in neurones in this region, including glutamate, substance P, galanin, thyrotropin–releasing hormone (TRH), dynorphin, dopamine (A11 group), and melanin-concentrating hormone (MCH) (Lin 2000).

Recently, two new transmitters have been implicated in these waking mechanisms, these being histamine and orexin/hypocretin, which are only synthesised in this part of the brain in neurones which send projections rostrally and caudally to the entire nervous system.

Histamine Network

Histamine-containing neurons are exclusively located in the tuberomammillary nucleus and closely adjacent areas. Histamine is synthesised by decarboxylation of L-histidine by histidine-decarboxylase, an enzyme only expressed in these neurones.

These neurones send projections rostrally to the cortex, anterior hypothalamus and basal forebrain and caudally to the MRn and cholinergic and aminergic neurones (DA, NA, and 5-HT).

The firing rate of histaminergic neurones is high during wake, then dramatically decreases at sleep onset and stops during deep SWS and PS. In vitro studies have shown that these neurones exhibit slow spontaneous firing at 2 Hz (Sakai *et al.* 1990, Steininger *et al.* 1999), whereas, in animals in which the brainstem has been sectioned just behind the diencephalon («cerveau isolé» preparation), cortical slow waves are continuously recorded (Sakai *et al.* 1990), suggesting that the histamine neurones are inactive and need stimulation from the brainstem.

Histamine binds to three types of receptors, H1, H2 and H3. H1 and H2 receptors act synergistically in the stimulation of target cells. Activation of post-synaptic H1 receptors causes a slow depolarisation and tonic firing pattern of target cells, while activation of H2 receptors causes an increase in neuronal excitability and thus facilitates the neuronal discharge rate (Reiner *et al.* 1989, Khateb *et al.* 1990). H3 receptors are mainly presynaptic (autoreceptors) and have an inhibitory effect on histamine release and synthesis (Schwartz *et al.* 1991).

Pharmacological studies support the role of histamine in wake regulation. Long-lasting wake results from injection of a H1-receptor agonist (phenylephrine), a H3-autoreceptor antagonist (ciproxifan), or substances increasing the availability of histamine by blocking its catabolism (Lin 2000). H3-receptor antagonists might prove of use in arousing human patients from trauma-induced coma.

Administration of a histamine synthesis inhibitor (α-fluoromethylhistidine) or a H1-receptor antagonist (mepyramine) causes a decrease in wake duration and an increase in slow saves (Lin *et al.* 2000). In man, classical antihistaminics are well known to decrease vigilance and cause sedation and drowsiness, effects previously considered as side-effects of their antiallergenic actions. These substances are now classified as H1-receptor antagonists and the sedative action is their main cerebral effect.

Orexin/Hypocretin Network

Two new hypothalamic peptides, discovered in 1998 by two independent laboratories, were named *orexin A* (Orx-A) and *orexin B* (Orx-B) by one team because of their effect on food intake (Sakurai *et al.* 1998) and *hypocretin 1* (Hcrt-1) and *hypocretin 2* (Hcrt-2) by the second team because their amino acid sequences were similar to that of the gastric hormone, secretin (*hypo*thalamic se*cretin*) (de Lecea *et al.* 1998, Peyron *et al.* 1998). These peptides bind to two G protein-coupled receptors, HcrtR1 and HcrtR2. HcrtR1 preferentially binds Orx-A/Hcrt-1, whereas HcrtR2 binds both peptides with equal affinity (Sakurai *et al.* 1998). Orexin-containing neurones are exclusively located in the perifornical area of the dorso-lateral part of the posterior hypothalamus and send ascending and descending fibres to all brain regions, which might explain the different roles of this peptide (Peyron *et al.* 1998). A year later, these peptides were found to be involved in narcolepsy, a sleep disorder characterised by the subject suddenly falling asleep, and cataplexy. Orexin gene KO transgenic mice present narcoleptic attacks during the active phase of the day (Chemelli *et al.* 1999) and genetic narcoleptic dogs have a mutation which causes the loss of the Hcrt-2 receptor gene (Lin *et al.* 1999).

Several orexin target neurones are involved in wake-sleep regulation. Numerous recent studies have implicated this peptide in wake regulation. Intracerebroventricular injection or in situ infusion of orexin A into different wake structures provokes a long lasting wakefulness (Hagan *et al.* 1999, Thakkar *et al.* 2001).

Orexin is now accepted as a member of the wake transmitter family. The question to be solved is how the lack of only one arousing peptide can provoke sleep during the wakefulness period (narcolepsy) and how this peptide is involved in wakefulness. Moreover, orexin neurones also contain dynorphin-A and prolactin (Griffond *et al.* 1994, Paut-Pagano *et al.* 1993) and their interaction with orexin remains to be elucidated.

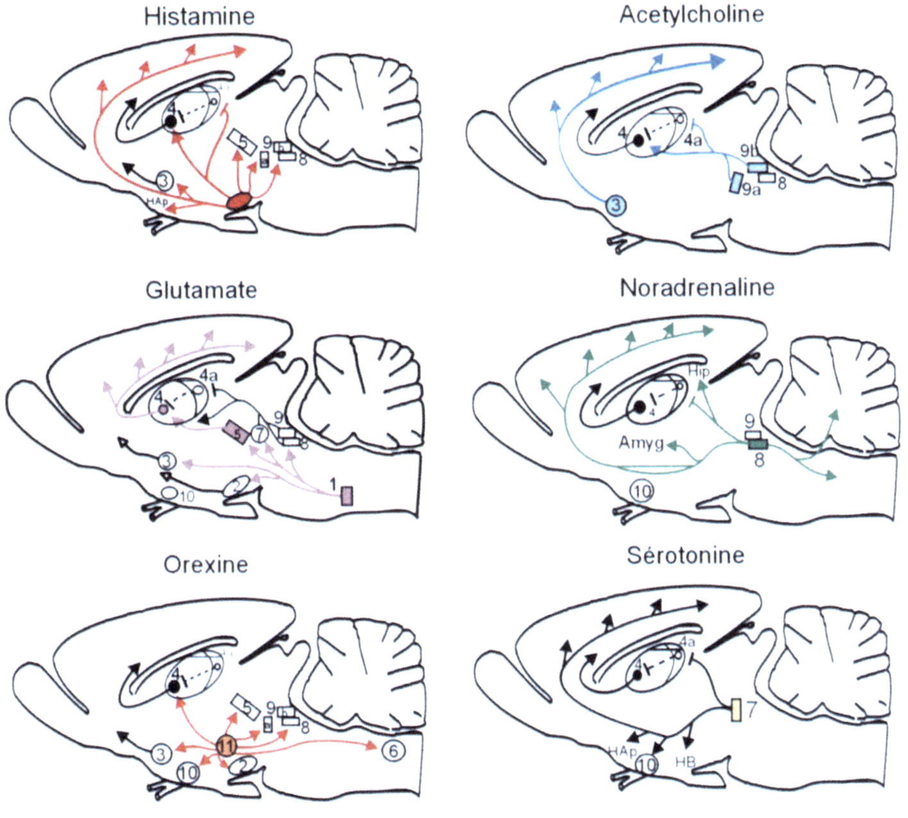

Fig. 3. Main neurotransmitters of the wake network. For structures see Fig. 2. *Hip* Dorsal hippocampus, *HAp* preoptic anterior hypothalamus, *HB* basal hypothalamus

Brainstem Acetylcholine and Amine Networks

Acetylcholine-containing neurones are located in the mesopontine area and the dorsolateral tegmental nuclei and send projections rostrally to the thalamic and hypothalamic nuclei and the basal forebrain. Ach is involved in the stimulation of forebrain waking structures via nicotinic receptors and in the blockade of sleep generation (thalamic reticular nucleus) via muscarinic receptors. Ach-containing neurones receive glutamatergic inputs from the bulbar magnocellular nucleus.

The catecholamine system is composed of three types of neurones, synthesising dopamine (DA), noradrenaline (NA) or adrenaline (A) from the amino-acid, L-tyrosine, via several sequential enzymatic steps, involving tyrosine hydroxylase (TH) and DOPA decarboxylase for DA production, followed by dopamine-β-hydroxylase (DBH) for NA production, then phenyl-N-methyl transferase (PNMT) for A production. These transmitters are catabolised by two monoamine oxidases (MAO), MAO-A for NA and MAO-B for DA and A. MAO inhibitors induce a long lasting wakefulness by increasing the amount of neurotransmitter in the synaptic cleft.

Dopamine-containing neurones, located in the substantia nigra (SN) and ventral tegmental area (VTA), send axons, respectively, to the striate nuclei (nigro-striatal tract) and accumbus nucleus and frontal cortex (meso-cortical tract). Their firing pattern fluctuates with locomotion and not with the sleep wake cycle. SN/VTA lesion results in akinesia, as in Parkinson's disease, with no changes in cortical EEG activation.

Pharmacological studies support the involvement of DA in behavioural motricity. Injection of substances, such as amphetamine, that cause release of DA into the synaptic cleft, results in increased locomotion and, consequently, an increased wake duration.

Noradrenaline-containing neurones are composed of 8 groups, the main one being the *locus coeruleus* (LC) situated in the dorso-lateral part of the pons at the angle of the fourth ventricle (Dalhstrom *et al.* 1964). This sends projections to almost the entire brain, caudally to the spinal cord, dorsally to the cerebellum and rostrally to the thalamus, amygdala, hippocampus, basal forebrain and cortex. LC neurones receive inputs from many forebrain and brainstem nuclei and from vegetative territories through the solitary tract nucleus. They fire at about 2 Hz during quiet waking; the rate of firing increases with stress and new somatic or vegetative stimuli, decreases during SWS and stops during PS. When a new stimulus is repeated several times, LC neurones no longer respond.

NA synthesis inhibitors or adrenergic α1-receptor antagonists cause slowing of the EEG, whereas re-uptake inhibitors induce prolonged wakefulness (see Jouvet 1972).

The Serotonin System

Serotonin (5-hydroxytryptamine, 5-HT) is synthesised from tryptophan by tryptophan-hydroxylase in 9 neuronal groups located along the raphe (midline) of the brainstem. The caudal nuclei send projections to the spinal cord as far down as the lumbar level. The anterior group, the dorsal raphe nucleus (DRN), innervates the mesencephalon, hypothalamus, amygdala, basal forebrain, septum, hippocampus and cortex.

Electrical stimulation of the DRN causes prolonged arousal. The firing rate is low during quiet waking, decreases during SWS and stops during PS. In vitro experiments have shown that 5-HT neurones are not spontaneously active, but need to be stimulated by different neuronal afferents containing histamine, orexine, or NA (Brown *et al.* 2002).

Since it binds to at least seven types of receptor, serotonin has many, sometimes opposing, effects, depending on the receptor type of the target cell.

Lesion of serotonin nuclei or inhibition of serotonin synthesis causing complete loss of this transmitter does not lead to slowing of the EEG, but to insomnia (Jouvet 1972), which can be explained by the non-stimulation of the so-called anti-wake system (see Fig. 1). Infusion of the serotonin precursor, 5-hydroxytryptophan (5-HTP), into the basal preoptic area, where the antiwake system is located, restores SWS and PS in insomniac animals (Denoyer *et al.* 1989).

Tentative Functions of the Wake System

We have seen that the organisation of the wake network is very complex, involving at least a dozen neuronal structures. This raises the question of whether all are necessary for the awake state or whether some are redundant. For example, we have seen that a neurotoxic lesion of the mesencephalic reticular nucleus disturbs the wake system for only a few days, and similar results are obtained after posterior hypothalamus lesion. Moreover, a few days after lesion of both structures, wake duration is almost identical to that in sham-operated animals (Denoyer *et al.* 1991). We may therefore ask whether these structures are unimportant for wakefulness. Taking into account some recent results (Parmentier 2002), we believe that, in wake analysis, it is necessary to examine factors other than the mere duration of waking based on EEG activation and to assess the quality of wakefulness by screening with different behavioural tests.

We hypothesise that each wake structure contributes to one or more aspects of wake quality and thus to cognitive performance, an idea that has not yet been systematically taken into account. For example, many studies have used various pathophysiological deficit models to examine the role of Ach in memory acquisition and information storage mechanisms without determining whether the observed change in memory is due to memory processing changes or to the decreased vigilance level caused by the Ach deficit, as seen in atropinised animals (Longo 1966).

We hypothesise that structures using glutamate as transmitter (mesencephalic reticular nucleus, intralaminar thalamic nuclei and bulbar magnocellular nucleus) are responsible for general non-specific mechanisms of arousal and its maintenance, while other wake structures are responsible both for this and for the quality of wakefulness.

Dopamine-Containing Neurones. Dopamine is indispensable for voluntary movements (locomotion and speech) and a lack of dopamine leads to a decrease or loss of movements (tremor and akinesia), while an overdose causes stereotypies and involuntary movements. Dopamine has been known for several years to be involved in mood regulation, pleasure, reward circuit and drug dependence mechanisms (Koob *et al.* 1998). One may imagine that a dopamine dysfunction or imbalance may alter cognition.

Histamine-Containing Neurones. A very recent study has shown that, in addition to its role in brain activation, histamine plays a special role when a subject encounters a new situation (Parmentier *et al.* 2002). Genetic molecular engineering in mice makes it possible to remove or add a single target gene and study the effects on behaviour. We have used this approach to generate mice lacking histamine by deleting the gene coding for its synthesis enzyme, histidine decarboxylase (HDC). Compared to wild-type (WT) mice, HDC-/- mice show no major change in the daily amount of waking, confirming the results of the lesion experiments described above, but do show an increase in PS duration, and power spectral analysis of the EEG shows an increase in the δ slow waves density during waking, suggesting weaker inhibitory control of synchronisation and PS mechanisms. However, the most striking difference is seen when the mice encounter different stimulating conditions, namely a litter change or transfer to a new cage with several objects to explore. In both of these situations, WT mice stay awake and explore the cage for at least 2 hours, whereas mutant mice (HDC-/-) are unable to maintain wakefulness and fall asleep within a few minutes of transfer. Thus, histaminergic neurones seem to be necessary for maintaining the brain awake when the animal is faced with novelty and behavioural challenges (Parmentier *et al.* 2002). In humans, it would be interesting to check whether patients receiving antihistamine treatment respond less to novelty.

Noradrenergic Locus Coeruleus (NA-LC). These neurones are activated by various sensory and vegetative stimuli, but cease to react when the stimuli become repetitive. Several studies have demonstrated that physiological and behavioural responses to stress factors require a functioning NA-LC. For example, after one hour of immobilisation stress at the beginning of the active period, rats sleep more (+100%) (Rampin *et al.* 1991) and this sleep increase is not seen in animals given a corticotrophin releasing hormone (CRH) receptor antagonist just before immobilisation (Gonzalez *et al.* 1997). Injection of CRH receptor antagonists is known to prevent almost all stress responses (Heinrichs *et al.* 1994, Morimoto *et al.* 1993). The same result is obtained after systemic administration of DSP-4, a neurotoxic drug which causes irreversible degeneration of NA-LC terminals. (Gonzalez *et al.* 1998b). Similarly, after 10 hours of sleep deprivation, a 54% decrease in sleep rebound is observed in animals treated with either DSP-4 or a CRH receptor antagonist (Gonzalez *et al.* 1998a). These results indicate that sleep rebound has two components, namely the recuperation of sleep debt and the reaction to the inherent stress of any sleep deprivation technique. Given the widespread nature of LC terminals, is it possible to focus on a particular target region? Preliminary results indicate that the hippocampal amygdala is the next link after the LC in the limbic reactive stress network, since DSP-4 lesion of NA-LC terminals restricted to the amygdala provokes the same rebound decrease as systemic DSP-4 injection (Charifi *et al.* 2000). Amygdala NA terminals are in contact with local CRH neurones, which may explain the subsequent steps involved in stress responses.

Furthermore, repetition of restraint stress results in a decreased response (sleep rebound), no statistically significant effect being seen after the fourth stress session (Rampin *et al.* 1991). We hypothesise that the NA-LC, via its connections to the limbic system (the prefrontal cortex and hippocampus) informs decision centres about the intensity of stress factors and situation novelty and plays a different, but complementary, role to that of histamine.

Effect of Sleep Deprivation on Cognitive Performances

Since the first description of PS, many total deprivation experiments have been carried out on animals and human volunteers (Dement *et al.* 1960, Jouvet 1962). The first sign to appear after staying up all night is increasing irritability, and the subject gets angry about small facts and has alternate periods of calm and euphoric agitation, resulting in an attention span of no more than five minutes. After two or three sleepless nights, memory disorders are observed, namely learning disability due to poor attention, inability to find the correct word and a very curious disorder called «amnesia of future» characterised by the subject being unable to think of a new project or to see the consequences of his own acts. Certain paranoid traits and a persecution syndrome are also noted after several days of sleep deprivation. Non-total deprivation for several weeks or months, as in shift or night work, may, in addition, trigger chronic fatigue and digestive tract disorders e.g. peptic ulcer or alternating periods of diarrhoea and constipation (Valatx 1988). During the daytime, these patients complain of sleepiness and decreased physical and mental performance.

Sleep deprivation is a common feature in medicine. In emergency departments and intensive care units, patient monitoring and hourly examinations result in total sleep deprivation. If this lasts more than three days, the above-described effects frequently appear, especially persecution syndrome, and a sleep-deprived patient feels convinced that a nurse or doctor does not give him appropriate care. In addition, after long-lasting deep anaesthesia, patients complain of insomnia for several days and ask for sleeping tablets, although the nurse often notices that they have been asleep. This classical post-surgical after-effect is a memory disorder probably linked to benzodiazepine anaesthetics, the patient not remembering that he has been asleep and only remembering the wake periods that occur normally at the end of each sleep cycle. These sleep-deprivation effects are also seen in physicians or surgeons on duty for several days, especially from the third day onwards, when errors may be made in diagnosis and treatment.

In conclusion, wakefulness is the basic component of consciousness and is a very complex phenomenon supported by a neural network, which is not yet well understood. The same transmitters are used in interneuronal communication in different networks, and human brain pharmacology is therefore often difficult to analyse. However, dysfunction of one or more elements of the wake network, demonstrated by very specific tests, may explain some consciousness disorders. Consciousness disorders may be caused by dysfunction of many other neuronal networks in addition to those involved in sleep.

References

1. Berger BD, Stein L (1969) An analysis of the learning deficits produced by scopolamine. Psychopharmacol14: 271–283
2. Borbely AA, Valatx JL (1984) Sleep mechanisms. Exp Brain Res Suppl 8, Springer-Berlin Heidelberg New York Tokyo
3. Born J, Hansen K, Marshall L, Molle M, Fehm HL (1999) Timing the end of nocturnal sleep. Nature 397: 29–30
4. Bremer F (1935) Cerveau isolé et physiologie du sommeil. CR Soc Biol (Paris) 18: 1235–1241
5. Bremer F (1970) Inhibitions intrathalamiques récurrentielles et physiologie du sommeil. Electroenceph Clin Neurophysiol 28:1–16
6. Brown RE, Sergeeva OA, Eriksson KS, Haas HL (2002) Convergent excitation of dorsal raphe serotonin neurons by multiple arousal systems (orexin/hypocretin, histamine and noradrenaline). J Neurosci 22: 8850–8859

7. Buda C, Sastre JP, Jouvet M (1994) Inhibition of the lateral periaqueductal gray induces a dramatic increase in paradoxical sleep. J Sleep Res 3: 34
8. Charifi C, Paut-Pagano L, Debilly G, Cespuglio R, Jouvet M, Valatx JL (2000) Effect of noradrenergic denervation of the amygdala upon recovery after sleep deprivation in the rat. Neurosci Lett 287: 41–44
9. Chemelli RM, Willie JT, Sinton CM, Elmquist JK, Scammell T, Lee C, Richardson JA, Williams SC, Xiong YM, Kisanuki Y, Fitch TE, Nakazato M, Hammer RE, Saper CB, Yanagisawa M (1999) Narcolepsy in orexin knockout mice: Molecular genetics of sleep regulation. Cell 98: 437–451
10. Dahlström A, Fuxe K (1964) Evidence for the existence of monoamine containing neurons in the central nervous system. Acta Physiol Scand 62 [Suppl] 230: 224–232
11. De Lecea L, Kilduff TS, Peyron C, Gao XB, Foye PE, Danielson PE, Fukuhara C, Battenberg ELF, Gaulvik VT, Bartlett FS, Franket WN, Van den Pol AN, Bloom FE, Gaulvik KM, Sutcliffe JG (1998) The hypocretins: hypothalamus-specific peptides with neuroexcitatory activity. Proc Natl Acad Sci USA 95: 322–327
12. Dement WC (1960) The effect of dream deprivation. Science 131: 1705–1707
13. Denoyer M, Sallanon M, Kitahama K, Aubert C, Jouvet M (1989) Reversibility of para-chlorophenylalanine-induced insomnia by intrahypothalamic micro-injection of 1–5-hydroxytryptophan. Neuroscience 28:83–94
14. Denoyer M, Sallanon M, Buda C, Kitahama K, Jouvet M (1991) Neurotoxic lesion of the mesencephalic reticular formation and/or the posterior hypothalamus does not alter waking in the cat. Brain Res 539: 287–303
15. Deutsch JA, Rocklin KW (1967) Amnesia induced by scopolamine and its temporal variations. Nature. 216: 89–90
16. von Economo C (1929) Schlaftheorie. Ergebn Physiol 28: 312–339
17. Freund TF, Meskenaite V (1992) γ-aminobutyric acid-containing basal forebrain neurons innervate inhibitory interneurons in the neocortex. Proc Natl Acad Sci USA 89:738–742
18. Gonzalez MM, Debilly G, Valatx JL, Jouvet M (1995) Sleep increase after immobilization stress: role of the noradrenergic locus coeruleus system in the rat. Neurosci Lett 202: 5–8
19. Gonzalez MM, Valatx JL (1997) Effect of intracerebroventricular administration of a-helical CRH (9-41) on the sleep/waking cycle in rats under normal conditions or after subjection to an acute stressful stimulus. J Sleep Res 6: 164–170
20. Gonzalez MM, Valatx JL (1998a) Involvement of stress in the sleep rebound mechanism induced by sleep deprivation in the rat: use of alpha-helical CRH (9-41). Behav Pharmacol 9: 655–662
21. Gonzalez MM, Debilly G, Valatx JL (1998b) Noradrenaline neurotoxin DSP-4 effects on sleep and brain temperature in the rat. Neurosci Lett 248: 93–96
22. Griffond B, Colard C, Deray A, Fellmann D, Bugnon C (1994) Evidence of the expression of dynorphin gene in the prolactin-immunoreactive neurons of the rat lateral hypothalamus. Neurosci Lett 165: 89–92

23. Hagan JJ, Leslie RA, Patel S, Evans ML, Wattam TA, Holmes S, Benham CD, Taylor SG, Routeledge C, Hemmati P, Munton RP, Ashmeade TE, Shah AS, Hatcher JP, Hatcher PD, Jones DN, Smith MI, Piper DC, Hunter AJ, Porter RA, Upton N (1999) Orexin-A activates locus coeruleus cell firing and increases arousal in the rat. Proc Natl Acad Sci USA 96:10911–10916
24. Heinrichs SC, Menzaghi F, Merlo Pich E, Baldwin HA, Rassnick S, Britton KT, Koob GF (1994) Anti-stress action of a corticotropin-releasing factor antagonist on behavioral reactivity to stressors of varying type and intensity. Neuropsychopharmacol 11: 179–186
25. Jones BE (1993) The organization of central cholinergic systems and their functional importance in sleep-waking states. In: Cuello AC (ed) Progress in Brain Research, Vol 98. Elsevier Science Publishers, Amsterdam, pp 61–71
26. Jones BE, Muhlethaler M (1999) Cholinergic and GAB Aergic neurons of the basal forebrain: role in cortical activation. In: Lydic R, Baghdoyan HA (eds) Handbook of behavioral state control: cellular and molecular mechanisms. CRC Press, Florida, pp 213–233
27. Jouvet M, Michel F, Courjon J (1959) Sur un stade d'activité électrique cérébrale rapide au cours du sommeil physiologique. CR Soc Biol Paris 153: 1024–1028
28. Jouvet M (1962) Recherches sur les structures nerveuses et les mécanismes responsables des différentes phases du sommeil physiologique. Arch Ital Biol 100: 125–206
29. Jouvet M (1967) Neurophysiology of the states of sleep. Physiol Rev 47:117–177
30. Jouvet M (1972) The role of monoamines and acetylcholine-containing neurons in the regulation of the sleep-waking cycle. Ergebn Physiol 64: 166–307
30. Jouvet M (1991) Le sommeil paradoxal est-il le gardien de l'individuation psychologique? Rev Canad Psychol 45: 148–168
31. Khateb A, Serafin M, Mühlethaler M (1990) Histamine excites pedunculopontine neurones in guinea pig brainstem slices. Neurosci Lett 112: 257–262
32. Kleitmann N (1939) Sleep and wakefulness, Chicago University Press
33. Koob GF, Sanna PP, Bloom FE (1998) Neuroscience of addiction. Neuron 21: 467–476
34. Lin JS, Sakai K, Jouvet M (1988) Evidence for histaminergic arousal mechanisms in the hypothalamus of cats. Neuropharmacology 27: 111–122
35. Lin JS, Sakai K, Vanni-Mercier G, Jouvet M (1989) A critical role of the posterior hypothalamus in the mechanisms of wakefulness determined by microinjection of muscimol in freely moving cats. Brain Res 479: 225–240
36. Lin JS (2000) Brain structures and mechanisms involved in the control of cortical activation and wakefulness, with emphasis on the posterior hypothalamus and histaminergic neurons. Sleep Med Rev 4: 471–503
37. Lin L, Faraco J, Li R, Kadotani H, Rogers W, Lin X, Qiu X, de Jong PJ, Nishino S, Mignot E (1998) The sleep disorder canine narcolepsy is caused by a mutation in the hypocretin (orexin) receptor 2 gene. Cell 98: 365–376
38. Longo VG (1966) Mechanisms of the behavioral and electroencephalographic effects of atropine and related compounds. Pharmacol Rev 18: 965–996

39. Morimoto A, Nakamori T, Morimoto K, Tan N, Murikami N (1993) The central role of corticotropin-releasing factor (CRF-41) in psychological stress in rats. J Physiol 460: 211–229
40. McGinty D, Sterman MB (1968) Sleep suppression after basal forebrain lesions in the cat. Science 160: 1253–1255
41. Moore RY (1982) Organization and function of a central nervous system oscillator: the suprachiasmatic nucleus. Fed Proc 42:2783–2789
42. Moore RY (1992) The organization of the human circadian timing system. In: Schwaab DF, Hofman MA, Mirmiran M, Ravid R, Van Leeuwen FW (eds) The human hypothalamus in health and disease. Progress in Brain Research. Elsevier, Amsterdam 93: 101–117
43. Morimoto A, Nakamori T, Morimoto K, Tan N, Murakami N (1993) The central role of corticotrophin-releasing factor (CRF-41) in psychological stress in rats. J Physiol 460: 221–229
44. Moruzzi G (1972) The sleep-waking cycle. Ergebn Physiol 64: 1–165
45. Moruzzi G, Magoun HW (1949) Brainstem reticular formation and activation of the EEG. Electroenceph. Clin Neurophysiol 1: 455–473
46. Nauta WJH (1946) Hypothalamic regulation of sleep in rats: an experimental study. J Neurophysiol 9: 285–316
47. Parmentier R, Hotsu H, Djebarra-Hannas Z, Valatx JL, Watanabe T, Lin JS (2002) Anatomical, physiological, and pharmacological characteristics of histidine decarboxylase knock-out mice: evidence for the role of brain histamine in behavioral and sleep-wake control. J Neurosci 22: 7695–7711
48. Paut Pagano L, Roky R, Valatx JL, Kitahama K, Jouvet M (1993) Anatomical distribution of prolactin-like immunoreactivity in the rat brain. Neuroendocrinol 58: 682–695
49. Peyron C, Tighe DK, Van den Pol AN, de Lecea L, Heller HC, Sutcliffe JG, Kilduff AJ (1998) Neurons containing hypocretin (orexin) project to multiple neuronal systems. J Neurosci 18: 9996–10015
50. Puizillout JJ, Gaudin-Chazal G, Bras H (1984) Vagal mechanisms in sleep regulation. In: Borbely AA, Valatx JL (eds) Sleep mechanisms. Springer, Berlin Heidelberg New York Tokyo, pp 19–38
51. Rampin C, Cespuglio R, Chastrette N, Jouvet M (1991) Immobilization stress induces a paradoxical rebound in rat. Neurosci Lett 126: 113–118
52. Ranson SW (1939) Somnolence caused by hypothalamic lesion in the monkey. Arch Neurol Psychiat 41: 1–23
53. Rechtschaffen A, Kales A (1968) A manual of standardized terminology, techniques and scoring system to sleep stages of human subjects. UCLA Brain Information Services
54. Rees G, Greiman G, Koch C (2002) Neural correlates of consciousness in humans. Nature Rev Neurosci 3: 261–270
55. Reiner PB, Kamondi A (1989) Mechanisms of antihistamine-induced sedation in the human brain: H1 receptor activation reduces a background leakage potassium current. Neuroscience 59: 579–588
56. Rougeul A, Verdeaux J, Letalle A (1969) Effects électrographiques et comportementaux de divers hallucinogènes chez le chat. Rev Neurol 120:391–394

57. Sakai K, El Mansari M, Lin JS, Zhang JG, Vanni-Mercier G (1990) The posterior hypothalamus in the regulation of wakefulness and paradoxical sleep. In: Mancia M, Marini G (eds) The diencephalon and sleep. Raven Press, New York, pp 171–198
58. Sakai K, Crochet S, Onoe H (2001) Pontine structures and mechanisms involved in the generation of paradoxical (REM) sleep. Arch Ital Biol 139: 93–107
59. Sakurai T, Amemiya A, Ishii M, Matsuzaki I, Chemelli RM, Tanaka H, Williams SC, Richardson JA, Kozlowski GP, Wilson S, Arch JR, Buckingham RE, Haynes AC, Carr SA, Annan RS, McNulty DE, Liu WS, Terrett JA, Elshourbagy, Bergsma DJ, Yanagisawa M (1998) Orexins and orexin receptors: a family of hypothalamix neuropeptides and G protein-coupled receptors that regulate feeding behavior. Cell 92: 573–585
60. Sallanon M, Denoyer M, Kitahama K, Aubert C, Gay N, Jouvet M (1989) Long-lasting insomnia induced by preoptic neuron lesions and its transient reversal by muscimol injection into the posterior hypothalamus in the cat. Neuroscience 32: 669–6839
61. Schwartz JC, Arrang JM, Garbarg M, Pollard H, Ruat M (1991) Histaminergic transmission in the mammalian brain. Physiol Rev 71: 1–51
62. Steininger LT, Alam MN, Gong H, Szymusiak R, McGinty D (1999) Sleep-waking discharge of neurons in the posterior lateral hypothalamus of the albino rat. Brain Res 840: 198–147
63. Stériade M, Buszacki G (1991) Parallel activation of thalamic and cortical neurons by brainstem and basal forebrain cholinergic systems. In: Steriade M, Biesold D (eds) Brain cholinergic systems. Oxford University Press, Oxford, pp 3–64
64. Thakkar MM, Ramesh V, Strecker RE, McCarley RW (2001) Microdialysis perfusion of orexin-A in the basal forebrain increases wakefulness in freely behaving rats. Arch Ital Biol 139: 313–328
65. Urbain N, Rentero N, Gervasoni D, Renaud B, Chouvet G (2002) The switch of subthalamic neurons from an irregular to a bursting pattern does not solely depend on their GABAergic inputs in the anesthetic-free rat. J Neurosci 22: 8665–8675
66. Valatx JL (1988) La privation de sommeil. Méd Hyg 46: 2248–2253
67. Valatx JL (1995a) Génétique du sommeil. In: Benoit O, Foret J (eds) Le sommeil humain. Bases expérimentales physiologiques et physiopathologiques. Masson, Paris, pp 17–24
68. Valatx JL (1995b) Régulation du cycle veille sommeil. In: Benoit O, Foret J (eds) Le sommeil humain. Bases expérimentales physiologiques et physiopathologiques. Masson, Paris, pp 25–37
69. Valatx JL (2001) Economie énergétique et sommeil. In: Barré H, Chatonnet J, Le Maho Y, Valatx JL (eds) Physiologie énergétique. BelinSup Sciences, Paris, pp 315–381

Advances in Craniosynostosis Research and Management

J. Guimarães-Ferreira[1], J. Miguéns[2], and C. Lauritzen[3]

[1] Departments of Human Anatomy and Plastic Surgery,
University of Lisbon School of Medicine, Lisbon, Portugal
[2] Department of Neurosurgery, University of Lisbon School of Medicine,
Lisbon, Portugal
[3] Head, Craniofacial Unit, Department of Plastic Surgery,
Sahlgrenska University Hospital, Göteborg, Sweden

With 26 Figures

Contents

Abstract

The purpose of the present paper is to analyze the most recent advances in the field of craniosynostosis basic and clinical research and management, and to give an overview of the more frequently adopted surgical strategies.

After reviewing some basic concepts regarding normal craniofacial embryology and growth, aetiopathogenesis of craniosynostosis and craniofacial dysostosis, classification and diagnosis and historical evolution of surgical treatment, the authors elaborate on a selection of topics that have modified our current understanding of and therapeutical approach to these disease processes. Areas covered include advances in molecular biology and genetics, imaging techniques and surgical planning, resorbable fixation technology, bone substitutes and tissue engineering, distraction osteogenesis and the spring-mediated cranioplasties, resorbable distractor devices, minimally invasive surgery and in utero surgery.

A review of the main subtypes of craniosynostosis and craniofacial dysostosis is presented, including their specific clinical features and a commentary on the presently available surgical options.

Keywords: Craniosynostoses; pathology; radiography; surgery; surgery; plastic/ methods.

Introduction

Human deformities, especially of the head, have always moved man's imagination. Craniofacial deformity has at times been regarded as a sign of the supernatural. The Taoist god of long life and luck, Shou-xing, is one typical example, being usually portrayed with an extremely high head. Fukurokuju, the Shinto god of wisdom, is another example also depicted with a very high forehead (Fig. 1). Certain cultures have used particular head shapes as marks of elitist distinction or divinity (Gaudier B *et al.* 1967). Examples of this are the Incas, Mayas (Fig. 2), Pueblos, Navajos and Apaches in the Americas, the Ethiopians in Africa, the Philippines and the French (Montaut J *et al.* 1977) (in Brittany, Normandy and especially the Toulouse region, where the practice of skull deformation through bandaging persisted until the end of the XIXth century).

At the same time pathological deformities historically and currently have branded a person with a stigma of inferiority. Homer (Homer *et al.* 1994) (ca 800 BC) wrote about the warrior Thersites, who "... *was the ugliest man who came before Troy ... his head ran up to a point ...*". This description would fit a man with craniosynostosis, and the expression "tête à la Thersite" has been used in French literature as a synonym for oxycephaly.

Fig. 1. Shou-xing, god of long life (left) and Fukurokuju, god of wisdom (right)

Homer's description was written in a disdainful tone toward an aesthetically handicapped person and it is a fact that social attitudes toward these individuals have varied substantially through the ages. Disposal of newborns by abandonment or killing by drowning, strangulation, live burial, etc are known to have happened in pre-Christian time (Sailer HF *et al.* 1995). Individuals with a craniofacial deformity are often met with prejudice and misconceptions about their intellectual capabilities. The medieval poet Deschamps, wrote about the misfortune of having malformed children (Huizinga J 1996):

That a malformed man
Is also deformed in his mind
Full of sins and full of vices

Normal Craniofacial Embryology and Growth

Formation of the human cranium begins between the 23rd and the 26th gestational days, at a time where the anterior neuropore is closing. The

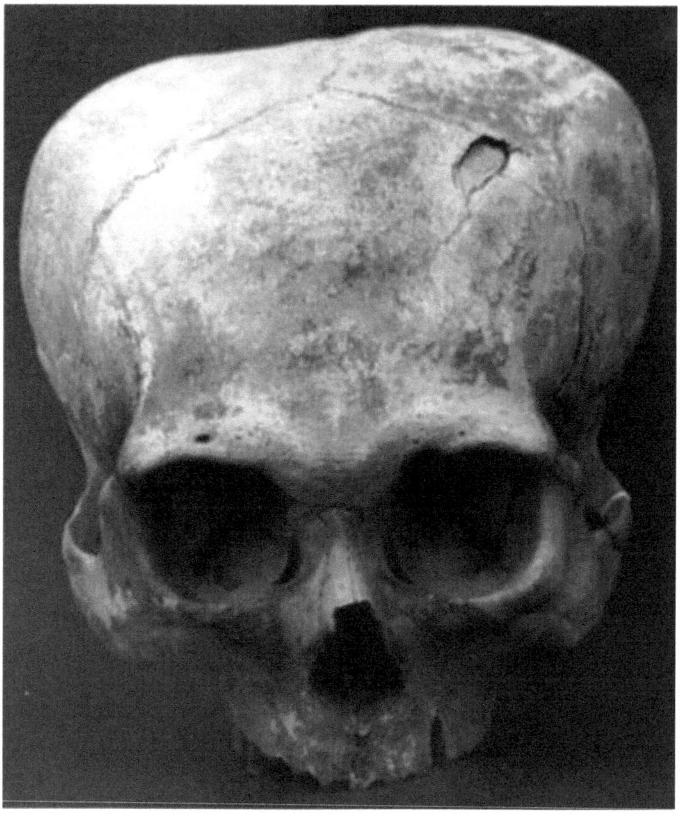

Fig. 2. Precolombian skull, showing the effects of intentional deformation

skull of the embryo and early foetus can be divided into the *neurocranium*, the capsule surrounding the brain, and the *viscerocranium*, from which the facial skeleton develops. The neurocranium can be further subdivided into *chondrocranium* (originating the cranial base bones that undergo endochondral ossification, i.e., the occipital, otic, orbitotemporal, and ethmoidal regions) and *membranous neurocranium* (which ossifies directly in the mesenchyme and will originate the cranial vault). The early stages of skull development can be viewed as a sequence of three phases: membrane formation, chondrification, and ossification (Lemire RJ 2000).

Dura mater appears between the 51st and 53rd days of gestation (Lemire RJ 2000). With the ongoing differential regional growth of the neural structures, several dural reflections appear, which delineate mesenchymal areas that start to ossify around the 10th gestational week. The dural reflections themselves do not ossify but instead become the sites where cranial sutures form (Friede H 1981; Smith DW *et al.* 1978).

Two different but closely integrated types of bone growth occur in the craniofacial skeleton. One is the *remodelling* type, caused by the continuous osteoblastic and osteoclastic activities in periosteal and endosteal surfaces. The other is the *displacement* type, through which whole bones move apart from each other, with a separating effect on the intervening articulations (sutures, synchondroses, condyles). Articular contact is maintained by the apposition of bone at the articular margins (Enlow DH 1999; Enlow DH 2000). When bone displacement cannot accommodate brain growth, as in cases of generalized premature sutural fusion, intracranial volume expansion must rely exclusively on the remodelling process. Since this is a slower mechanism, elevations of intracranial pressure may occur, particularly during phases of rapid brain growth.

Cerebral and cerebellar expansion are considered to be the main engine underlying calvarial growth (Enlow DH 2000). Approximately 68% of adult brain weight is achieved at one year of age, 79% at two years, 84% at three years, and 92% at five years (Coppoletta JM *et al.* 1933). During the first 3 years of life, the value quadruples over that at birth (Dekaban AS 1978) (Fig. 3).

The postnatal increase in intracranial volume does not follow a linear pattern, but rather closely matches the curve of brain weight. At two and five years of age, intracranial volume reaches 77% and 90%, respectively, of the volume at 15 years (Sgouros S *et al.* 1999a).

Aetiology, Pathogenesis and Function

Craniosynostosis is the process of premature cranial sutural fusion, which may lead to *craniostenosis*, a deformity of the skull (Tessier P 1971b).

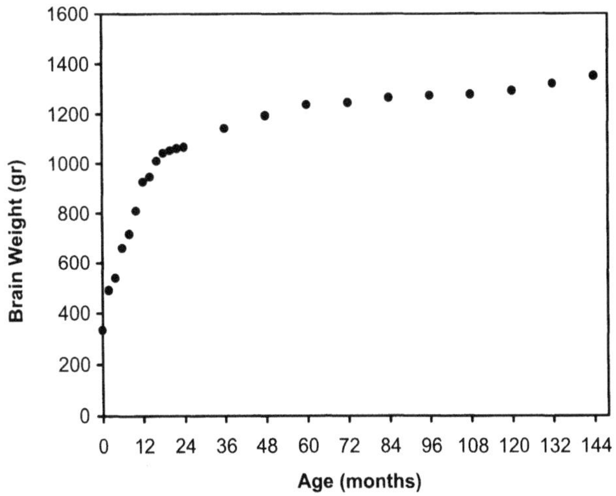

Fig. 3. The postnatal evolution of mean brain weight, from birth until twelve years of age. Based on data from Coppoletta and Wolbach. (Coppoletta JM *et al.* 1933)

Table 1. *Some Conditions Associated with Secondary Craniosynostosis*

Metabolic	*Hematologic*
– Hyperthyroidism	– sickle cell anemia
– Rickets	– polycythemia vera
– Mucopolysaccharidoses	*Malformations*
Teratogens	– holoprosencephaly
– Diphenylhydantoin	– microcephaly
– Valproate	*Iatrogenic*
– Fluconazole	– shunted hydrocephalus

Technicalities aside, both expressions may be used interchangeably, although most authors currently favour the first. *Craniofacial dysostosis* is a term, which appears to have gained generalized acceptance to designate the condition that results from premature closure of *both* cranial *and* facial sutures, as in cases of certain craniofacial syndromes (Lauritzen C *et al.* 1999; McCarthy JG *et al.* 1978; Posnick JC 1994; Tessier P 1971a).

Most cases of craniosynostosis are *primary*. A number of conditions are associated with the development of premature synostosis, which, in such a setting is termed *secondary* (Table 1). The following sections deal essentially with *primary* craniosynostosis.

Sutures and Craniosynostosis

"*Human heads are not at all similar to each other. Nor are the sutures constant in number or in location*". These are the opening words of "*De Vulneribus Capitis Liber*" (Fig. 4), written by Hippocrates (460–367 BC). The hint of a possible relationship between sutures and skull morphology becomes apparent as he writes on: "*Thus, whoever has a projection in the forehead (...), his sutures have the appearance of the letter tau, T. (...) But whoever has the prominence in the posterior part of the head, his sutures are opposite to the former. (...) But whoever has prominences both in the front and the back of the head, in him the sutures resemble the letter eta, H.*"

Other ancient physicians have stated the importance of cranial sutures. Galen (129–216 AD) mentioned the association of exophthalmus and headache with the reduction in the number of sutures (Cohen MM 2000b), and Celsus (first century AD) described skulls without sutures (Laitinen L 1956). In Vesalius' "*De Humani Corporis Fabrica*", published in 1543, there are drawings of deformed skulls, some of which appear clearly craniosynostotic (Fig. 5). In 1557 Lycosthenes wrote the following description of the birth of a child: "*Natus est infans manibus ac pedibus anserinis,*

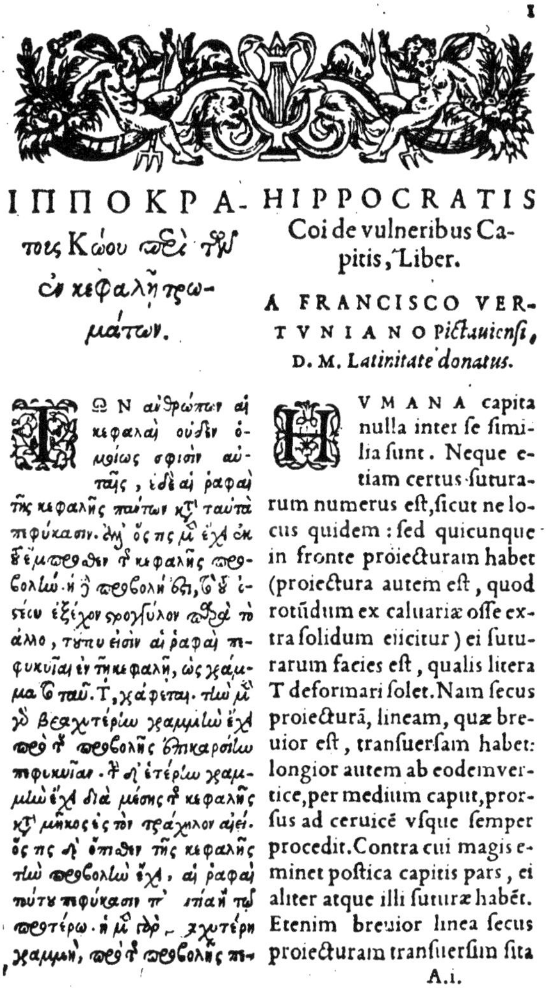

Fig. 4. Frontpage of Hippocrates' (460–367 BC) "De Vulneribus Capitis Liber"

horrendo capite ..." ("*An infant was born with goose hands and feet, a horrible head ...*") This could correspond to an early description of a craniofacial dysostosis syndrome, with its associated syndactyly (Laitinen L *et al.* 1955).

Sömmerring (Sömmerring ST 1800) was the first modern age investigator to study the structure of cranial sutures, and to assign them an essential role in cranial growth. He also observed that premature sutural synostosis would lead to deformity. Otto (Otto AW 1830) studied the phenomenon extensively and concluded that premature sutural closure resulted in skull deformity through a combination of localized growth restriction and compensatory expansion elsewhere. His ideas were an obvious precursor of the well-known Virchow's law (Virchow R 1851), which

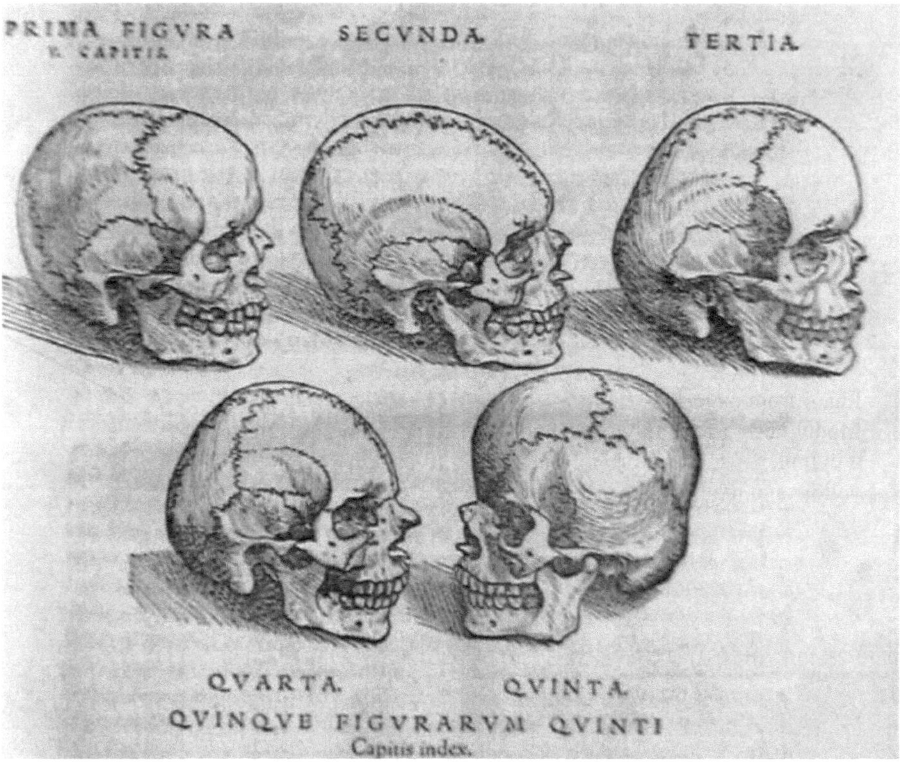

Fig. 5. The two skulls on the right have a craniosynostotic configuration. The drawings of sutures are not accurate, however. From Andreas Vesalius' "De Humani Corporis Fabrica", 1543 (Central Library, University of Lisbon School of Medicine, with permission)

states that when a suture is prematurely closed, growth becomes restricted in the perpendicular direction. This leads to overexpansion of adjacent open sutures, resulting in an exaggeration of cranial growth in a direction parallel to the closed suture.

In opposition to Virchow's, who has put strong emphasis on the primary role of sutural anomalies as the basis for the skull deformities that accompany premature craniosynostosis, is Moss' theory of the "*functional matrix*" (Moss ML 1959; Moss ML 1975). This American orthodontist postulated that the primary defect that leads to craniosynostosis lies in the cranial base, rather than locally at the sutural level. The primary cranial base abnormality would initiate a sequence of events leading to premature sutural closure and skull deformity, mediated by the transmission of abnormal dural tensions to an otherwise normal cranial vault. The author also proposed that the development and final morphology of the craniofacial complex are determined by the "*functional matrix*", i.e. the combined mechanical interactions of the various muscles, aponeurosis and organs of the head and neck (Moss ML *et al.* 1969). In an interesting experiment involving the microvascular heterotopic transplantation of rat heads, Hirabayashi showed that this is not the case, at least in this rodent model of craniofacial growth (Hirabayashi S *et al.* 1988).

Whatever the sequence of events leading to premature suture closure might be, the initial determinants of the phenomenon remain unclear in most cases. There is however, general agreement over the assumption that a heterogeneous group of pathological mechanisms does exist that may lead to this common end result (Cohen MM 2000c).

Mechanical Factors

Mechanical forces, e.g. those related to intrauterine constraint, have been implicated as possible causes for craniosynostosis, both clinically (Graham JM, Jr. *et al.* 1980a; Graham JM, Jr. *et al.* 1979; Graham JM, Jr. *et al.* 1980b; Higginbottom MC *et al.* 1980; Koskinen-Moffett LK *et al.* 1982) and in experimental settings (Koskinen-Moffett LK *et al.* 1989). One indication of mechanical factors causing craniosynostosis is the premature fusion of sutures seen in patients with ventricular (over-)shunting for hydrocephalus (Kloss JL 1968; Roberts JR *et al.* 1970). One such case is presented in Fig. 6.

Genetics

Approximately 8% of all cases of craniosynostosis are familial, with an autosomal dominant inheritance pattern (Cohen MM *et al.* 2000). The advent of molecular genetics has allowed the identification of several mutations which are implicated in premature sutural fusion, either as an isolated anomaly or, more frequently, in the context of a craniofacial syndrome (Table 2).

Many of these mutations involve the genes that code for three of the four fibroblast growth factor receptors (*FGFR1*, *FGFR2* and *FGFR3*), members of the receptor tyrosine kinase family (Jabs EW *et al.* 1994; Meyers GA *et al.* 1995; Mulliken JB *et al.* 1999; Paznekas WA *et al.* 1998; Reardon W *et al.* 1994; Rutland P *et al.* 1995; Wilkie AO *et al.* 1995a). The *TWIST* (Howard TD *et al.* 1997) and *MSX2* genes have also been implicated in craniosynostosis. *MSX2*, a member of the highly conserved and widely distributed homeobox gene family, was in fact the first gene to be found responsible for a human form of hereditary craniosynostosis, the so called Boston-type, in 1993 (Jabs EW *et al.* 1993). Also, transgenic mice expressing increased levels of *Msx2* exhibit premature fusion of cranial sutures (Liu YH *et al.* 1995; Liu YH *et al.* 1999).

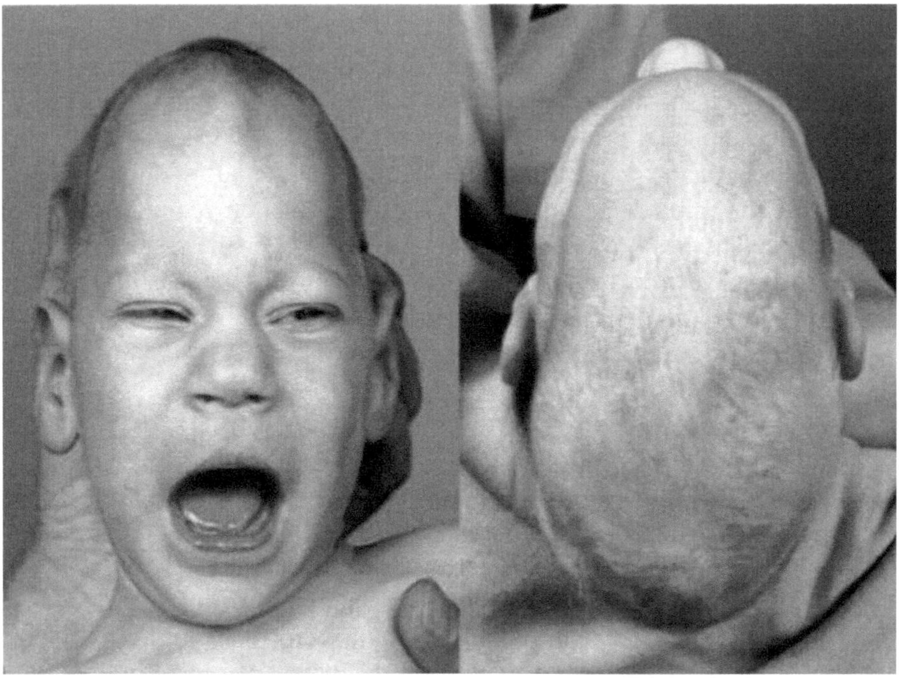

Fig. 6. Sagittal synostosis and extreme scaphocephalic head shape in a 3-month-old boy previously subjected to ventriculo-peritoneal shunting

Table 2. *Known Mutations in Some Craniofacial Syndromes*

Syndrome	Chromosome	Gene	Number
Apert	10q25.3-q26	*FGFR2*	6
Crouzon	10q25.3-q26	*FGFR2*	30
Pfeiffer	10q25.3-q26	*FGFR2*	29
	8p11.2-p12	*FGFR1*	1
Saethre-Chotzen	7p21	*TWIST*	40
	10q25.3-q26	*FGFR2*	2
	4p16	*FGFR3*	1
Jackson-Weiss	10q25.3-q26	*FGFR2*	1
Beare-Stevenson cutis gyrata	10q25.3-q26 4p16	*FGFR2*	2
		FGFR3	1
Muenke craniosynostosis	4p16	*FGFR3*	1
Crouzon syndrome with acantosis nigricans	4p16	*FGFR3*	1
Boston-type craniosynostosis	5q34-q35	*MSX2*	2

A great deal of overlap exists between the phenotypic expressions of different mutations. Conversely, it is also possible that individuals sharing the same mutation be assigned to different clinical diagnostic groups. One probable exception to this are the two *FGFR2* mutations causative of more than 98% of the cases of Apert syndrome, which present the most specific mutation-phenotype correlations known to date.

Functional Aspects

It has been assumed that craniosynostosis would lead to a smaller than normal intracranial volume. This is probably not true in most situations of single suture involvement since volume expansion may still occur through compensational growth at unaffected sutures and by appositional bone growth (Gault DT *et al.* 1990; Posnick JC *et al.* 1994). A recent study using computerized tomography has shown that although children with craniosynostosis are born with slightly reduced intracranial volumes, normal values are reached at six months of age without any surgical manipulation (Sgouros S *et al.* 1999b). Two exceptions were children with complex pansynostoses, who had smaller preoperative volumes, and patients with Apert syndrome, in whom greater than normal volumes were found (Sgouros S *et al.* 1999b).

Although raised intracranial pressure (ICP) is more frequent in multiple sutural synostosis and is generally considered to be rare in single suture cases, several reports of this have nevertheless been published (Gault DT *et al.* 1992; Renier D 1989; Renier D *et al.* 1982; Thompson DN *et al.* 1995). The finding of elevated ICP cannot be blamed entirely on intracranial volume restriction (Fok H *et al.* 1992; Gault DT *et al.* 1992; Posnick JC 1999). Other factors like hydrocephalus (Cinalli G *et al.* 1998; Collmann H *et al.* 1988; Golabi M *et al.* 1987; Hanieh A *et al.* 1989) and altered intracranial venous haemodynamics may also play a role (Mursch K *et al.* 1999).

One of the most devastating (yet frequently preventable) effects of long-standing elevated ICP is visual impairment (Stavrou P *et al.* 1997). In most cases, this is thought to result from papilloedema and subsequent optic nerve atrophy (Miller MT 2000). Other factors may be involved, such as direct optic nerve compression or kinking (Wood-Smith D *et al.* 1976), exophthalmus-related corneal exposure, and amblyopia due to strabismus, ametropia, or ptosis (Hertle RW *et al.* 1991).

Mental handicaps associated with craniosynostosis may, in theory, be caused by long-standing elevations in ICP, distorted brain growth due to the skull deformity and primary brain abnormalities (Camfield PR *et al.* 2000). It is an infrequent finding in single suture synostosis and, when present, it can usually be attributed to an underlying brain malformation or to perinatal factors (Hunter AG *et al.* 1976; Hunter AG *et al.* 1977; Noetzel MJ *et al.* 1985). Among the syndromic craniosynostoses, the incidence of mental handicap is highest in Apert syndrome (Patton MA *et al.* 1988; Renier D *et al.* 1996a; Renier D *et al.* 1996b), although some affected individuals have normal intelligence. Psychomotor and intellectual retardation have also been reported in patients with Pfeiffer (Cohen MM 1993) and Carpenter (Cohen DM *et al.* 1987) syndromes, and whenever the cloverleaf skull deformity is present (Angle CR *et al.* 1967). Mental retardation is infrequent in Crouzon syndrome (Kreiborg S 1981).

In cases of craniofacial dysostosis, additional problems appear, in relation to the deficient growth of the orbits and midface. Insufficient eye protection, respiratory obstruction – sometimes life-threatening –, hearing and speech disorders, and malocclusion may require specific interventions (Alonso N *et al.* 1998; Britto JA *et al.* 1998; Bruce DA 1996; Epstein F *et al.* 1981; Jarund M *et al.* 1996; Lauritzen C *et al.* 1999; Lauritzen C *et al.* 1986; Lauritzen C *et al.* 1998; Marchac D *et al.* 1994; McCarthy JG *et al.* 1995; McCarthy JG *et al.* 1984; McCarthy JG *et al.* 1990; Ortiz-Monasterio F *et al.* 1978; Peterson-Falzone SJ *et al.* 1981; Posnick JC 1991; Posnick JC 1994; Raulo Y *et al.* 1981; Sculerati N *et al.* 1998; Tessier P 1967a; Tessier P 1967b; Tessier P 1971a; Tessier P 1971b; Toth BA *et al.* 1998).

Classification and Diagnosis

Virchow published the first classification of craniosynostosis in 1851 (Virchow R 1851) (Table 3). Several other classifications have since been

Table 3. *Virchow's Classification (Abbreviated)*

Simple macrocephaly	
Simple microcephaly	
Dolichocephaly	Dolichocephaly
	Sphenocephaly
	Leptocephaly
	Clinocephaly
	Trigonocephaly
Brachycephaly	Pachycephaly
	Oxycephaly
	Platycephaly
	Trochocephaly
	Plagiocephaly
	Simple Brachycephaly

Table 4. *Anatomic Classification of Craniosynostosis*

Monosutural synostosis	sagittal
	unicoronal
	metopic
	unilambdoid
Multiple suture synostosis	bicoronal
	bilambdoid
	other combinations
	cloverleaf skull deformity
Craniofacial dysostosis syndromes	apert
	crouzon
	Pfeiffer
	Saethre-Chotzen
	other

proposed, some of them based upon recent knowledge from the field of molecular genetics (Cohen MM 2000b).

Possibly, the most useful classifications for the clinical setting are the anatomic ones, based on the specific sutural involvement. One such classification is presented in Table 4. The incidence of craniosynostosis, all diagnoses combined, was recently estimated to be of 476 per million live births (Lajeunie E *et al.* 1995).

Diagnosis

The most important diagnostic tool in craniofacial pathology is clinical observation. An experienced observer can establish a correct diagnosis in the majority of cases without the need for any diagnostic studies. Frequently however, specific studies are requested to document a condition, to aid in surgical planning or to evaluate the functional status of a patient. Table 5 lists the main diagnostic studies used in craniofacial surgery.

Evolution of Surgical Treatment

"*Can you not unlock my poor child's brain and let it grow?*" was the request of a mother to L. C. Lane, Professor of Surgery at the Copper Medical College, San Francisco. The child, who had a premature sutural synostosis, was operated on with a cross-shaped craniectomy in August 1888, and died 14 hours later. The same surgeon performed a second craniectomy in 1892, with a better result (Lane LC 1892). The first report in the literature of a double parasagittal linear craniectomy for the treatment of

Table 5. *Diagnostic Studies*

Photography
Roentgencephalometry
Computerized tomography
Magnetic resonance imaging
3D-imaging studies
Stereolithography
ICP monitoring
Fundoscopy
Sleep studies
Visual evoked potentials
Psychometric studies
Genetic studies

scaphocephaly was published in 1890 by Lannelongue. The procedure was performed in Paris on a four-year old girl with a severe psychomotor handicap. Reportedly, the patient had an almost immediate and complete neurologic recovery (Lannelongue M 1890).

Harvey Cushing, the most influential neurosurgeon of his time, was very critical of these early attempts: "The introduction in 1891 of linear craniotomy, which has led to innumerable operations said to have been followed by an improvement in mentality, is a lamentable instance of the furor operandi running away with surgical judgement" (Cushing H 1908). Despite this and other voices of disbelief, decompressive surgery for craniosynostosis became an accepted treatment modality mainly due to the concerns of visual loss and intellectual impairment (Dandy WE 1943; Faber HK 1924; Faber HK *et al.* 1927; Ingraham F *et al.* 1948; King J 1942; Laitinen L *et al.* 1955; McLaurin RL *et al.* 1952). Operations were viewed not only as preventive, but in some cases also as potentially corrective of an already established dysfunction. Walter Dandy, in a paper describing his method of cranioplasty for scaphocephaly (release of two large parietal bone segments) wrote the following: "*Immediately after the first operation the child became alert, active and responsive and showed a remarkable change from the preoperative listlessness and rather stupid appearance*" (Dandy WE 1943).

A shift away from this trend was initiated in 1954 by Ingraham and Matson, who included cosmetic benefits as indications for surgery in their textbook on paediatric neurosurgery (Ingraham FD *et al.* 1954). After Shillito's study of a series of 519 cases of craniosynostosis operated on with minimal mortality and morbidity (Shillito J, Jr. *et al.* 1968), aesthetical correction became generally accepted as an indication for surgery.

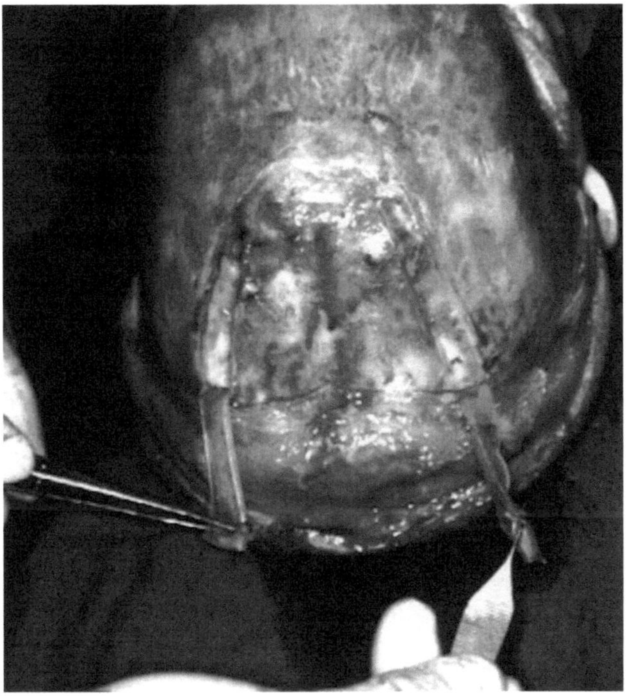

Fig. 7. Reoperation for sagittal craniosynostosis. Silicone sheet from a previous procedure is being removed, showing complete reossification of the craniectomy lines

Regardless of the indications, treatment for craniosynostosis largely relied on linear craniectomies (Ingraham F *et al.* 1948). Electrocautery and caustic agents were used trying to diminish the osteogenic potential of the dura mater (the use of caustic agents was largely discontinued due to evidence of dura permeation and brain toxicity) (Marlin AE *et al.* 1980; Pawl RP *et al.* 1972). In addition, a variety of materials implanted along the osteotomies was used to prevent reossification (Duff TA *et al.* 1991; Ingraham F *et al.* 1948). Although still widely used, the effectiveness of ossification barriers was never convincingly demonstrated. Figure 7 shows an example of a patient with sagittal craniosynostosis at the time of reoperation for residual deformity. In recent years, a trend has developed toward more extensive reshaping operations and most studies show the superiority of these procedures when compared to the simpler linear craniectomies (Boop FA *et al.* 1996; Cohen MM 2000a; Joos U *et al.* 1993; Kaiser G 1988; Marsh JL *et al.* 1991; Maugans TA *et al.* 1997; Panchal J *et al.* 1999).

The first attempt at complete surgical correction of the facial deformity of a craniofacial dysostosis syndrome was reported by Sir Harold Gillies in

1950 (Gillies H *et al.* 1950). The patient was a woman with a recessed middle third of the face with exorbitism, and the treatment was a subcranial LeFort III-type operation, carried out in 1942. No bone grafts were used, and despite the excellent initial result, an almost complete relapse eventually occurred (Fernando Ortiz-Monasterio, Göteborg 2001, personal communication).

The foundation of modern craniofacial surgery was laid down by Paul Tessier in the 1960's. His extensive cadaveric dissection studies of the orbits and midface culminated with the presentation of a series of Crouzon and Apert patients operated on with sub cranial LeFort III-type osteotomies, at the IV International Congress of Plastic and Reconstructive Surgery in 1967 (Tessier P 1967a). His presentation had an enormous impact (Wolfe SA 1999), and was shortly followed by a paper on the correction of orbital hypertelorism (Tessier P *et al.* 1967). The latter was the first report of a transcranial procedure for the correction of an orbital deformity. Most of the fundamental principles of modern craniofacial surgery had been defined: thorough exposure of the osteotomy sites; the use of bone grafting to secure the remodelled anatomy; and the close cooperation between a plastic (craniofacial) surgeon and a neurosurgeon for the combined access to the craniofacial skeleton. Major advances in the management of craniosynostosis emerged from this cooperation. Procedures involving extensive remodelling and repositioning of bone segments were conceptualized and put into practice (Anderson FM 1981; Converse JM *et al.* 1971; David DJ *et al.* 1982; Delashaw JB *et al.* 1986; Jane JA *et al.* 1978; Lauritzen C *et al.* 1996; Marchac D *et al.* 1979; Marchac D *et al.* 1981; McCarthy JG 1979; Ortiz-Monasterio F *et al.* 1978; Persing JA *et al.* 1987; Tessier P 1971a; Tessier P 1971b; Whitaker LA *et al.* 1977).

The technique of bone distraction, pioneered by Codivilla (Codivilla A 1905) in the early 1900's and popularized by Ilizarov (Ilizarov GA *et al.* 1980) for long bone lengthening, was successfully applied to the entire craniofacial skeleton (Chin M *et al.* 1997; Cohen SR *et al.* 1998; Hirabayashi S *et al.* 1998a; Kobayashi S *et al.* 1999; Matsumoto K *et al.* 2000; McCarthy JG *et al.* 1992; Molina F *et al.* 1995; Polley JW *et al.* 1995; Shwyrkow MB *et al.* 1989; Sugawara Y *et al.* 1998; Toth BA *et al.* 1998).

The goals of surgery for craniosynostosis are aesthetical, functional and psychological (Tessier P 1979). Besides the coordinated interactions of the craniofacial/neurosurgical/anaesthetic teams, the collaboration of other specialists is therefore often required. These include intensive care specialists, paediatricians, orthodontists, neuroradiologists, ophthalmologists, otolaryngologists, pneumologists, psychologists, geneticists and speech therapists, among others (Hockley A 1993).

Recent Advances

Molecular Studies

The molecular mechanisms regulating sutural fusion have been extensively investigated in recent years. Opperman demonstrated that tissue interactions between the rat coronal suture and the underlying dura mater are required for the maintenance of suture patency (Opperman LA *et al.* 1998; Opperman LA *et al.* 1995; Opperman LA *et al.* 1993). The same author showed that local periosteum does not contribute significantly to the regulation of suture fate (Opperman LA *et al.* 1994).

Transforming growth factors (*TGF*) $\beta1$, $\beta2$ and $\beta3$, a family of molecules which have been strongly implicated in the regulation of cellular proliferation and osteogenesis, were identified as the potential mediators of those interactions (Mehrara BJ *et al.* 1999; Roth DA *et al.* 1997b). Other components of growth factor signalling systems like *insulin-like growth factors* (*IGF*) and *fibroblast growth factors* (*FGF*) have been shown to play important regulatory roles in the maintenance of suture patency and in the process of sequential sutural fusion in experimental animal models (Mehrara BJ *et al.* 1998; Most D *et al.* 1998). Evidence of the involvement of these factors in human premature synostosis has also been presented (Bresnick S *et al.* 1995; Bresnick S *et al.* 1998; Lin KY *et al.* 1997; Pensler JM *et al.* 1998; Roth DA *et al.* 1997a).

Imaging Techniques and Surgical Planning

The first applications of three-dimensional computerized tomography imaging (3D-CT) of the craniofacial skeleton are now over 20 years old (Hemmy DC *et al.* 1983). The continuous development of computer hardware and reconstruction algorithms has brought us from the initial rough-looking and undetailed imagery to the high-definition renderings now routinely available at most modern units. 3D-CT is now a well established and valuable tool for the diagnosis and surgical planning in craniofacial surgery (Cutting C *et al.* 1986; Lo LJ *et al.* 1994; Vannier MW *et al.* 1984).

Stereolithography, a relatively new technique originally developed for industrial use, allows the creation of acrylic skull replicas from patient's CT data. The method involves computerized LASER-induced polymerization of a plastic, guided from CT-slice data (Anderl H *et al.* 1994; Klein HM *et al.* 1992). The replicas thus obtained are very precise reproductions of the skull and may be used for preoperative surgical planning and also intraoperatively after sterilization (Hirabayashi S *et al.* 1998b). The main indication of the technique seems to be as an aid to surgical planning in complex craniofacial reconstruction (Sailer HF *et al.* 1998).

A direct consequence of the evolution of imaging technologies was the production of computer-generated patient-matched implants (Chang PS *et al.* 2003; Eppley BL 2002). Although intellectually appealing, for the moment, except in some carefully selected cases, this rather sophisticated technology probably adds more to the costs than to the results.

Resorbable Fixation

Titanium plate and screw fixation techniques are widely used in craniofacial surgery, both for trauma and in congenital deformities. This is due to the ease of application, reliability and biocompatibility of titanium implants. Some concern exists over the use of metallic fixation in paediatric patients, however, including the passive intracranial migration of the implants (Goldberg DS *et al.* 1995; Papay FA *et al.* 1995; Stelnicki EJ *et al.* 1998a) (Fig. 8), the restriction of calvarial and facial growth (Connelly SM *et al.* 1998; Wong L *et al.* 1993), the unknown effects of prolonged interaction with tissues over several decades and the interference with imaging techniques.

To overcome these problems, a range of resorbable fixation systems has been developed, based on polylactic acid (PLA), polyglycolic acid (PGA) or a combination thereof. Plates and screws made of this type of

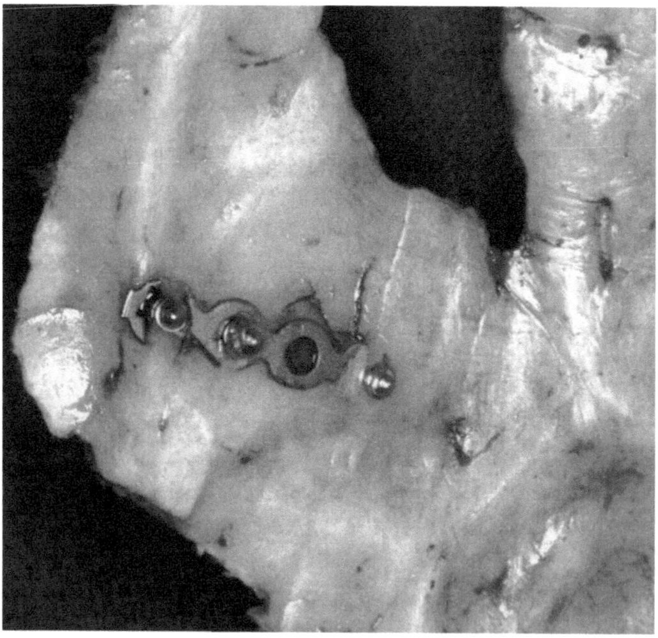

Fig. 8. Titanium plate and screws "migrated" intracranially through the temporal bone of a reoperated patient

material undergo three sequential phases when implanted in a living organism: hydrolysis, fragmentation and phagocytosis with disappearance of the material. The rate at which these phases succeed and, consequently, the rate of mechanical strength loss and the total resorption time, are highly dependant on the chemical and isomeric composition and on several aspects of the manufacturing process (Heino H 1999; Peltoniemi HH *et al.* 1999a).

Several experimental and clinical studies have shown that modern polymeric fixation systems are comparable to titanium-based systems for fixation strength and healing of osteotomy lines (Peltoniemi HH *et al.* 1998; Peltoniemi HH *et al.* 1999b; Wiltfang J *et al.* 2000). Some problems related to excessively long resorption time, inflammatory reactions and granuloma formation were reported initially (Bergsma EJ *et al.* 1993; Bergsma JE *et al.* 1995; Bos RR *et al.* 1991) but seem to be rare with currently employed systems (Kumar AV *et al.* 1997; Kurpad SN *et al.* 2000; Losken A *et al.* 2001; Pensler JM 1997; Wiltfang J *et al.* 2000). Despite the current degree of technological evolution of the commercially available products, problems may still be encountered such as prolonged, exuberant inflammatory reactions or occasional granuloma and fistula formation (Fig. 9), and caution must be exercised when placing an indication for the use of resorbable fixation materials. Also, the current plates are still too bulky to use outside hair bearing scalp. A recent survey on paediatric cranial fixation techniques among paediatric neurosurgeons showed that only 29% use absorbable plating for routine craniotomies (Dickerman RD *et al.* 2002).

Bone Substitutes and Tissue Engineering

Bone defects are a common finding in craniosynostosis and craniofacial dysostosis patients, especially after multiple surgical procedures. The use of autologous bone grafts is possibly the best solution for this problem,

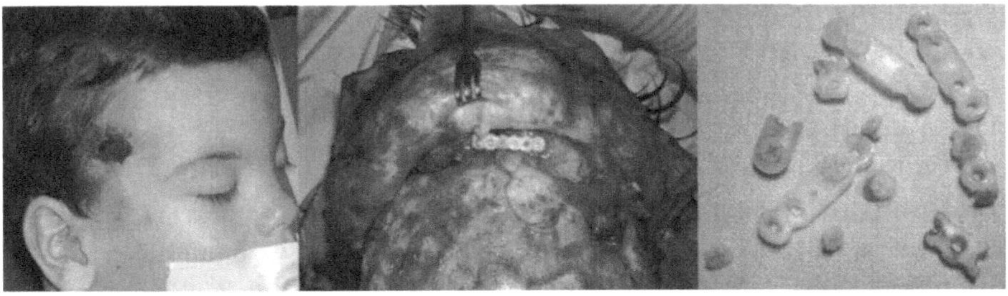

Fig. 9. Fistula and extensive granuloma formation in relation to PLA plates and screws implanted two years previously

since transplanted bone tends to revascularize and acquire the biomechanical features and remodelling potential of the lost tissue. The availability of donor sites, however, is limited and the procedure of bone harvesting adds to the overall morbidity and cost of treatment. This justifies the search for adequate bone substitutes and several materials have been developed which are currently in clinical use with varying degrees of success.

Synthetic materials like silicone, polymethylmetacrylate and porous polyethylene have long been used as bone graft substitutes, mainly for contour restoration. Since they are non-absorbable and intrinsically different from bone, they completely lack the ability to integrate functionally in a growing skeleton. They are unable to grow and remodel, their function being merely structural. As with any foreign body, infection and extrusion are ever-present dangers. Polymethylmetacrylate, nowadays armed with gentamycin, has been used a long time, but at one point of craniofacial euphoria considered obsolete. Today at certain centres (such as ours) the material has regained popularity and been frequently used over the last 7 years virtually without any problems.

Calcium phosphate-based materials (hydroxyapatite, tricalcium phosphate) share chemical similarities with the mineral component of bone and some initial studies fuelled the hope that they would gradually be replaced with the living tissue, once implanted (Byrd HS *et al.* 1996; Byrd HS *et al.* 1993; Philippe B *et al.* 1993; Salyer KE *et al.* 1989). This process is, however, very slow and incomplete and complications such as infection, extrusion and implant loss have been reported (Matic D *et al.* 2002; Moreira-Gonzalez A *et al.* 2003).

The emerging field of tissue engineering combines knowledge from biological and biomaterials sciences, to develop biological substitutes that restore, maintain and, if possible, improve the function of lost tissue and organs.

Periosteum and bone marrow have osteoprogenitor cells with the capability to form bone. The discovery of the osteoinductive potential of demineralised bone (Urist MR 1965) prompted the search for the growth factors responsible, many of which have been identified, along with their coding genes. Most of these factors belong to the transforming growth factor beta (TGF-β) superfamily, bone morphogenetic protein 7 (BMP-7) being one of the most potent.

Cells and growth factors are now being added to scaffolds made of bioabsorbable materials, like hydroxyapatite, poly-lactic and poly-glycolic acids, rendering them osteoinductive, i.e., conferring them the ability to transform into biologically functioning bone (Boyan BD *et al.* 1999; Breitbart AS *et al.* 1998; Gosain AK 1997; Kuznetsov SA *et al.* 1997; Ripamonti U 1991; Ripamonti U *et al.* 2001; Ripamonti U *et al.* 1992a;

Ripamonti U *et al.* 1992b; Sweeney TM *et al.* 1995). Gene therapy technology is also starting to play a role, and one study has been published where critical size cranial defects in rabbits have been repaired with PGA implants seeded with genetically modified cultured periosteal cells (Breitbart AS *et al.* 1999). In this experiment, the cultured periosteal cells were retrovirally transduced with the human gene coding BMP-7 and thus forced to produce high amounts of this protein.

Nevertheless, despite these fascinating prospects it seems clear that, to date, no bone substitute has consistently performed as well clinically as the fresh bone autograft, provided the cardinal principle for bone graft survival is respected – totally rigid fixation.

Distraction Osteogenesis and the Spring-Mediated Cranioplasties

Distraction osteogenesis (DOG) of the craniofacial skeleton has evolved as a spin-off of long bone distraction techniques, pioneered by Codivilla (Codivilla A 1905) in the early 1900's and was later developed by Ilizarov (Ilizarov GA *et al.* 1980) into an effective and safe procedure. Mandibular DOG was first reported by Snyder (Snyder CC *et al.* 1973) in an experimental setting and it was subsequently used for the reconstruction of traumatic and surgical defects in humans (Shwyrkow MB *et al.* 1989). The publication of McCarthy's paper on mandibular distraction (McCarthy JG *et al.* 1992) and the presentation of a large series by Molina (Molina F *et al.* 1995) prompted its diffusion in the western world, and the technique became an established clinical modality. The method is now routinely applied to the entire craniofacial skeleton, both for midface advancement (Chin M *et al.* 1997; Cohen SR *et al.* 1998; Polley JW *et al.* 1995; Rachmiel A *et al.* 1993; Talisman R *et al.* 1997; Toth BA *et al.* 1998) and for calvarial reshaping (Fukuta K *et al.* 1998; Hirabayashi S *et al.* 1998a; Kobayashi S *et al.* 1999; Lalikos JF *et al.* 1995; Losken HW *et al.* 1999; Losken HW *et al.* 1998; Persing JA *et al.* 1986; Sugawara Y *et al.* 1998; Tschakaloff A *et al.* 1994) and appears to be effective morphologically, as well as from a functional standpoint (skull expansion, eye and airway protection, see Fig. 10).

The advantages of these methods include (1) generation of new bone, eliminating the need for bone grafting and intermaxillary fixation, (2) avoidance of bone dissection from the dura mater with creation of dead space, thus limiting the risk of infection, transfusion requirements and bone resorption, and (3) simultaneous expansion of surrounding soft tissues, which is of particular relevance when the midface is concerned. The majority, however, still requires cyclic manual activation of the distractor devices and the use of external hardware (skin-penetrating pins or activator screws) which cause scarring and may lead to infection.

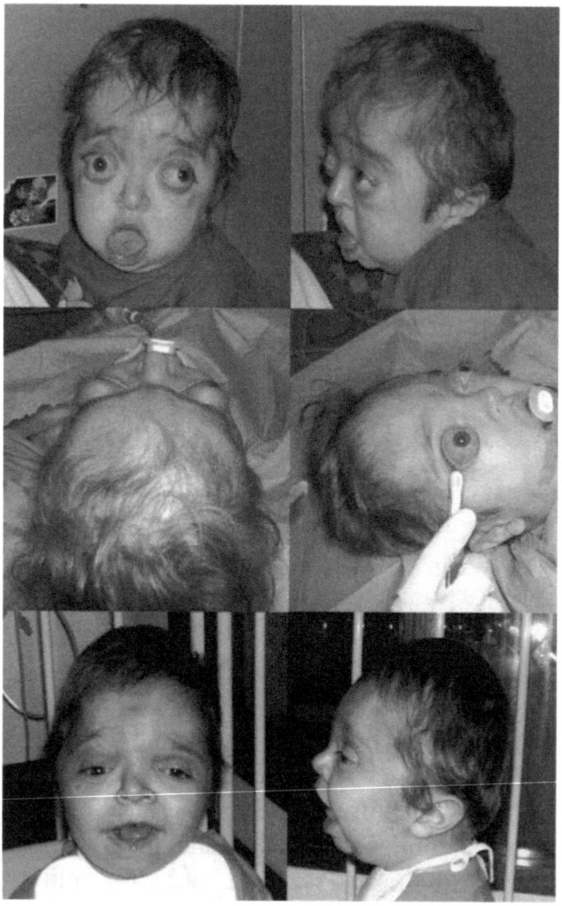

Fig. 10. Pre- and postoperative views of a three-year old patient with Pfeiffer syndrome operated on with a monobloc frontofacial advancement with bone distraction. Preoperatively there was recurrent spontaneous luxation of the eyballs and severe obstructive sleep apnoea. Both problems were corrected

In a recent report, Kremer (Kremer M *et al.* 2001) presented the preliminary experimental results of internal mandibular distraction with a miniaturized mechatronic plate. The device is fully implantable and the movement is generated by the transcutaneous cyclic activation of shape memory alloy wires through the external application of an electrical current.

The spring-mediated cranioplasties were reported by Lauritzen *et al.* (Guimarães-Ferreira J *et al.* 1999; Lauritzen C *et al.* 1998). These are less extensive procedures than "traditional" cranioplasty techniques, and consist of simple linear craniectomies powered by implantable dynamic elements in the form of compressive or expansive (or both) stainless steel springs. With this technique, reshaping of the skull does not rely exclusively on the action of residual brain growth, contrary to what happens in patients operated upon with linear craniectomies alone. These procedures combine the benefits of "conventional" distraction techniques, with those of fully implantable elastic devices that apply continuous force to the bone segments being mobilized, avoiding the risk of prolonged skin penetration by pins or activator screws. Furthermore, the method does not require cyclic activation of the distractor.

The procedure roughly consists of the following steps: (1) Scalp incision followed by limited supraperiostal dissection and exposure of the synostosed suture(s). (2) Strip craniectomies parallel to or excising the synostosed suture(s). (3) Minimal lateral dissection of the dura mater. (4) Insertion of the expander (or compressive) elements which are manufactured at the operating table with sterilized pliers and cutters from medical grade stainless steel wire. (5) Closure of the incision.

During the last five years the technique has been shown to be safe and effective in experimental settings (David LR *et al.* 2002b; Gewalli F *et al.* 2001a) and it has been used clinically in different kinds of craniosynostoses and craniofacial dysostoses, both for skull reshaping (Fig. 11) and midface advancement (Fig. 12), with good results (Guimarães-Ferreira J *et al.* 1999; Lauritzen C *et al.* 1999; Lauritzen C *et al.* 1998).

Intense research on the histological, biomechanical and molecular mechanisms of DOG is currently under way, along with efforts to improve distractor hardware and to optimize distraction protocols (Bouletreau P

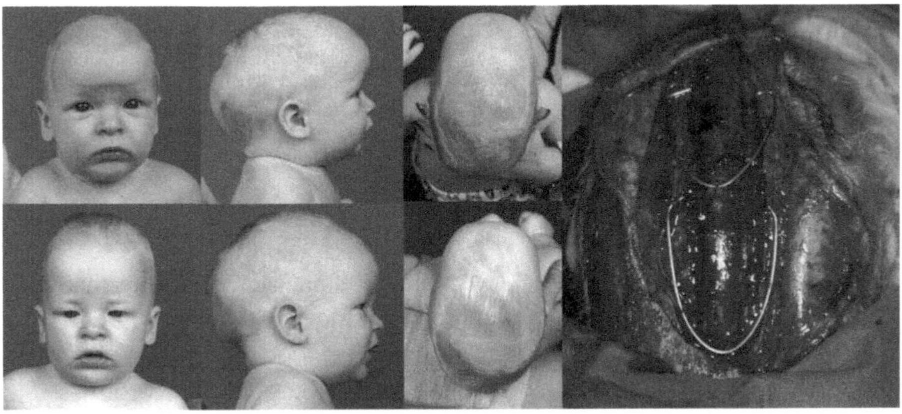

Fig. 11. Pre- and postoperative views of a three-month old patient with sagittal synostosis operated on with a spring-mediated cranioplasty (post-op at one year of age). Drawings: 1- Outline of the scalp "lazy-S" incision (the patient is in the prone position). 2- A linear craniectomy is made with rongeur forceps and two spring expanders are placed between the bone margins to achieve progressive transverse skull expansion

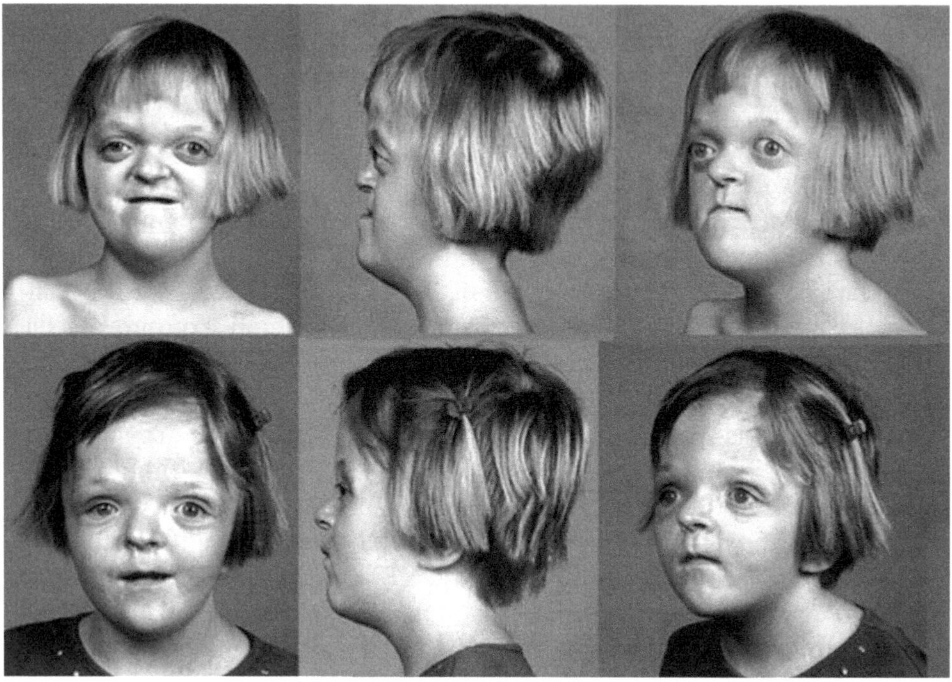

Fig. 12. Pre- and postoperative views of a seven-year old patient with Apert syndrome operated on with a spring-mediated monobloc frontofacial advancement

et al. 2001; Cohen SR 1999; Gosain AK *et al.* 2000; Haluck RS *et al.* 1999; Karp NS *et al.* 1992; Kusnoto B *et al.* 2001; Mehrara BJ *et al.* 1999; Meyer U *et al.* 1999; Nott RL *et al.* 2002; Robinson RC *et al.* 2001; Rowe NM *et al.* 1999; Smith SW *et al.* 1999; Stewart KJ *et al.* 1999; Tavakoli K *et al.* 1999; Tong L *et al.* 2003; Warren SM *et al.* 2001). Biological enhancement of osteogenesis at the distraction zone is also being tried through the local delivery of growth factors (Grayson BH *et al.* 2001).

Resorbable Distractor Devices

While the "traditional" DOG techniques and the newer spring-mediated cranial vault and midface remodelling techniques allow the progressive reshaping of bone and surrounding soft tissues, both technologies still require at least two operations: one for performing the osteotomies and placing the distractors, the second for hardware removal. The search for a totally implantable, self-activated resorbable dynamic element (which would avoid scarring and contamination from transcutaneous pins or activators and spare the patient the discomfort and the risk of additional surgery) is currently under way.

Some hybrid solutions have been developed, such as the one devised by Cohen by substituting the metallic fixation plates of a specific distraction system, with a resorbable fixation mesh (Cohen SR *et al.* 2001). This obviates for the secondary removal of the plates, but not for the removal of the distractor screw and cable drive, which requires a second operative procedure. Burstein devised a partially resorbable distraction device that does not require a second operative procedure (Burstein FD *et al.* 2002). In this variant the metallic drive screw is simply pulled out at the end of the consolidation period during an office visit. External activation is still needed during the distraction phase with both devices, and both imply the presence of skin-penetrating activator screws.

In a recent experimental study a totally implantable and bioresorbable device was successfully used to mobilize calvarial bone in rabbits (Guimarães-Ferreira J *et al.* 2002). The device consisted of bioresorbable polymer plates made of specially processed 70L/30D,L PLA, which were shown to undergo contraction after implantation. Resorbable polymers also may serve as carriers for the delivery of osteogenesis promoting factors (Boyan BD *et al.* 1999; Breitbart AS *et al.* 1998; Breitbart AS *et al.* 1999). At the end of their intended period of mechanical activity, the release of such mediators could help to prevent recurrence of the deformity, by promoting a strong osseous union between the approaching bone flaps. The authors concluded that polymer contractility will likely constitute the basis of a new generation of bioresorbable distractors for utilization in craniofacial surgery.

Minimally Invasive Surgery

Advances in anaesthesia and intensive care have allowed the safe performance of very extensive surgical procedures in infants and small children. It is desirable, however, particularly in the paediatric population, to minimize the surgical aggression and the blood replacement requirements. This explains the recent trend towards minimally invasive surgery typified by the endoscopic approach to craniosynostosis management.

Endoscopic techniques have been used in various fields of head and neck surgery, such as ENT surgery, aesthetic surgery (Marchac D *et al.* 2001; Rubin JP *et al.* 1998), post-traumatic reconstruction (Chen CT *et al.* 1999a; Chen CT *et al.* 1999b; Forrest CR 1999; Hammer B *et al.* 1999), and skull base surgery (Casiano RR *et al.* 2001; Thaler ER *et al.* 1999).

The potential of endoscopic and minimally invasive procedures in the field of craniosynostosis appears to be great, especially if used in the early stages of the deformity (Rubin JP *et al.* 1998; Tutino M *et al.* 1998), or when combined with postoperative external skull-moulding devices (Jimenez DF *et al.* 2002; Stelnicki EJ 2002), spring-assisted skull remodelling

(David LR *et al.* 2002a) and distraction osteogenesis techniques (Hollier L *et al.* 2002; Levine JP *et al.* 1998; Tutino M *et al.* 2000).

In Utero Surgery

It has long been known that up to a certain stage in pregnancy foetal skin wounds heal without the inflammation, contraction, and scarring that follow most wounds in adult individuals (Burrington JD 1971; Goss AN 1977; Idriss FS 1968; Rowlatt U 1979). The cellular and molecular mechanisms underlying this phenomenon have been extensively investigated but remain largely unknown (Longaker MT *et al.* 2001). It appears that the foetal fibroblast, with its highly specific response to growth factors and cytokines, plays a central role in the process (Moulin V *et al.* 2001). Understanding the intimate mechanisms regulating scarless foetal wound healing will open the way for biological manipulation of wound healing postnatally and, hopefully, lead to better outcomes in many fields of surgery. For the moment, most attempts to pharmacologically modulate the process (e.g., topical application of neutralizing antibodies, antisense oligonucleotides, and growth factors) have met with only limited success (Dang C *et al.* 2003).

The prospect of in utero therapy is appealing to craniofacial surgeons because it would allow: (1) a scarless repair, (2) correction of the primary deformity at an early stage (surgically or through pharmacological manipulation), (3) prevention of secondary deformities, and, finally, (4) possibility of giving the parents a normal-appearing child at birth.

Several experimental studies have shown the feasibility and advantages of prenatal surgical correction of craniofacial deformities like cleft lip and palate (Boon L *et al.* 1992; Canady JW *et al.* 1994; Estes JM *et al.* 1992; Hallock GG 1985; Kaban LB *et al.* 1993; Longaker MT *et al.* 1990; Longaker MT *et al.* 1992; Stelnicki EJ *et al.* 1999; Weinzweig J *et al.* 1999) and craniosynostosis (Stelnicki EJ *et al.* 1998b; Stelnicki EJ *et al.* 1998c).

Foetal surgery in humans is currently limited to situations that carry high mortality or morbidity risks if left untreated prior to birth. Advances in foetal monitoring techniques and development of more effective tocolytic agents may further widen its field of application to craniofacial deformities.

Surgical Techniques

There are no well defined technical standards in surgery for craniosynostosis and craniofacial dysostosis. This is due to a combination of factors that include (a) the relatively young age of the discipline, (b) the proliferation of craniofacial centres with different preferences, (c) the lack of detailed studies of surgical outcome, (d) the variable age at which patients are referred to specialized care, and (e) the very nature of the deformities, which are highly variable in intensity. An individualized approach must therefore be adopted, in order to effectively deal with each specific situation.

Sagittal Synostosis

Premature synostosis of the sagittal suture is the most frequent of the isolated craniosynostoses (Cohen MM 2000a; Marchac D *et al.* 1994; Montaut J *et al.* 1977). Lajeunie estimated an incidence of 190 in 1,000,000 live births for this condition (Lajeunie E *et al.* 1996). About three fourths of the cases are sporadic with males more often affected than females (Shillito J, Jr. *et al.* 1968).

Although premature sagittal craniosynostosis has been observed in a number of syndromes (Cohen MM 2000a), most cases represent isolated anomalies.

Secondary to premature synostosis of the sagittal suture the skull becomes elongated and narrowed with a marked increase in head circumference (Fig. 13). This head shape has been termed scaphocephaly, by analogy with a hull of a boat (Greek skafos = craft, vessel). If left untreated the deformity does not tend to improve (Lin KY *et al.* 1999) and may even be progressive in some cases (Barritt J *et al.* 1981).

Early surgery has been recommended to improve head shape and to prevent increased intracranial pressure (Friede H *et al.* 1996; Jane JA *et al.* 2000; Renier D *et al.* 1982). As already mentioned the first attempts to surgically treat the condition date from the late XIXth century and consisted of simple linear craniectomies (Lannelongue M 1890).

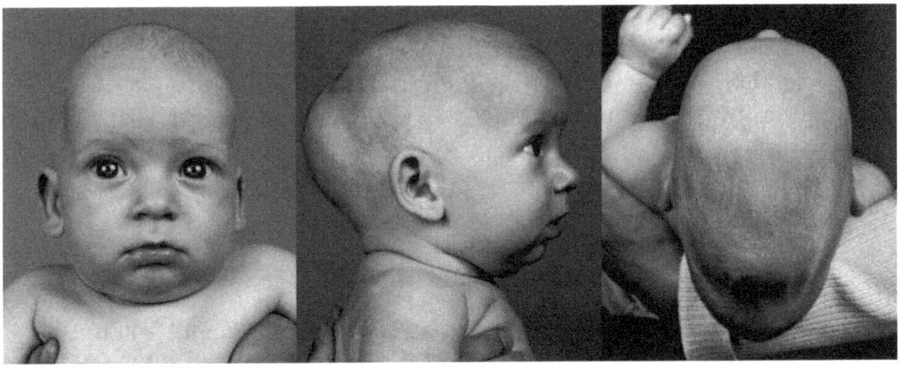

Fig. 13. Patient with sagittal synostosis. Long and narrow skull, absence of parietal bosses, occipital step-like deformity

The inadequacies of the results, due to early reossification of the osteotomy lines, led to modifications of this technique toward more extensive reshaping operations, shown to be more effective than the simpler linear craniectomies (Boop FA *et al.* 1996; Cohen MM 2000a; Joos U *et al.* 1993; Kaiser G 1988; Marsh JL *et al.* 1991; Maugans TA *et al.* 1997; Panchal J *et al.* 1999).

In 1978 John Jane (Jane JA *et al.* 1978) introduced the first dynamic cranioplasty, (defined as a technique that uses force, either compressive or expansive, to produce both immediate and progressive reshaping of the skull. From the Greek dynamikos = of or pertaining to force producing motion). This technique was further developed at our unit (Friede H *et al.* 1996) and it provided the basis for the creation of another dynamic surgical procedure, aimed at the correction of brachycephaly (Gewalli F *et al.* 1999; Guimarães-Ferreira J *et al.* 2001b; Lauritzen C *et al.* 1996).

The spring-mediated cranioplasty recently reported by Lauritzen (Lauritzen C *et al.* 1998) represents a return to less extensive surgery. This is a surgical technique where strip craniectomies are supplemented by springs inserted in the bone gaps. There is little dissection of calvaria from the dura mater (Guimarães-Ferreira J *et al.* 1999).

Although extended linear craniectomies may effectively correct mild to moderate deformities when performed during the first few weeks of life, the outcome is not completely predictable and we prefer not to rely exclusively on residual brain growth to achieve the correction.

Whenever possible, we choose to perform a spring-mediated cranioplasty (Fig. 11) at three months of age. The rationale for this age choice is that in younger patients the bone is not strong enough to withstand the pressure of the springs and in older patients the skull may be too elongated to achieve an adequate correction with transverse expansion alone. When the skull is very long already at three months of age, a longitudinally compressive spring may be added to reduce the occipital bulge after performing a semi-circular osteotomy at the approximate location of the lambdoid suture (Fig. 14).

For patients aged between six months and one year, we perform a modified pi-plasty (Guimarães-Ferreira J *et al.* 2001a) (Fig. 15). In older patients, depending on the severity of the deformity, more extensive calvarial reshaping procedures are usually required (Fig. 16).

Unicoronal Synostosis

The birth prevalence of coronal synostosis (unilateral and bilateral) has been estimated as 94 in 1,000,000 (Lajeunie E *et al.* 1995), unilateral involvement being far more frequent than bilateral. Females are affected more often than males at reported male : female ratios of 1 : 2 (Lajeunie E

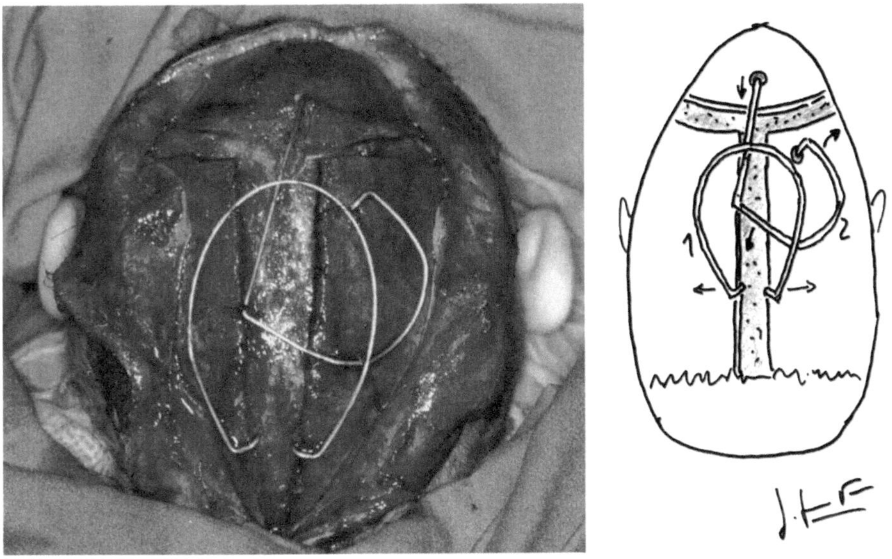

Fig. 14. Spring-mediated cranioplasty with a posterior compressive spring to shorten the skull, for patients whose skulls are already too long to treated with transverse expansion alone. In addition to the longitudinal craniectomy, a semi-circular osteotomy is added at the approximate location of the lambdoid suture. Two springs are then placed, one for transverse expansion *(1)* and another to pull the occiput forwards *(2)*. Both springs are secured to bone by the pressure exerted at their hook-shaped extremities. The arrows in the schema show the direction of the forces generated by the spring expanders

et al. 1995), 2:3 (Shillito J, Jr. *et al.* 1968) and 3:4 (Hunter AG *et al.* 1977).

Unilateral coronal synostosis results in a complex three-dimensional, asymmetric calvarial, orbital and facial deformity called anterior plagiocephaly (Fig. 17). Ipsilateral to the fused suture, growth restriction causes the frontal bone to be flatter, the upper orbital rim to be higher and posterior to the normal position and the anterior cranial fossa to be shorter. Compensational growth at the contralateral coronal and spheno-frontal sutures produces an excessive convexity in the forehead and lowers the supraorbital rim. There is frequently an apparent lateral bulge of the anterior portion of the temporal fossa on the affected side and the floor of the middle cranial fossa is pushed down to a lower level than that of the non-fused side. Facial scoliosis of variable intensity is also present, and the mid point of the chin deviates towards the non-fused side.

Although the lower-face deformity seldom requires correction, the orbitocranial components need to be addressed surgically. Simple excision of the area of fusion is not enough to correct the constellation of deform-

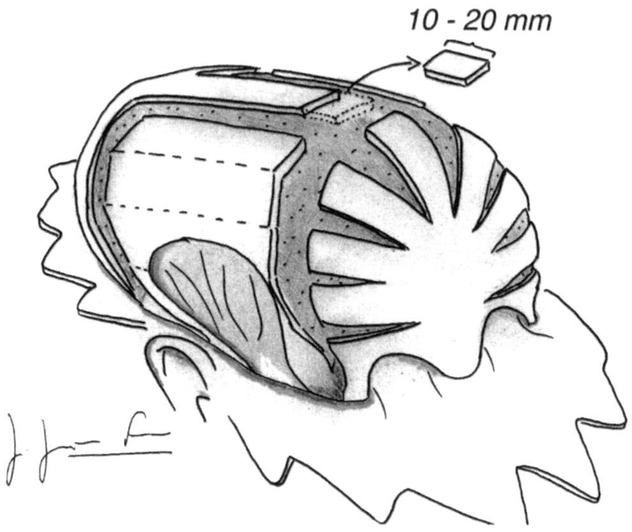

Fig. 15. The modified pi-plasty procedure is based on the procedure described by Jane in 1978. The technique includes the intraoperative use of longitudinal compressive force in order to dynamically shorten the skull. The modifications consist of the following: 1. Access through a coronal zigzag incision to reduce scar visibility. 2. Addition of vertical posterior parietal osteotomies, thus creating symmetric, approximately rectangular squamo-parietal flaps. The inferior third of the osteotomy is directed obliquely downwards and anteriorly, towards the squamous suture. The flaps are then outfractured at the level of this suture and three parallel longitudinal scores are carved in the inner table. These are used to guide the production of "green-stick" fractures in order to increase the convexity of the parietal regions. Since the attachments of the temporal muscles and periosteum are left undisturbed, they will bridge the fractures. 3. Reshaping of the forehead flap with radial osteotomies. 4. Active shortening of the skull to a greater extent than in previously reported series. In most cases this was achieved by wiring the forehead bone flap to the sagittal strip of bone, after excising the foremost 10 to 20 mm of the latter

ities that compose the clinical picture and treatment has evolved through unilateral bone surgery to the bilateral orbitocranial recontouring procedures widely used today.

To achieve a satisfactory result with surgical correction several points must be addressed (Fig. 18). First, the supraorbital bar and its tenon extensions must be removed and changed into a symmetrical piece. Sometimes this implies adding a small interpositional bone graft to increase the horizontal dimension of the supraorbital rim on the synostosed side. Repositioning of this segment has to be done in such a way as to symmetrize the orbital vertical and horizontal dimensions, i.e., advancing and placing it somewhat lower on the fused side and somewhat higher on the contralateral side. To adequately judge the amount of correction needed on

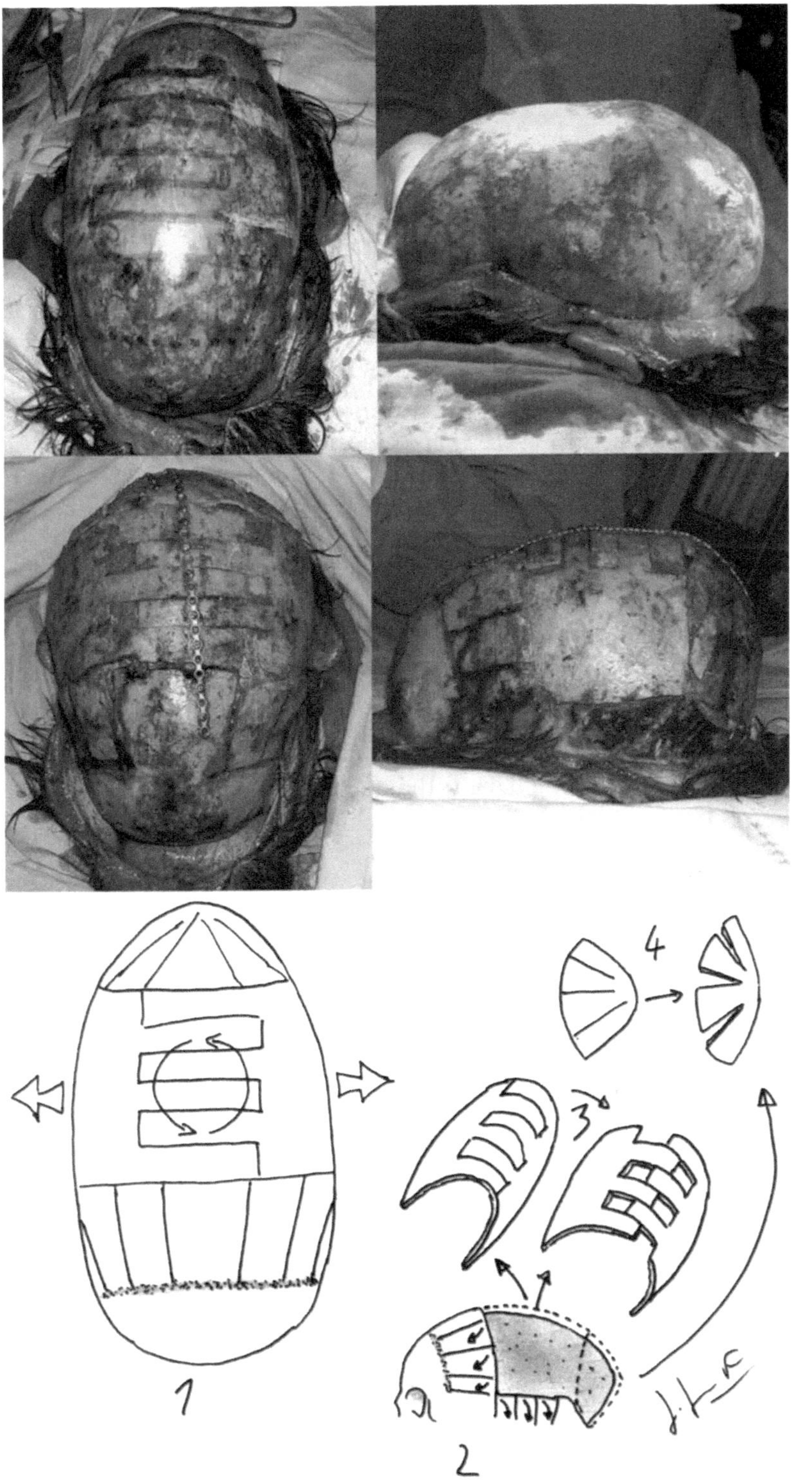

Fig. 16. Total calvarial cranioplasty for older children with sagittal synostosis

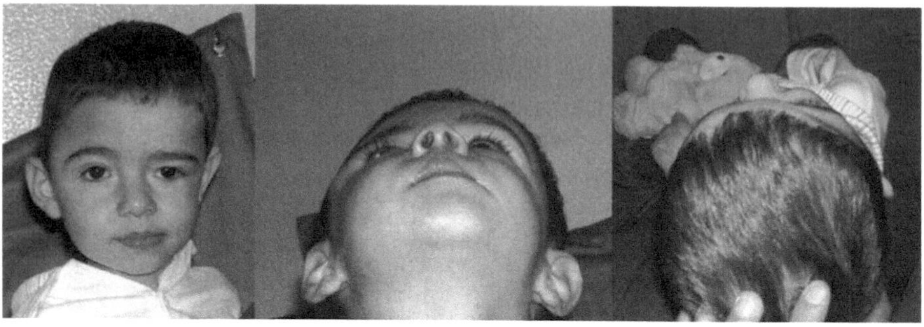

Fig. 17. Right unicoronal synostosis (anterior plagiocephaly). Ipsilateral to the fused suture: forehead flattening, higher and posteriorly placed upper orbital rim, lateral bulge of the anterior portion of the temporal fossa. Contralateral to the fused suture: forehead bulge, lowered superior orbital rim. Facial scoliosis with mid point of the chin deviating towards the non-fused side

each orbit it is imperative that the nose, infraorbital rims and malar eminences be accessible to direct intraoperative inspection and palpation.

The bulge in the ipsilateral temporal fossa must be corrected if it is marked. This can be achieved by removing the deformed bone as a circular piece and repositioning it with the inner, concave, table outwards or, if the overall convexity is too marked it can be flattened with radial osteotomies.

To obtain a nicely contoured and symmetrical forehead we prefer to choose a piece of bone from the temporo-parietal region and transplant it as a single unit to the appropriate position, over the remodelled supraorbital bar.

After the fronto-supraorbital complex is remodelled and fixed in the desired position with wires and/or microplates, the temporo-parietal regions are adapted, as needed, with "barrel-stave" osteotomies and any remaining gaps are filled with available bone fragments.

Fig. 18. Unicoronal synostosis – surgical correction in an eight month-old girl. Above: preoperative. Middle: intraoperative views. Bottom: postoperative. Drawings: *1* Placement of osteotomies. *2* The supraorbital bar and its tenon extensions are removed and changed into a symmetrical piece with the aid of kerfs in the inner table made with a rotating instrument. *3* The completed operation after repositioning of the supraorbital bar, which is advanced and lowered on the fused side and, if necessary, placed at a higher position on the contralateral side. The resulting gaps are filled with the remaining bone fragments which are fixed with wires or plates. The bulge of the anterior portion of the temporal fossa on the side of fusion may be corrected with "barrel-stave" osteotomies which are green-stick fractured inwards

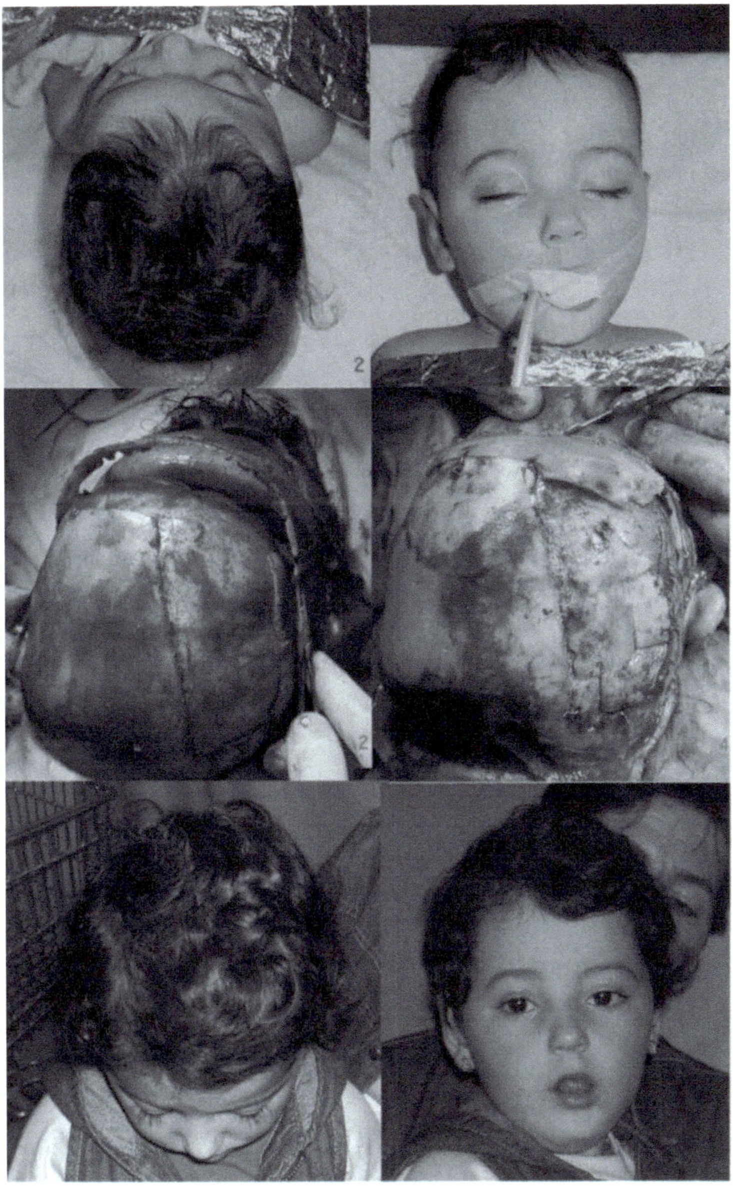

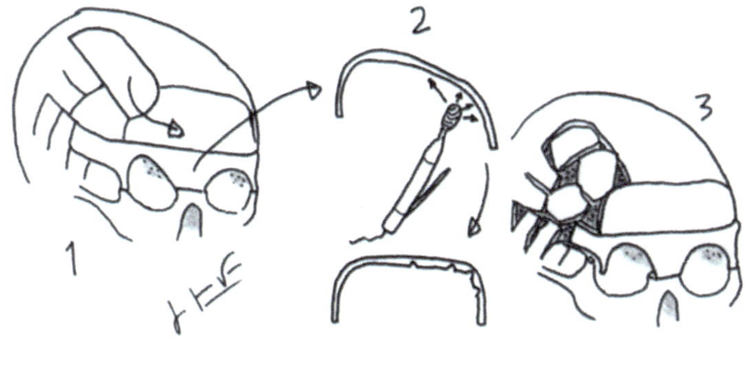
2
1
3

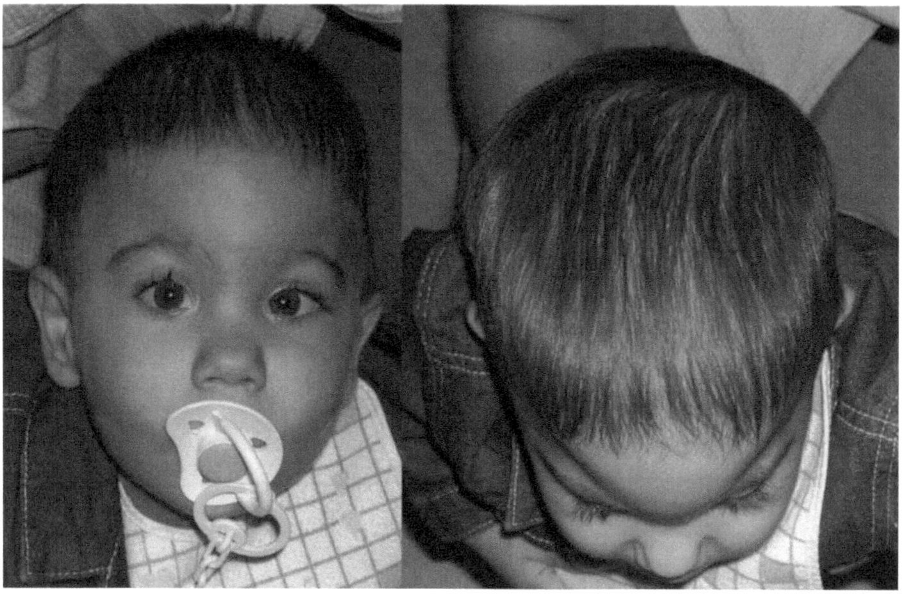

Fig. 19. Metopic synostosis. Median forehead ridge, mild orbital hypotelorism, narrow bitemporal distance, triangular shape of the anterior portion of the skull when viewed from above

Metopic Synostosis

Metopic synostosis has an estimated birth prevalence of 67 in 1,000,000 and it affects males 3.3 times more often than females (Lajeunie E *et al.* 1998).

The clinical spectrum of the deformity is quite wide, ranging from an isolated forehead median ridge to the complete trigonocephaly picture characterized by a triangular shaped forehead with a protruding median ridge, a reduced bi-temporal width and a mild to moderate orbital hypotelorism (Fig. 19).

Patients with isolated prominent metopic ridging may be treated effectively by simply reducing the prominence with a rotating burr, either through a coronal incision or an endoscopic approach.

When the forehead deformity is mild to moderate and hypotelorism correction is not judged necessary, we perform what we call the dynamic cranioplasty for trigonocephaly (DCT) (Fig. 20). This technique is somewhat reminiscent of the pi-plasty procedure, since intra-operative longitudinally compressive force is also applied to the bregma area. After a standard coronal zig-zag incision, the forehead, parietal and temporal regions are approached subperiosteally and a 10 mm wide craniectomy is placed parallel to and slightly in front of the coronal suture. A series of four ra-

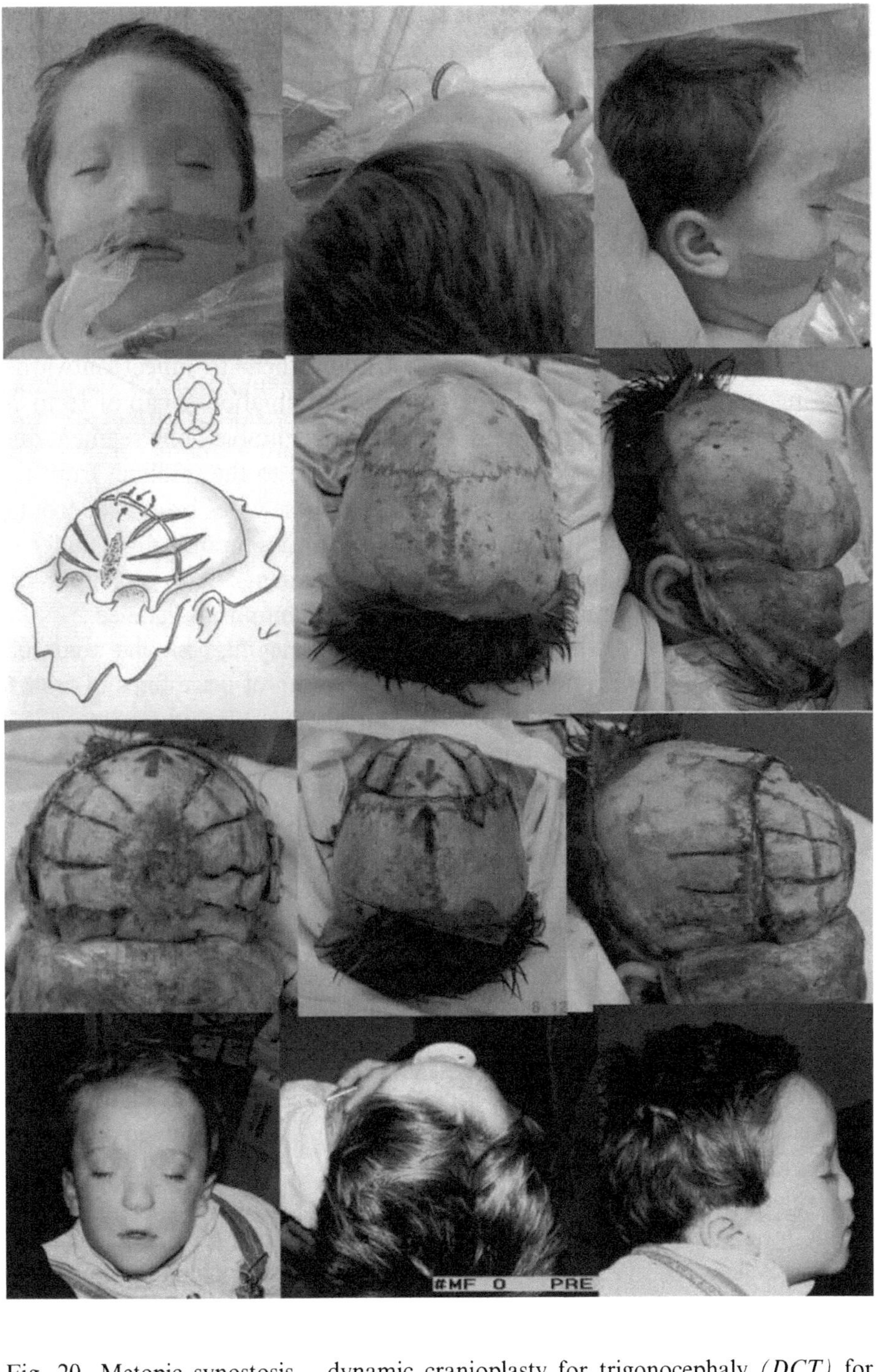

Fig. 20. Metopic synostosis – dynamic cranioplasty for trigonocephaly *(DCT)* for mild to moderate deformities with no need for hypotelorism correction. This eleven month-old patient has a diagnosis of Opitz C-syndrome. Above: preoperative. Middle: intraoperative views and drawing. Bottom: Postoperative. See text for details

dial osteotomies are then made with the craniotome on each side of the frontal bone, converging to the glabella area, the lowest of which is horizontal, the uppermost being at an angle of approximately 45 degrees. The craniotome is again used to cut three or four horizontal "barrel-stave"-type osteotomies in the parieto-temporal regions, extending from the coronal osteotomy backwards to the area of maximum calvarial width. A shaping burr is used to reduce the metopic ridge and to carve a grove in the outer table of the frontal radial segments, at mid distance between their extremities and perpendicular to their longer axis. These two steps will lessen the rigidity of the frontal bone and allow it to curve when pressure is applied. In older patients a similar groove is then cut at the posterior base of the "barrel-staves", to allow them to inflect outwards, creating a wider bi-temporal diameter. A longitudinal bone gap of 15 to 20 mm is created at the bregma area by resecting a suitable bone segment and two metal wires are passed across this gap, close to the midline, from the frontal to the parietal bones. Slow twisting of these wires brings the frontal and parietal bones together and causes the fronto-temporal dura mater to bulge laterally, pushing the osteotomized segments. This reshapes the forehead, making it rounder, and widens the bi-temporal distance.

The advantages of this procedure over the ones that involve remodelling of the orbital rims and removal and reshaping of large flaps of frontal bone include (1) a greater simplicity and shorter operative time, (2) a smaller compromise to the bone vascular supply and (3) the absence of postoperative extra-dural dead space.

When the deformity is moderate to severe and orbital hypotelorism has to be addressed, the entire frontal bone must be removed together with the supraorbital bar and its tenon extensions to allow for complete reshaping of the fronto-supraorbital regions (Fig. 21).

Unilambdoid Synostosis

This is the rarest of the isolated craniosynostoses, corresponding to about 1% of cases in most published series. Morphologically, it is characterized by ipsilateral flattening of the occiput and an inferior and posterior displacement of the ipsilateral ear. There is some degree of forehead bossing contralateral to the fused suture, which, together with the ipsilateral occipital flattening gives the head a trapezoidal contour when viewed from above. When observed from behind, the head has a parallelogram shape due to an upward and outward compensatory bossing in the contralateral parietal region, coupled with an occipito-mastoid outward and downward bulge on the affected side. These features usually allow for a straightforward clinical distinction from the far more common deformational posterior plagiocephaly.

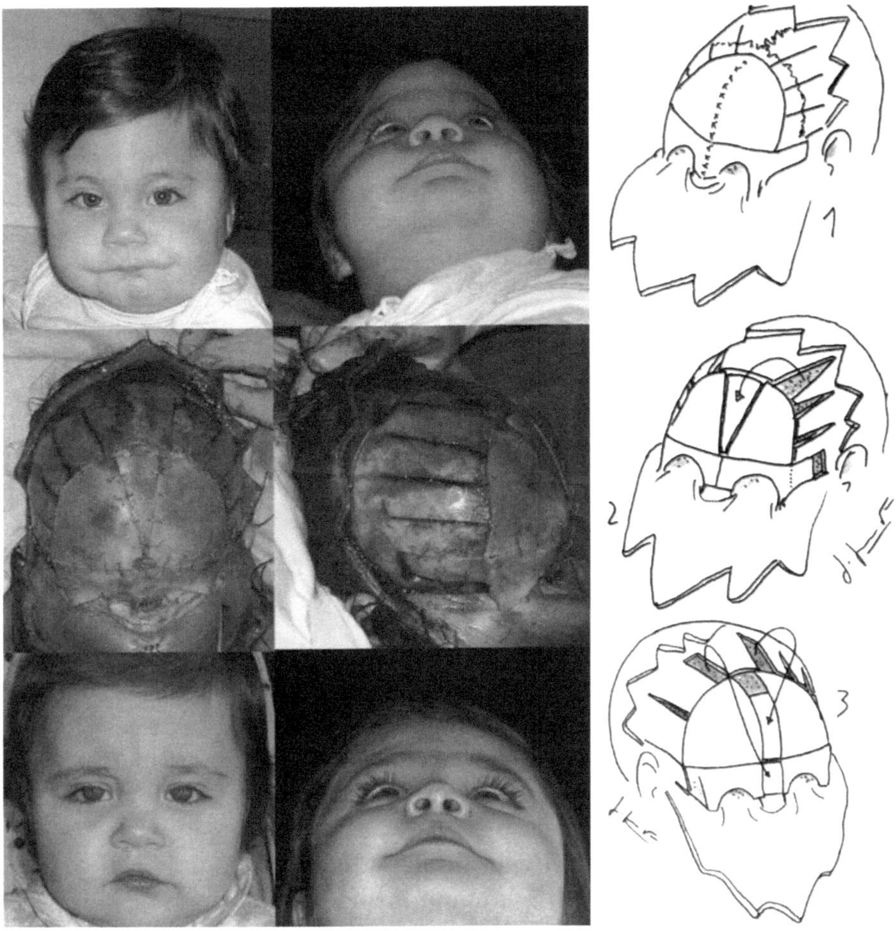

Fig. 21. Metopic synostosis – total fronto-supraorbital cranioplasty for more marked deformities. Drawings: *1* Design of the osteotomies. *2* The completed operation. *3* When there is the need for a greater increase in the interorbital distance due to the existence of a more marked hypotelorbitism, this is achieved with the interposition of a bone graft between the two halves of the supraorbital bar

When detected early, a wide strip craniectomy excising the affected suture may be all that is needed to prevent the possible consequences of increased regional intracranial pressure as well as the deformity of the occipital and, to a lesser extent, frontal regions. For more advanced deformities, complete reshaping of a parieto-occipital bone flap with radial osteotomies in combination with "barrel-stave" osteotomies will be necessary.

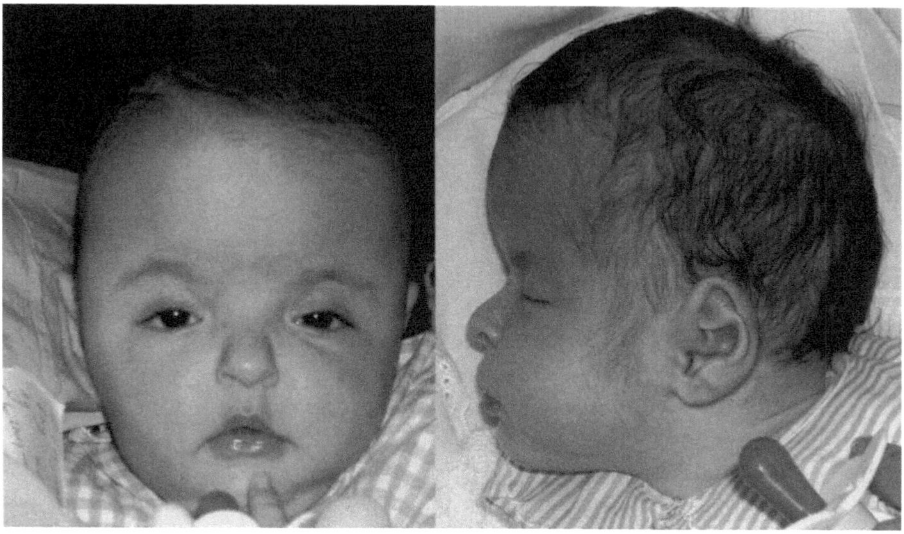

Fig. 22. Bicoronal synostosis (brachycephaly). Short, broad and high skull

Bicoronal Synostosis

Bilateral synostosis of the coronal suture results in an abnormally short, wide and high skull termed brachycephalic or brachyturricephalic, each of these dimensions being affected to a variable extent in different cases (Fig. 22).

The cranial vault deformity has been addressed by a variety of surgical techniques (Fig. 23), the most common being orbito-frontal advancement

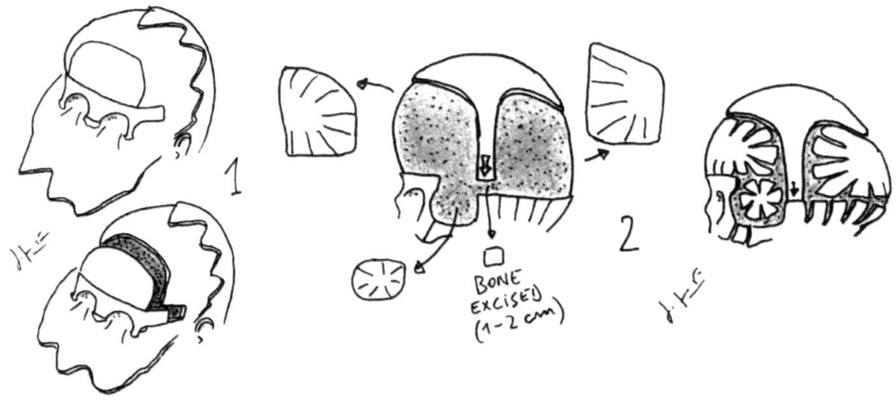

Fig. 23. Bicoronal synostosis – surgical techniques. *1* Forehead advancement. *2* "Barrel-stave" osteotomy

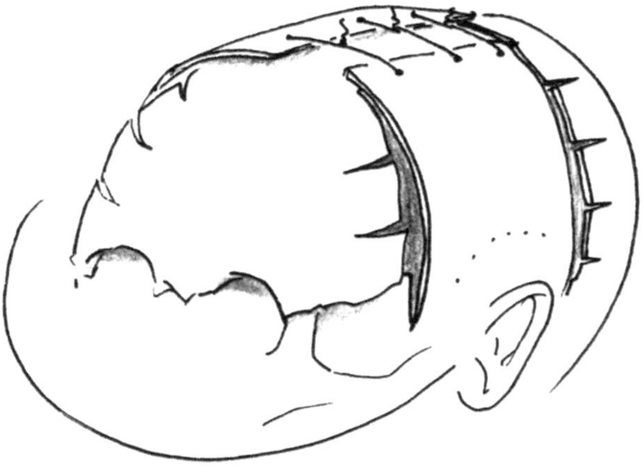

Fig. 24. Bicoronal synostosis – dynamic cranioplasty for brachycephaly *(DCB)*. The procedure consists of the creation and reshaping of a forehead bone flap, which includes the upper portion of the orbital rims, hinged on a superiorly located strip of bone, partially fractured across the upper midline. A caudally hinged occipital flap is also created and reshaped with radial osteotomies. Two squamo-parietal bone flaps are developed and approximated over the midline by slow twisting of transverse wires

with fixation (Tessier P 1971b), the floating forehead (Marchac D *et al.* 1979) and the barrel-stave osteotomy (Persing JA *et al.* 1987). In 1996, a technique named the dynamic cranioplasty for brachycephaly (DCB) was reported by Lauritzen (Lauritzen C *et al.* 1996), and it became the preferred technique at our Units, since it was shown to produce consistently good morphological results both in isolated and in syndromic brachyturricephaly (Gewalli F *et al.* 2001b; Guimarães-Ferreira J *et al.* 2001b). The procedure (Fig. 24) consists of the creation and reshaping of a forehead bone flap, including the upper portion of the orbital rims, hinged on a superiorly located strip of bone, partially fractured across the upper midline. A caudally hinged occipital flap is also created and reshaped with radial osteotomies. Two squamo-parietal bone flaps are developed and approximated over the midline strip of bone by gentle twisting of transverse wires. In the DCB procedure all three dimensions of the calvarium are addressed. Skull width is directly reduced by the use of intraoperative inter-parietal wire compression. The advancement of the squamo-parietal flaps over the convex midline strip of bone, applies downward pressure on the later, leading to a reduction in skull height. An increase in skull length is obtained by the action of the expanding brain on the released and reshaped fronto-supraorbital and occipital flaps. Since the upper portion of the orbital rims and the supraorbital bar region are included in the forehead bone flap, these structures are advanced as well.

Fig. 25. Father and son with Saethre-Chotzen syndrome. Unicoronal synostosis with plagiocephaly, ptosis of the eyelid, low frontal hairline

Saethre-Chotzen Syndrome

Saethre-Chotzen syndrome was first described in 1931 and it consists of a variable pattern of malformations that may include craniosynostosis, most frequently unicoronal, eyelid ptosis, low-set anterior hairline, brachydactily (of the first fingers and toes) and incomplete soft tissue syndactily (usually between the second and third fingers and toes). Most patients have a mutation in the TWIST gene located on the short arm of chromosome 7 and the inheritance pattern is autosomal dominant (Fig. 25).

Although the craniofacial features of Saethre-Chotzen patients tend to be more marked than in non-syndromic unicoronal synostosis, the same general surgical techniques are applicable.

Crouzon and Pfeiffer Syndromes

The birth prevalence of Crouzon syndrome is 15 to 16 in 1,000,000 live births (Cohen MM, Jr. *et al.* 1992a), while that of Pfeiffer syndrome is unknown. Both syndromes share many craniofacial features, which include craniosynostosis of multiple calvarial and cranial base sutures, midface hypoplasia, orbital hypertelorism and proptosis in variable degrees

(Fig. 26). Pfeiffer syndrome is further defined by the presence of hands and feet anomalies in the form of broad, sometimes deviated thumbs and great toes, and variable syndactyly. More than 30 different mutations have been identified so far as causes for each of these syndromes, most of them located in the fibroblast growth factor receptor 2 genes. For the purposes of craniofacial reconstruction the two syndromes present much the same problems and challenges.

Pure brachycephaly is very rare in these patients, in marked opposition to what we see in Apert syndrome. Most patients have small, pansynostotic skulls, typically pointed at the bregma area, showing generalized convolutional markings in radiographs. A rare Crouzon patient may present with a scaphocephalic calvarium, if the sagittal suture is affected some time before pansynostosis appear (Fig. 26). The cloverleaf skull configuration may, very rarely, be part of any of these two syndromes.

Patients must be closely monitored for signs of increased intracranial pressure which, if detected, must be aggressively managed with shunt placement, skull expansion or both when indicated, in order to prevent visual loss. If the constriction is judged to be anterior, the forehead will be advanced. If it is the transverse dimension that is lacking, a transverse skull expansion (spring-mediated for patients up to one year of age) will be performed. For posterior fossa and parieto-occiput crowding, a posterior skull expansion will be elected. Repeat skull expansion procedures may be needed at any time if signs of increased intracranial pressure attributable to insufficient volume appear.

Eye protection may require an early forehead advancement or even midface advancement surgery. Correction of the midface deformity is ideally carried out after completion of facial growth, in adolescent patients. This allows the definitive correction in a single surgical procedure (usually a LeFort III or monobloc advancement combined with a LeFort I to adapt the occlusion) and avoids the obligatory relapse into a class III malocclusion caused by physiological mandibular growth that supervenes when midface surgery is carried out at an earlier age. Unfortunately, one is frequently forced to perform a midface advancement during infancy or childhood as an alternative to permanent tracheotomy, due to progressive respiratory obstruction which is not adequately compensated by tonsillectomy and continuous positive airway pressure appliances. In these cases we prefer to use spring- or distractor-mediated procedures, since they allow for greater degrees of overcorrection than the traditional techniques.

Apert Syndrome

This malformative disorder was described in 1906 by E. Apert (Apert E 1906), under the name acrocephalosyndactylie. It is characterized by the

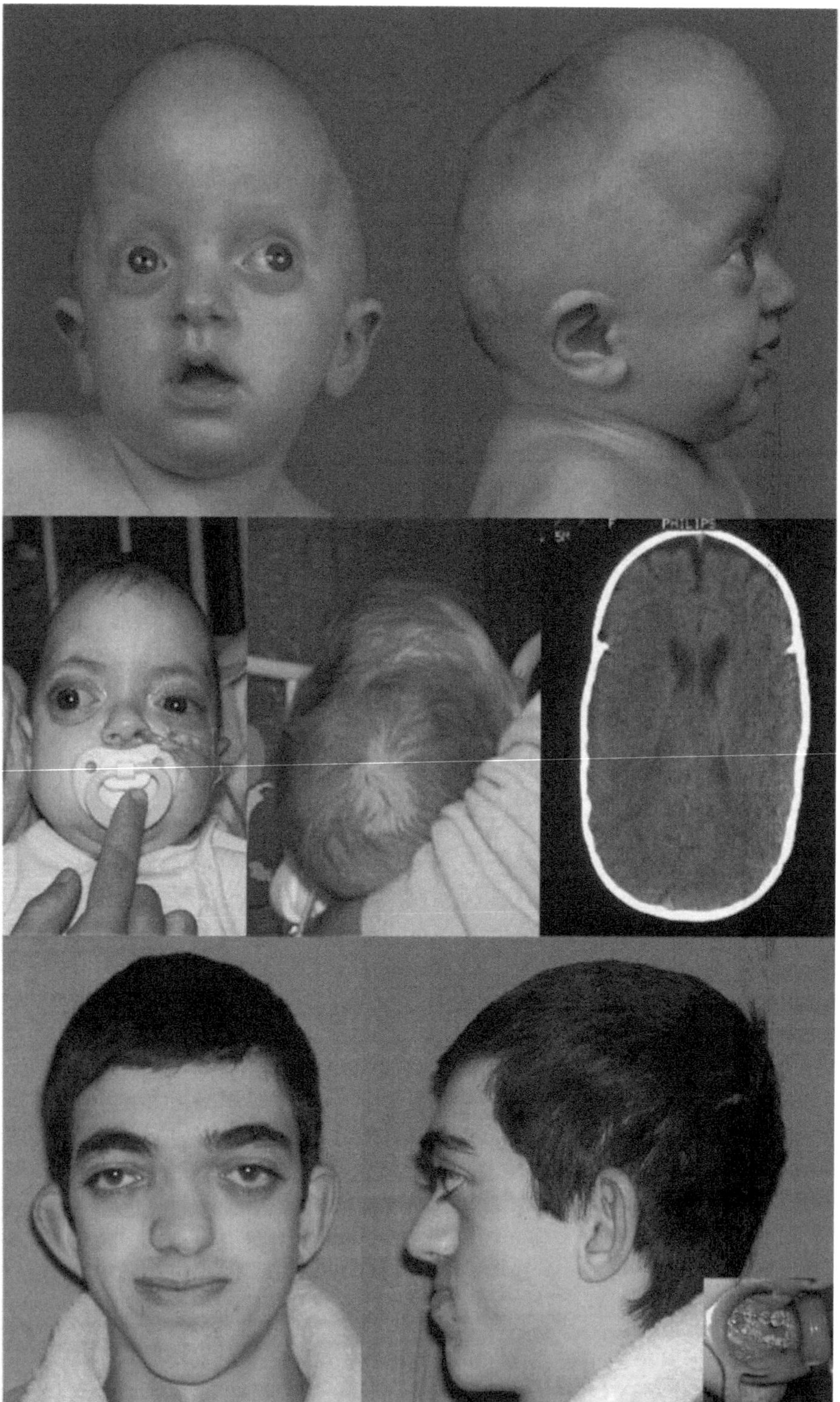
PHILIPS

combination of craniosynostosis (most frequently in the form of bicoronal and cranial base synostosis leading to brachycephaly, midface retrusion, typical nose deformity, mild to moderate exophthalmus and orbital hypertelorism), symmetrical soft-tissue and bone syndactyly of the hands and feet, among many other possible skeletal and visceral malformations. Two mutations in the fibroblast growth factor receptor 2 gene have been shown to cause the disease (Wilkie AO *et al.* 1995b). Birth prevalences of 15.5 (Cohen MM, Jr. *et al.* 1992b) and 12.4 (Tolarova MM *et al.* 1997) cases per million live births have been calculated for this condition.

As in Crouzon and Pfeiffer syndromes, surgical treatment must be staged flexibly, so that emerging problems are dealt with in a timely and appropriate fashion. We usually perform a DCB between six months and one year of age to correct brachyturricephaly and provide some degree of eye protection. This will be followed by a spring- or distractor-mediated midface advancement during infancy or childhood, if so dictated by the need to enlarge the airway and/or protect the eyes. The choice of the procedure (LeFort III, modified monobloc or facial bipartition) will be a function of the specific deformity: if it is only the midface that needs advancement, a LeFort III osteotomy will be made. When there is residual forehead retrusion, a modified monobloc may be indicated. Any of these two procedures can be combined with a facial bipartition to correct orbital hypertelorism and the maxillary occlusal plane dysmorphology. The final correction of the skull and face deformities is usually deferred until adolescence, when facial growth is completed, and typically will include a LeFort III advancement and a LeFort I osteotomy to adjust the occlusion, possibly coupled with a facial bipartition.

Acknowledgements

This work was supported in part by scholarships granted to the author by Forskningsstiftelsen Livskraft (Sweden), Göteborg University (LUA E06101), Fundação Calouste Gulbenkian (Portugal) and Subprograma Ciência e Tecnologia do 2° Quadro Comunitário de Apoio (PRAXIS XXI/BD/21406/99) (Portugal).

Fig. 26. Crouzon syndrome. Above: Three month-old boy with turricephaly and anterior calvarial bulge related to pansynostosis. There is moderate exophthalmus due to the combined effect of forehead and midface retrusion. Middle: One month old girl with Crouzon syndrome presenting with a scaphocephalic shape. Below: young adult operated upon with a forehead advancement in infancy, showing exophthalmus, mild orbital hipertelorism, marked midface hypoplasia and class III malocclusion

References

1. Alonso N, Munhoz AM, Fogaca W, Ferreira MC (1998) Midfacial advancement by bone distraction for treatment of craniofacial deformities. J Craniofac Surg 9: 114–118; discussion 119–122
2. Anderl H, Zur Nedden D, Muhlbauer W, Twerdy K, Zanon E, Wicke K, Knapp R (1994) CT-guided stereolithography as a new tool in craniofacial surgery. Br J Plast Surg 47: 60–64
3. Anderson FM (1981) Treatment of coronal and metopic synostosis: 107 cases. Neurosurgery 8: 143–149
4. Angle CR, McIntire MS, Moore RC (1967) Cloverleaf skull: Kleeblattschadel-deformity syndrome. Am J Dis Child 114: 198–202
5. Apert E (1906) L'acrocephalosyndactylie. Bull Soc Med 23: 1310–1313
6. Barritt J, Brooksbank M, Simpson D (1981) Scaphocephaly: aesthetic and psychosocial considerations. Dev Med Child Neurol 23: 183–191
7. Bergsma EJ, Rozema FR, Bos RR, de Bruijn WC (1993) Foreign body reactions to resorbable poly(L-lactide) bone plates and screws used for the fixation of unstable zygomatic fractures. J Oral Maxillofac Surg 51: 666–670
8. Bergsma JE, de Bruijn WC, Rozema FR, Bos RR, Boering G (1995) Late degradation tissue response to poly(L-lactide) bone plates and screws. Biomaterials 16: 25–31
9. Boon L, Manicourt D, Marbaix E, Vandenabeele M, Vanwijck R (1992) A comparative analysis of healing of surgical cleft lip corrected in utero and neonates. Plast Reconstr Surg 89: 11–17; discussion 18–20
10. Boop FA, Shewmake K, Chadduck WM (1996) Synostectomy versus complex cranioplasty for the treatment of sagittal synostosis. Childs Nerv Syst 12: 371–375
11. Bos RR, Rozema FR, Boering G, Nijenhuis AJ, Pennings AJ, Verwey AB, Nieuwenhuis P, Jansen HW (1991) Degradation of and tissue reaction to biodegradable poly(L-lactide) for use as internal fixation of fractures: a study in rats. Biomaterials 12: 32–36
12. Bouletreau P, Warren S, Paccione M, Greenwald J, Nijher N, Kummer F, LeGeros R, Grayson B, Loboa E, Carter D, Longaker M (2001) New developments in craniofacial distraction. In: Arnaud E, Diner P (eds) III International Congress on Cranial and Facial Bone Distraction Processes Monduzzi, Paris, France
13. Boyan BD, Lohmann CH, Romero J, Schwartz Z (1999) Bone and cartilage tissue engineering. Clin Plast Surg 26: 629–645, ix
14. Breitbart AS, Grande DA, Kessler R, Ryaby JT, Fitzsimmons RJ, Grant RT (1998) Tissue engineered bone repair of calvarial defects using cultured periosteal cells. Plast Reconstr Surg 101: 567–574; discussion 575–576
15. Breitbart AS, Grande DA, Mason JM, Barcia M, James T, Grant RT (1999) Gene-enhanced tissue engineering: applications for bone healing using cultured periosteal cells transduced retrovirally with the BMP-7 gene. Ann Plast Surg 42: 488–495

16. Bresnick S, Schendel S (1995) Crouzon's disease correlates with low fibroblastic growth factor receptor activity in stenosed cranial sutures. J Craniofac Surg 6: 245–248
17. Bresnick S, Schendel S (1998) Apert's syndrome correlates with low fibroblast growth factor receptor activity in stenosed cranial sutures. J Craniofac Surg 9: 92–95
18. Britto JA, Evans RD, Hayward RD, Jones BM (1998) Maxillary distraction osteogenesis in Pfeiffer's syndrome: urgent ocular protection by gradual midfacial skeletal advancement. Br J Plast Surg 51: 343–349
19. Bruce DA (1996) Consensus: craniofacial synostoses. Apert and Crouzon syndromes. Childs Nerv Syst 12: 734–736
20. Burrington JD (1971) Wound healing in the fetal lamb. J Pediatr Surg 6: 523–528
21. Burstein FD, Williams JK, Hudgins R, Graham L, Teague G, Paschal M, Simms C (2002) Single-stage craniofacial distraction using resorbable devices. J Craniofac Surg 13: 776–782
22. Byrd HS, Hobar PC (1996) Alloplastic nasal and perialar augmentation. Clin Plast Surg 23: 315–326
23. Byrd HS, Hobar PC, Shewmake K (1993) Augmentation of the craniofacial skeleton with porous hydroxyapatite granules. Plast Reconstr Surg 91: 15–22; discussion 23–26
24. Camfield PR, Camfield CS, Cohen MM (2000) Neurologic aspects of craniosynostosis. In: Cohen MM, MacLean R (eds) Craniosynostosis: diagnosis, evaluation, and management. Oxford University Press, New York, pp 177–183
25. Canady JW, Landas SK, Morris H, Thompson SA (1994) In utero cleft palate repair in the ovine model. Cleft Palate Craniofac J 31: 37–44
26. Casiano RR, Numa WA, Falquez AM (2001) Endoscopic resection of esthesioneuroblastoma. Am J Rhinol 15: 271–279
27. Chang PS, Parker TH, Patrick CW Jr, Miller MJ (2003) The accuracy of stereolithography in planning craniofacial bone replacement. J Craniofac Surg 14: 164–170
28. Chen CT, Chen YR, Tung TC, Lai JP, Rohrich RJ (1999a) Endoscopically assisted reconstruction of orbital medial wall fractures. Plast Reconstr Surg 103: 714–720; quiz 721
29. Chen CT, Lai JP, Tung TC, Chen YR (1999b) Endoscopically assisted mandibular subcondylar fracture repair. Plast Reconstr Surg 103: 60–65
30. Chin M, Toth BA (1997) Le Fort III advancement with gradual distraction using internal devices. Plast Reconstr Surg 100: 819–830; discussion 831–832
31. Cinalli G, Sainte-Rose C, Kollar EM, Zerah M, Brunelle F, Chumas P, Arnaud E, Marchac D, Pierre-Kahn A, Renier D (1998) Hydrocephalus and craniosynostosis. J Neurosurg 88: 209–214
32. Codivilla A (1905) On the means of lengthening, in the lower limbs, the muscles and tissues which are shortened through deformity. Am J Orthop Surg 2: 353

33. Cohen DM, Green JG, Miller J, Gorlin RJ, Reed JA (1987) Acrocephalopolysyndactyly type II—Carpenter syndrome: clinical spectrum and an attempt at unification with Goodman and Summit syndromes. Am J Med Genet 28: 311–324
34. Cohen MM (1993) Pfeiffer syndrome update, clinical subtypes, and guidelines for differential diagnosis. Am J Med Genet 45: 300–307
35. Cohen MM (2000a) Craniosynostosis: diagnosis, evaluation, and management, 2nd edn. Oxford University Press, New York
36. Cohen MM (2000b) History, terminology, and classification of craniosynostosis. In: Cohen MM, MacLean R (eds) Craniosynostosis: diagnosis, evaluation, and management. Oxford University Press, New York, pp 103–111
37. Cohen MM (2000c) Sutural pathology. In: Cohen MM, MacLean R (eds) Craniosynostosis: diagnosis, evaluation, and management. Oxford University Press, New York, pp 53
38. Cohen MM, Jr, Kreiborg S (1992a) Birth prevalence studies of the Crouzon syndrome: comparison of direct and indirect methods. Clin Genet 41: 12–15
39. Cohen MM, Jr, Kreiborg S, Lammer EJ, Cordero JF, Mastroiacovo P, Erickson JD, Roeper P, Martinez-Frias ML (1992b) Birth prevalence study of the Apert syndrome [see comments]. Am J Med Genet 42: 655–659
40. Cohen MM, MacLean RE (2000) Anatomic, genetic, nosologic, diagnostic and psychosocial considerations. In: Cohen MM, MacLean R (eds) Craniosynostosis: diagnosis, evaluation, and management. Oxford University Press, New York, pp 119
41. Cohen SR (1999) Craniofacial distraction with a modular internal distraction system: evolution of design and surgical techniques. Plast Reconstr Surg 103: 1592–1607
42. Cohen SR, Boydston W, Burstein FD, Hudgins R (1998) Monobloc distraction osteogenesis during infancy: report of a case and presentation of a new device. Plast Reconstr Surg 101: 1919–1924
43. Cohen SR, Holmes RE (2001) Internal Le Fort III distraction with biodegradable devices. J Craniofac Surg 12: 264–272
44. Collmann H, Sorensen N, Krauss J, Muhling J (1988) Hydrocephalus in craniosynostosis. Childs Nerv Syst 4: 279–285
45. Connelly SM, Smith RJ (1998) Effects of rigid plate fixation and subsequent removal on craniofacial growth in rabbits. Arch Otolaryngol Head Neck Surg 124: 444–447
46. Converse JM, Telsey D (1971) The tripartite osteotomy of the mid-face for orbital expansion and correction of the deformity in craniostenosis. Br J Plast Surg 24: 365–374
47. Coppoletta JM, Wolbach SB (1933) Body lengths and organ weights of infants and children. A study of the body length and normal weights of the more important vital organs of the body between birth and twelve years of age. Am J Pathol 9: 55–70
48. Cushing H (1908) Surgery of the head. In: Keen WW, Da Costa JC (eds) Surgery, its principles and practice. W.B. Saunders, Philadelphia London, pp 6 v

49. Cutting C, Bookstein FL, Grayson B, Fellingham L, McCarthy JG (1986) Three-dimensional computer-assisted design of craniofacial surgical procedures: optimization and interaction with cephalometric and CT-based models. Plast Reconstr Surg 77: 877–887
50. Dandy WE (1943) An operation for scaphocephaly. Arch Surg 47: 247–249
51. Dang C, Ting K, Soo C, Longaker MT, Lorenz HP (2003) Fetal wound healing current perspectives. Clin Plast Surg 30: 13–23
52. David DJ, Poswillo DE, Simpson D (1982) The craniosynostoses: causes, natural history, and management. Springer, Berlin Heidelberg New York Tokyo
53. David LR, Gewalli F, Guimarães-Ferreira J, Sanger C, Glazier S, Argenta LC (2002a) Dynamic spring-mediated cranioplasty in a rabbit model. J Craniofac Surg 13: 794–801
54. David LR, Gewalli F, Guimarães-Ferreira J, Sanger C, Glazier S, Argenta LC (2002b) Dynamic spring-mediated cranioplasty in a rabbit model. J Craniofac Surg 13: 794–801
55. Dekaban AS (1978) Changes in brain weights during the span of human life: relation of brain weights to body heights and body weights. Ann Neurol 4: 345–356
56. Delashaw JB, Persing JA, Park TS, Jane JA (1986) Surgical approaches for the correction of metopic synostosis. Neurosurgery 19: 228–234
57. Dickerman RD, McConathy WJ, Pearl NA, Stevens QE, Cohen A, Schneider SJ (2002) Pediatric cranial fixation: a survey of pediatric neurosurgeons. J Craniofac Surg 13: 769–771
58. Duff TA, Mixter RC (1991) Midline craniectomy for sagittal suture synostosis: comparative efficacy of two barriers to calvarial reclosure. Surg Neurol 35: 350–354
59. Enlow DH (1999) Craniofacial growth and development: normal and deviant patterns. In: Posnick JC (ed) Craniofacial and maxillofacial surgery in children and young adults. W.B. Saunders, Philadelphia, pp 22–36
60. Enlow DH (2000) Normal craniofacial growth. In: Cohen MM, MacLean R (eds) Craniosynostosis: diagnosis, evaluation, and management. Oxford University Press, New York, pp 35
61. Eppley BL (2002) Craniofacial reconstruction with computer-generated HTR patient-matched implants: use in primary bony tumor excision. J Craniofac Surg 13: 650–657
62. Epstein F, McCarthy J (1981) Neonatal craniofacial surgery. Scand J Plast Reconstr Surg 15: 217–219
63. Estes JM, Whitby DJ, Lorenz HP, Longaker MT, Szabo Z, Adzick NS, Harrison MR (1992) Endoscopic creation and repair of fetal cleft lip. Plast Reconstr Surg 90: 743–746; discussion 747–749
64. Faber HK (1924) The importance of early diagnosis in oxycephaly and allied cranial deformities, with reference to the prevention of blindness and other sequelae. Calif West Med 22: 542–545
65. Faber HK, Towne EB (1927) Early craniectomy as a preventive measure in oxycephaly and allied conditions. With special reference to the prevention of blindness. Am J Med Sci 173: 701–711

66. Fok H, Jones BM, Gault DG, Andar U, Hayward R (1992) Relationship between intracranial pressure and intracranial volume in craniosynostosis. Br J Plast Surg 45: 394–397
67. Forrest CR (1999) Application of endoscope-assisted minimal-access techniques in orbitozygomatic complex, orbital floor, and frontal sinus fractures. J Craniomaxillofac Trauma 5: 7–12; discussion 13–14
68. Friede H (1981) Normal development and growth of the human neurocranium and cranial base. Scand J Plast Reconstr Surg 15: 163–169
69. Friede H, Lauritzen C, Figueroa AA (1996) Roentgencephalometric follow-up after early osteotomies in patients with scaphocephaly. J Craniofac Surg 7: 96–101
70. Fukuta K, Saito K, Potparic Z (1998) A comparison of single-stage versus gradual fronto-parietal advancement in terms of extradural dead space and bone deposition. Br J Plast Surg 51: 169–175
71. Gaudier B, Laine E, Fontaine G, Castier C, Farriaux JP (1967) Les crâniosynostoses (étude de vingt observations). Arch Fr Pediatr 24: 775–792
72. Gault DT, Renier D, Marchac D, Ackland FM, Jones BM (1990) Intracranial volume in children with craniosynostosis. J Craniofac Surg 1: 1–3
73. Gault DT, Renier D, Marchac D, Jones BM (1992) Intracranial pressure and intracranial volume in children with craniosynostosis. Plast Reconstr Surg 90: 377–381
74. Gewalli F, Guimarães-Ferreira J, Maltese G, Örtengren U, Lauritzen C (2001a) Expander elements in craniofacial surgery: an experimental study in rabbits. Scand J Plast Reconstr Surg Hand Surg 35: 149–156
75. Gewalli F, Guimarães-Ferreira J, Owman-Moll P, Sahlin P, Lauritzen C (1999) The dynamic cranioplasty for brachycephaly – five year follow-up. In: Ray-Chen Y (ed) VIII International Congress of the International Society of Craniofacial Surgery Monduzzi, Taipei, Taiwan
76. Gewalli F, Guimarães-Ferreira J, Sahlin P, Friede H, Owman-Moll P, Olsson R, David L, Lauritzen C (2001b) Long-term follow-up of dynamic cranioplasty for brachycephaly—non-syndromal bicoronal synostosis. Scand J Plast Reconstr Surg Hand Surg 35: 157–164
77. Gillies H, Harrison SH (1950) Operative correction by osteotomy of recessed malar maxillary compound in a case of oxycephaly. Br J Plast Surg 3: 123–127
78. Golabi M, Edwards MS, Ousterhout DK (1987) Craniosynostosis and hydrocephalus. Neurosurgery 21: 63–67
79. Goldberg DS, Bartlett S, Yu JC, Hunter JV, Whitaker LA (1995) Critical review of microfixation in pediatric craniofacial surgery [published erratum appears in J Craniofac Surg 1996 Jan; 7(1):83]. J Craniofac Surg 6: 301–307; discussion 308
80. Gosain AK (1997) Hydroxyapatite cement paste cranioplasty for the treatment of temporal hollowing after cranial vault remodeling in a growing child. J Craniofac Surg 8: 506–511
81. Gosain AK, Song LS, Santoro T, Weihrauch D, Bosi BO, Corrao MA, Chilian WM (2000) Effects of transforming growth factor-beta and me-

chanical strain on osteoblast cell counts: an in vitro model for distraction osteogenesis. Plast Reconstr Surg 105: 130–136; discussion 137–139
82. Goss AN (1977) Intra-uterine healing of fetal rat oral mucosal, skin and cartilage wounds. J Oral Pathol 6: 35–43
83. Graham JM, Jr, Badura RJ, Smith DW (1980a) Coronal craniostenosis: fetal head constraint as one possible cause. Pediatrics 65: 995–999
84. Graham JM, Jr, deSaxe M, Smith DW (1979) Sagittal craniostenosis: fetal head constraint as one possible cause. J Pediatr 95: 747–750
85. Graham JM, Jr, Smith DW (1980b) Metopic craniostenosis as a consequence of fetal head constraint: two interesting experiments of nature. Pediatrics 65: 1000–1002
86. Grayson BH, Rowe NM, Hollier LH, Jr, Williams JK, McCormick S, Longaker MT, McCarthy JG (2001) Development of a device for the delivery of agents to bone during distraction osteogenesis. J Craniofac Surg 12: 19–25
87. Guimarães-Ferreira J, Gewalli F, David L, Maltese G, Heino H, Lauritzen C (2002) Calvarial bone distraction with a contractile bioresorbable polymer. Plast Reconstr Surg 109: 1325–1331; discussion 1332
88. Guimarães-Ferreira J, Gewalli F, David L, Olsson R, Friede H, Lauritzen C (2001a) Clinical outcome of the modified pi-plasty procedure for sagittal synostosis. J Craniofac Surg 12: 218–224; discussion 225–226
89. Guimarães-Ferreira J, Gewalli F, David L, Olsson R, Friede H, Lauritzen C (In press) Spring-mediated cranioplasty compared with the modified pi-plasty for sagittal synostosis. Scand J Plast Reconstr Surg Hand Surg (in press)
90. Guimarães-Ferreira J, Gewalli F, Lauritzen C (1999) Spring assisted craniofacial surgery: 2 years experience. In: Ray-Chen Y (ed) VIII International Congress of the International Society of Craniofacial Surgery Monduzzi, Taipei, Taiwan
91. Guimarães-Ferreira J, Gewalli F, Sahlin P, Friede H, Owman-Moll P, Olsson R, Lauritzen C (2001b) Dynamic cranioplasty for brachycephaly in Apert syndrome – long term follow-up study. J Neurosurg 94: 757–764
92. Hallock GG (1985) In utero cleft lip repair in A/J mice. Plast Reconstr Surg 75: 785–790
93. Haluck RS, Mackay DR, Gorman PJ, Saggers GC, Manders EK (1999) A comparison of gradual distraction techniques for modification of the midface in growing sheep. Ann Plast Surg 42: 476–480
94. Hammer B, Kunz C, Schramm A, deRoche R, Prein J (1999) Repair of complex orbital fractures: technical problems, state-of-the-art solutions and future perspectives. Ann Acad Med Singapore 28: 687–691
95. Hanieh A, Sheen R, David DJ (1989) Hydrocephalus in Crouzon's syndrome. Childs Nerv Syst 5: 188–189
96. Heino H (1999) Biaxial orientation of lactide copolymers. Tampere University of Technology, Tampere
97. Hemmy DC, David DJ, Herman GT (1983) Three-dimensional reconstruction of craniofacial deformity using computed tomography. Neurosurgery 13: 534–541

98. Hertle RW, Quinn GE, Minguini N, Katowitz JA (1991) Visual loss in patients with craniofacial synostosis. J Pediatr Ophthalmol Strabismus 28: 344–349
99. Higginbottom MC, Jones KL, James HE (1980) Intrauterine constraint and craniosynostosis. Neurosurgery 6: 39–44
100. Hirabayashi S, Harii K, Sakurai A, Takaki EK, Fukuda O (1988) An experimental study of craniofacial growth in a heterotopic rat head transplant. Plast Reconstr Surg 82: 236–243
101. Hirabayashi S, Sugawara Y, Sakurai A, Harii K, Park S (1998a) Fronto-orbital advancement by gradual distraction. Technical note. J Neurosurg 89: 1058–1061
102. Hirabayashi S, Sugawara Y, Sakurai A, Nakatsuka T, Takato T, Harii K (1998b) Clinical value of computer-generated acrylic skull replicas produced by laser lithography. J Craniofac Surg 9: 222–227
103. Hockley A (1993) Craniofacial surgery [editorial]. Br J Neurosurg 7: 339–341
104. Hollier L, Kelly P, Babigumira E, Potochny J, Taylor T (2002) Minimally invasive Le Fort III distraction. J Craniofac Surg 13: 44–48
105. Homer, Reck M (1994) The Iliad. Icon Editions, New York
106. Howard TD, Paznekas WA, Green ED, Chiang LC, Ma N, Ortiz de Luna RI, Garcia Delgado C, Gonzalez-Ramos M, Kline AD, Jabs EW (1997) Mutations in TWIST, a basic helix-loop-helix transcription factor, in Saethre-Chotzen syndrome. Nat Genet 15: 36–41
107. Huizinga J (1996) O Declínio da Idade Média. Ulisseia, Lisboa
108. Hunter AG, Rudd NL (1976) Craniosynostosis. I. Sagittal synostosis: its genetics and associated clinical findings in 214 patients who lacked involvement of the coronal suture(s). Teratology 14: 185–193
109. Hunter AG, Rudd NL (1977) Craniosynostosis. II. Coronal synostosis: its familial characteristics and associated clinical findings in 109 patients lacking bilateral polysyndactyly or syndactyly. Teratology 15: 301–309
110. Idriss FS (1968) Experimental intrauterine surgery: studies in fetal sin healing. Proc Inst Med Chic 27: 93–94
111. Ilizarov GA, Devyatov AA, Kamerin VK (1980) Plastic reconstruction of longitudinal bone defects by means of compression and subsequent distraction. Acta Chir Plast 22: 32–41
112. Ingraham F, Alexander E, Matson D (1948) Clinical studies in craniosynostosis. Surgery 24: 518–541
113. Ingraham FD, Matson DD (1954) Neurosurgery of infancy and childhood. Thomas, Springfield, Ill
114. Jabs EW, Li X, Scott AF, Meyers G, Chen W, Eccles M, Mao JI, Charnas LR, Jackson CE, Jaye M (1994) Jackson-Weiss and Crouzon syndromes are allelic with mutations in fibroblast growth factor receptor 2 [published erratum appears in Nat Genet 1995 Apr; 9(4):451]. Nat Genet 8: 275–279
115. Jabs EW, Muller U, Li X, Ma L, Luo W, Haworth IS, Klisak I, Sparkes R, Warman ML, Mulliken JB, *et al* (1993) A mutation in the homeodomain of the human MSX2 gene in a family affected with autosomal dominant craniosynostosis. Cell 75: 443–450

116. Jane JA, Edgerton MT, Futrell JW, Park TS (1978) Immediate correction of sagittal synostosis. J Neurosurg 49: 705–710
117. Jane JA, Persing JA (2000) Neurosurgical treatment of craniosynostosis. In: Cohen MM, MacLean R (eds) Craniosynostosis: diagnosis, evaluation, and management. Oxford University Press, New York, pp 209–227
118. Jarund M, Lauritzen C (1996) Craniofacial dysostosis: airway obstruction and craniofacial surgery. Scand J Plast Reconstr Surg Hand Surg 30: 275–279
119. Jimenez DF, Barone CM, Cartwright CC, Baker L (2002) Early management of craniosynostosis using endoscopic-assisted strip craniectomies and cranial orthotic molding therapy. Pediatrics 110: 97–104
120. Joos U, Weingart D, Hoevels S (1993) Strip ostectomy and radical ostectomy in the treatment of craniosynostosis. An experimental animal study. J Craniomaxillofac Surg 21: 89–92
121. Kaban LB, Dodson TB, Longaker MT, Stern M, Umeda H, Adzick S (1993) Fetal cleft lip repair in rabbits: long-term clinical and cephalometric results. Cleft Palate Craniofac J 30: 13–21
122. Kaiser G (1988) Sagittal synostosis—its clinical significance and the results of three different methods of craniectomy. Childs Nerv Syst 4: 223–230
123. Karp NS, McCarthy JG, Schreiber JS, Sissons HA, Thorne CH (1992) Membranous bone lengthening: a serial histological study. Ann Plast Surg 29: 2–7
124. King J (1942) Oxycephaly. Ann Surg 115: 488–506
125. Klein HM, Schneider W, Alzen G, Voy ED, Gunther RW (1992) Pediatric craniofacial surgery: comparison of milling and stereolithography for 3D model manufacturing. Pediatr Radiol 22: 458–460
126. Kloss JL (1968) Craniosynostosis secondary to ventriculoatrial shunt. Am J Dis Child 116: 315–317
127. Kobayashi S, Honda T, Saitoh A, Kashiwa K (1999) Unilateral coronal synostosis treated by internal forehead distraction. J Craniofac Surg 10: 467–471
128. Koskinen-Moffett LK, Moffett BC (1989) Sutures and intrauterine deformation. In: Persing JA, Jane JA, Edgerton MT (eds) Scientific foundations and surgical treatment of craniosynostosis. Williams & Wilkins, Baltimore, pp 96–106
129. Koskinen-Moffett LK, Moffett BC, Graham JM (1982) Cranial synostosis and intrauterine compression: a developmental study of human sutures. In: Dixon AD, Sarnat BG (eds) Factors and mechanisms influencing bone growth. AR Liss, New York, pp xxiv, 657
130. Kreiborg S (1981) Crouzon syndrome. A clinical and roentgencephalometric study. Scand J Plast Reconstr Surg [Suppl] 18: 1–198
131. Kremer M, Butsch M, Schnell M, Genecov D, Salyer K (2001) Miniaturized mechatronic distraction plate for unidirectional internal distraction —technical concepts, design features and preliminary experimental results. In: Arnaud E, Diner P (eds) III International Congress on Cranial and Facial Bone Distraction Processes Monduzzi, Paris, France

132. Kumar AV, Staffenberg DA, Petronio JA, Wood RJ (1997) Bioabsorbable plates and screws in pediatric craniofacial surgery: a review of 22 cases. J Craniofac Surg 8: 97–99
133. Kurpad SN, Goldstein JA, Cohen AR (2000) Bioresorbable fixation for congenital pediatric craniofacial surgery: a 2-year follow-up. Pediatr Neurosurg 33: 306–310
134. Kusnoto B, Figueroa AA, Polley JW (2001) Radiographic evaluation of bone formation in the pterygoid region after maxillary distraction with a rigid external distraction (RED) device. J Craniofac Surg 12: 109–117; discussion 118
135. Kuznetsov SA, Krebsbach PH, Satomura K, Kerr J, Riminucci M, Benayahu D, Robey PG (1997) Single-colony derived strains of human marrow stromal fibroblasts form bone after transplantation in vivo. J Bone Miner Res 12: 1335–1347
136. Laitinen L (1956) Craniosynostosis. Premature fusion of the cranial sutures. An experimental, clinical and histological investigation with particular reference to the pathogenesis and etiology of the disease. Ann Paediatr Fenn 2 [Suppl] 7: 18
137. Laitinen L, Sulamaa M (1955) Craniosynostosis. Symptoms and results of treatment. Ann Paediatr Fenn 1: 283–296
138. Lajeunie E, Arnaud E, Le Merrer M, Cinalli G, Marchac D, Renier D (1998) Syndromal and nonsyndromal primary trigonocephaly: analysis of a series of 237 patients. Am J Med Genet 75: 211–215
139. Lajeunie E, Le Merrer M, Bonaiti-Pellie C, Marchac D, Renier D (1995) Genetic study of nonsyndromic coronal craniosynostosis. Am J Med Genet 55: 500–504
140. Lajeunie E, Le Merrer M, Bonaiti-Pellie C, Marchac D, Renier D (1996) Genetic study of scaphocephaly. Am J Med Genet 62: 282–285
141. Lalikos JF, Tschakaloff A, Mooney MP, Losken HW, Siegel MI, Losken A, Reitz P, Wright M (1995) Internal calvarial bone distraction in rabbits with experimental coronal suture immobilization: effects of overdistraction. Plast Reconstr Surg 96: 689–698
142. Lane LC (1892) Pioneer craniectomy for relief of mental imbecility due to premature sutural closure and microcephalus. JAMA 18: 49–50
143. Lannelongue M (1890) De la craniectomie dans la microcephalie. CR Acad Sci 110: 1382–1385
144. Lauritzen C, Friede H, Elander A, Olsson R, Jensen P (1996) Dynamic cranioplasty for brachycephaly. Plast Reconstr Surg 98: 7–14; discussion 15–16
145. Lauritzen C, Guimarães-Ferreira J, Gewalli F, Sahlin P, Owman-Moll P (1999) Implantable springs in reconstructive surgery for craniofacial dysostosis. In: Ray-Chen Y (ed) VIII International Congress of the International Society of Craniofacial Surgery Monduzzi, Taipei, Taiwan
146. Lauritzen C, Lilja J, Jarlstedt J (1986) Airway obstruction and sleep apnea in children with craniofacial anomalies. Plast Reconstr Surg 77: 1–6
147. Lauritzen C, Sugawara Y, Kocabalkan O, Olsson R (1998) Spring mediated

dynamic craniofacial reshaping. Case report. Scand J Plast Reconstr Surg Hand Surg 32: 331–338
148. Lemire RJ (2000) Embriology of the skull. In: Cohen MM, MacLean R (eds) Craniosynostosis: diagnosis, evaluation, and management. Oxford University Press, New York, pp 24–25
149. Levine JP, Rowe NM, Bradley JP, Williams JK, Mackool RJ, Longaker MT, McCarthy JG (1998) The combination of endoscopy and distraction osteogenesis in the development of a canine midface advancement model. J Craniofac Surg 9: 423–432
150. Lin KY, Jane JA Sr (1999) The natural history of sagittal craniosynostosis. In: Ray-Chen Y (ed) VIII International Congress of the International Society of Craniofacial Surgery Monduzzi, Taipei, Taiwan
151. Lin KY, Nolen AA, Gampper TJ, Jane JA, Opperman LA, Ogle RC (1997) Elevated levels of transforming growth factors beta 2 and beta 3 in lambdoid sutures from children with persistent plagiocephaly. Cleft Palate Craniofac J 34: 331–337
152. Liu YH, Kundu R, Wu L, Luo W, Ignelzi MA, Snead ML, Maxson RE (1995) Premature suture closure and ectopic cranial bone in mice expressing Msx2 transgenes in the developing skull. Proc Natl Acad Sci USA 92: 6137–6141
153. Liu YH, Tang Z, Kundu RK, Wu L, Luo W, Zhu D, Sangiorgi F, Snead ML, Maxson RE (1999) Msx2 gene dosage influences the number of proliferative osteogenic cells in growth centers of the developing murine skull: a possible mechanism for MSX2-mediated craniosynostosis in humans. Dev Biol 205: 260–274
154. Lo LJ, Marsh JL, Vannier MW, Patel VV (1994) Craniofacial computer-assisted surgical planning and simulation. Clin Plast Surg 21: 501–516
155. Longaker MT, Dodson TB, Kaban LB (1990) A rabbit model for fetal cleft lip repair. J Oral Maxillofac Surg 48: 714–719
156. Longaker MT, Peled ZM, Chang J, Krummel TM (2001) Fetal wound healing: progress report and future directions. Surgery 130: 785–787
157. Longaker MT, Stern M, Lorenz P, Whitby DJ, Dodson TB, Harrison MR, Adzick NS, Kaban LB (1992) A model for fetal cleft lip repair in lambs. Plast Reconstr Surg 90: 750–756
158. Losken A, Williams JK, Burstein FD, Cohen SR, Hudgins R, Boydston W, Reisner A, Simms C (2001) Outcome analysis for correction of single suture craniosynostosis using resorbable fixation. J Craniofac Surg 12: 451–455; discussion 456–457
159. Losken HW, Mooney MP, Zoldos J, Tschakaloff A, Burrows AM, Smith TD, Arnott R, Cano G, Sherwood C, Dechant J, Cooper GM, Kapucu MR, Siegel MI (1999) Coronal suture response to distraction osteogenesis in rabbits with delayed-onset craniosynostosis. J Craniofac Surg 10: 27–37
160. Losken HW, Mooney MP, Zoldos J, Tschakaloff A, Burrows AM, Smith TD, Cooper GM, Kapucu MR, Siegel MI (1998) Internal calvarial bone distraction in rabbits with delayed-onset coronal suture synostosis. Plast Reconstr Surg 102: 1109–1119; discussion 1120–1121

161. Marchac D, Goni S (2001) Endoscopic forehead lift. Acta Chir Belg 101: 210–217
162. Marchac D, Renier D (1979) Le front flottant, traitement précoce des faciocraniosynostoses. Ann Chir Plast 24: 121–126
163. Marchac D, Renier D (1981) Cranio-facial surgery for cranio-synostosis. Scand J Plast Reconstr Surg 15: 235–243
164. Marchac D, Renier D, Broumand S (1994) Timing of treatment for craniosynostosis and facio-craniosynostosis: a 20-year experience. Br J Plast Surg 47: 211–222
165. Marlin AE, Brown WE, Jr, Huntington HW, Epstein F (1980) Effect of the dural application of Zenker's solution on the feline brain. Neurosurgery 6: 45–48
166. Marsh JL, Jenny A, Galic M, Picker S, Vannier MW (1991) Surgical management of sagittal synostosis. A quantitative evaluation of two techniques. Neurosurg Clin N Am 2: 629–640
167. Matic D, Phillips JH (2002) A contraindication for the use of hydroxyapatite cement in the pediatric population. Plast Reconstr Surg 110: 1–5
168. Matsumoto K, Nakanishi H, Seike T, Shinno K, Hirabayashi S (2000) Application of the distraction technique to scaphocephaly. J Craniofac Surg 11: 172–176
169. Maugans TA, McComb JG, Levy ML (1997) Surgical management of sagittal synostosis: a comparative analysis of strip craniectomy and calvarial vault remodeling. Pediatr Neurosurg 27: 137–148
170. McCarthy JG (1979) New concepts in the surgical treatment of the craniofacial synostosis syndromes in the infant. Clin Plast Surg 6: 201–226
171. McCarthy JG, Coccaro PJ, Eptstein F, Converse JM (1978) Early skeletal release in the infant with craniofacial dysostosis: the role of the sphenozygomatic suture. Plast Reconstr Surg 62: 335–346
172. McCarthy JG, Glasberg SB, Cutting CB, Epstein FJ, Grayson BH, Ruff G, Thorne CH, Wisoff J, Zide BM (1995) Twenty-year experience with early surgery for craniosynostosis: II. The craniofacial synostosis syndromes and pansynostosis—results and unsolved problems. Plast Reconstr Surg 96: 284–295; discussion 296–298
173. McCarthy JG, Grayson B, Bookstein F, Vickery C, Zide B (1984) Le Fort III advancement osteotomy in the growing child. Plast Reconstr Surg 74: 343–354
174. McCarthy JG, La Trenta GS, Breitbart AS, Grayson BH, Bookstein FL (1990) The le fort III advancement osteotomy in the child under 7 years of age. Plast Reconstr Surg 86: 633–646; discussion 647–649
175. McCarthy JG, Schreiber J, Karp N, Thorne CH, Grayson BH (1992) Lengthening the human mandible by gradual distraction. Plast Reconstr Surg 89: 1–8; discussion 9–10
176. McLaurin RL, Matson DD (1952) Importance of early surgical treatment of craniosynostosis. Pediatrics 10: 637–652
177. Mehrara BJ, Mackool RJ, McCarthy JG, Gittes GK, Longaker MT (1998) Immunolocalization of basic fibroblast growth factor and fibroblast growth

factor receptor-1 and receptor-2 in rat cranial sutures. Plast Reconstr Surg 102: 1805–1817; discussion 1818–1820

178. Mehrara BJ, Rowe NM, Steinbrech DS, Dudziak ME, Saadeh PB, McCarthy JG, Gittes GK, Longaker MT (1999) Rat mandibular distraction osteogenesis: II. Molecular analysis of transforming growth factor beta-1 and osteocalcin gene expression. Plast Reconstr Surg 103: 536–547
179. Meyer U, Wiesmann HP, Kruse-Losler B, Handschel J, Stratmann U, Joos U (1999) Strain-related bone remodeling in distraction osteogenesis of the mandible. Plast Reconstr Surg 103: 800–807
180. Meyers GA, Orlow SJ, Munro IR, Przylepa KA, Jabs EW (1995) Fibroblast growth factor receptor 3 (FGFR3) transmembrane mutation in Crouzon syndrome with acanthosis nigricans. Nat Genet 11: 462–464
181. Miller MT (2000) Ocular findings in craniosynostosis. In: Cohen MM, MacLean R (eds) Craniosynostosis: diagnosis, evaluation, and management. Oxford University Press, New York, pp 184–194
182. Molina F, Ortiz Monasterio F (1995) Mandibular elongation and remodeling by distraction: a farewell to major osteotomies. Plast Reconstr Surg 96: 825–840; discussion 841–842
183. Montaut J, Stricker M (1977) Les dysmorphies cranio-faciales. Les synostoses prématurées (crâniosténoses et faciosténoses). Neurochirurgie 23: 1–299
184. Moreira-Gonzalez A, Jackson IT, Miyawaki T, Barakat K, DiNick V (2003) Clinical outcome in cranioplasty: critical review in long-term follow-up. J Craniofac Surg 14: 144–153
185. Moss ML (1959) The pathogenesis of premature cranial synostosis in man. Acta Anat (Basel) 37: 351–370
186. Moss ML (1975) Functional anatomy of cranial synostosis. Childs Brain 1: 22–33
187. Moss ML, Salentijn L (1969) The primary role of functional matrices in facial growth. Am J Orthod 55: 566–577
188. Most D, Levine JP, Chang J, Sung J, McCarthy JG, Schendel SA, Longaker MT (1998) Studies in cranial suture biology: up-regulation of transforming growth factor-beta1 and basic fibroblast growth factor mRNA correlates with posterior frontal cranial suture fusion in the rat. Plast Reconstr Surg 101: 1431–1440
189. Moulin V, Tam BY, Castilloux G, Auger FA, O'Connor-McCourt MD, Philip A, Germain L (2001) Fetal and adult human skin fibroblasts display intrinsic differences in contractile capacity. J Cell Physiol 188: 211–222
190. Mulliken JB, Steinberger D, Kunze S, Muller U (1999) Molecular diagnosis of bilateral coronal synostosis. Plast Reconstr Surg 104: 1603–1615
191. Mursch K, Enk T, Christen HJ, Markakis E, Behnke-Mursch J (1999) Venous intracranial haemodynamics in children undergoing operative treatment for the repair of craniosynostosis. A prospective study using transcranial colour-coded duplex sonography. Childs Nerv Syst 15: 110–116; discussion 117–118
192. Noetzel MJ, Marsh JL, Palkes H, Gado M (1985) Hydrocephalus and mental retardation in craniosynostosis. J Pediatr 107: 885–892

193. Nott RL, Stelnicki EJ, Mack JA, Ben Y, Mitchell R, Mooney MP (2002) Changes in the protein expression of hedgehog and patched-1 in perisutural tissues induced by cranial distraction. Plast Reconstr Surg 110: 523–532
194. Opperman LA, Chhabra A, Nolen AA, Bao Y, Ogle RC (1998) Dura mater maintains rat cranial sutures in vitro by regulating suture cell proliferation and collagen production. J Craniofac Genet Dev Biol 18: 150–158
195. Opperman LA, Passarelli RW, Morgan EP, Reintjes M, Ogle RC (1995) Cranial sutures require tissue interactions with dura mater to resist osseous obliteration in vitro. J Bone Miner Res 10: 1978–1987
196. Opperman LA, Persing JA, Sheen R, Ogle RC (1994) In the absence of periosteum, transplanted fetal and neonatal rat coronal sutures resist osseous obliteration. J Craniofac Surg 5: 327–332
197. Opperman LA, Sweeney TM, Redmon J, Persing JA, Ogle RC (1993) Tissue interactions with underlying dura mater inhibit osseous obliteration of developing cranial sutures. Dev Dyn 198: 312–322
198. Ortiz-Monasterio F, del Campo AF, Carrillo A (1978) Advancement of the orbits and the midface in one piece, combined with frontal repositioning, for the correction of Crouzon's deformities. Plast Reconstr Surg 61: 507–516
199. Otto AW (1830) Lehrbuch der Pathologischen Anatomie. Rücker, Berlin
200. Panchal J, Marsh JL, Park TS, Kaufman B, Pilgram T, Huang SH (1999) Sagittal craniosynostosis outcome assessment for two methods and timings of intervention. Plast Reconstr Surg 103: 1574–1584
201. Papay FA, Hardy S, Morales L, Jr, Walker M, Enlow D (1995) "False" migration of rigid fixation appliances in pediatric craniofacial surgery. J Craniofac Surg 6: 309–313
202. Patton MA, Goodship J, Hayward R, Lansdown R (1988) Intellectual development in Apert's syndrome: a long term follow up of 29 patients. J Med Genet 25: 164–167
203. Pawl RP, Sugar O (1972) Zenker's solution in the surgical treatment of craniosynostosis. J Neurosurg 36: 604–607
204. Paznekas WA, Cunningham ML, Howard TD, Korf BR, Lipson MH, Grix AW, Feingold M, Goldberg R, Borochowitz Z, Aleck K, Mulliken J, Yin M, Jabs EW (1998) Genetic heterogeneity of Saethre-Chotzen syndrome, due to TWIST and FGFR mutations. Am J Hum Genet 62: 1370–1380
205. Peltoniemi HH, Hallikainen D, Toivonen T, Helevirta P, Waris T (1999a) SR-PLLA and SR-PGA miniscrews: biodegradation and tissue reactions in the calvarium and dura mater. J Craniomaxillofac Surg 27: 42–50
206. Peltoniemi HH, Tulamo RM, Pihlajamaki HK, Kallioinen M, Pohjonen T, Tormala P, Rokkanen PU, Waris T (1998) Consolidation of craniotomy lines after resorbable polylactide and titanium plating: a comparative experimental study in sheep. Plast Reconstr Surg 101: 123–133
207. Peltoniemi HH, Tulamo RM, Toivonen T, Hallikainen D, Tormala P, Waris T (1999b) Biodegradable semirigid plate and miniscrew fixation compared with rigid titanium fixation in experimental calvarial osteotomy. J Neurosurg 90: 910–917

208. Pensler JM (1997) Role of resorbable plates and screws in craniofacial surgery. J Craniofac Surg 8: 129–134
209. Pensler JM, Ivescu AS, Radosevich JA (1998) Scaphocephaly: premature closure of the sagittal suture: a localized disorder of cellular metabolism? Ann Plast Surg 40: 48–52
210. Persing JA, Babler WJ, Nagorsky MJ, Edgerton MT, Jane JA (1986) Skull expansion in experimental craniosynostosis. Plast Reconstr Surg 78: 594–603
211. Persing JA, Edgerton MT, Park TS, Jane JA (1987) Barrel stave osteotomy for correction of turribrachycephaly craniosynostosis deformity. Ann Plast Surg 18: 488–493
212. Peterson-Falzone SJ, Pruzansky S, Parris PJ, Laffer JL (1981) Nasopharyngeal dysmorphology in the syndromes of Apert and Crouzon. Cleft Palate J 18: 237–250
213. Philippe B, Besins T (1993) Augmentation of craniofacial skeleton. Plast Reconstr Surg 92: 1415–1416
214. Polley JW, Figueroa AA, Charbel FT, Berkowitz R, Reisberg D, Cohen M (1995) Monobloc craniomaxillofacial distraction osteogenesis in a newborn with severe craniofacial synostosis: a preliminary report. J Craniofac Surg 6: 421–423
215. Posnick JC (1991) Craniofacial dysostosis. Staging of reconstruction and management of the midface deformity. Neurosurg Clin N Am 2: 683–702
216. Posnick JC (1994) The craniofacial dysostosis syndromes. Current reconstructive strategies. Clin Plast Surg 21: 585–598
217. Posnick JC (1999) Anterior plagiocephaly: unilateral coronal synostosis and skull moulding. Craniofacial and maxillofacial surgery in children and young adults. WB Saunders, Philadelphia, pp 128
218. Posnick JC, Lin KY, Chen P, Armstrong D (1994) Metopic synostosis: quantitative assessment of presenting deformity and surgical results based on CT scans. Plast Reconstr Surg 93: 16–24
219. Rachmiel A, Potparic Z, Jackson IT, Sugihara T, Clayman L, Topf JS, Forte RA (1993) Midface advancement by gradual distraction. Br J Plast Surg 46: 201–207
220. Raulo Y, Tessier P (1981) Fronto-facial advancement for Crouzon's and Apert's syndromes. Scand J Plast Reconstr Surg 15: 245–250
221. Reardon W, Winter RM, Rutland P, Pulleyn LJ, Jones BM, Malcolm S (1994) Mutations in the fibroblast growth factor receptor 2 gene cause Crouzon syndrome. Nat Genet 8: 98–103
222. Renier D (1989) Intracranial pressure in craniosynostosis: pre- and postoperative recordings – correlation with functional results. In: Persing JA, Jane JA, Edgerton MT (eds) Scientific foundations and surgical treatment of craniosynostosis. Williams & Wilkins, Baltimore
223. Renier D, Arnaud E, Cinalli G, Marchac D, Brunet L, Sebag G, Sainte-Rose C, Zerah M (1996a) Pronostic mental du syndrome d'Apert. Arch Pediatr 3: 752–760
224. Renier D, Arnaud E, Cinalli G, Sebag G, Zerah M, Marchac D (1996b) Prognosis for mental function in Apert's syndrome. J Neurosurg 85: 66–72

225. Renier D, Sainte-Rose C, Marchac D, Hirsch JF (1982) Intracranial pressure in craniostenosis. J Neurosurg 57: 370–377
226. Ripamonti U (1991) The induction of bone in osteogenic composites of bone matrix and porous hydroxyapatite replicas: an experimental study on the baboon (Papio ursinus). J Oral Maxillofac Surg 49: 817–830
227. Ripamonti U, Crooks J, Rueger DC (2001) Induction of bone formation by recombinant human osteogenic protein-1 and sintered porous hydroxyapatite in adult primates. Plast Reconstr Surg 107: 977–988
228. Ripamonti U, Ma SS, Reddi AH (1992a) Induction of bone in composites of osteogenin and porous hydroxyapatite in baboons. Plast Reconstr Surg 89: 731–739; discussion 740
229. Ripamonti U, Ma SS, van den Heever B, Reddi AH (1992b) Osteogenin, a bone morphogenetic protein, adsorbed on porous hydroxyapatite substrata, induces rapid bone differentiation in calvarial defects of adult primates. Plast Reconstr Surg 90: 382–393
230. Roberts JR, Rickham PP (1970) Craniostenosis following Holter valve operation. Dev Med Child Neurol [Suppl] 22: 145
231. Robinson RC, O'Neal PJ, Robinson GH (2001) Mandibular distraction force: laboratory data and clinical correlation. J Oral Maxillofac Surg 59: 539–544; discussion 544–555
232. Roth DA, Gold LI, Han VK, McCarthy JG, Sung JJ, Wisoff JH, Longaker MT (1997a) Immunolocalization of transforming growth factor beta 1, beta 2, and beta 3 and insulin-like growth factor I in premature cranial suture fusion. Plast Reconstr Surg 99: 300–309; discussion 310–316
233. Roth DA, Longaker MT, McCarthy JG, Rosen DM, McMullen HF, Levine JP, Sung J, Gold LI (1997b) Studies in cranial suture biology: Part I. Increased immunoreactivity for TGF-beta isoforms (beta 1, beta 2, and beta 3) during rat cranial suture fusion. J Bone Miner Res 12: 311–321
234. Rowe NM, Mehrara BJ, Luchs JS, Dudziak ME, Steinbrech DS, Illei PB, Fernandez GJ, Gittes GK, Longaker MT (1999) Angiogenesis during mandibular distraction osteogenesis. Ann Plast Surg 42: 470–475
235. Rowlatt U (1979) Intrauterine wound healing in a 20 week human fetus. Virchows Arch A Pathol Anat Histol 381: 353–361
236. Rubin JP, Posnick JC, Yaremchuk MJ (1998) Role of endoscopic and distraction techniques in facial and reconstructive surgery: new technology or improved results. J Craniofac Surg 9: 285–299
237. Rutland P, Pulleyn LJ, Reardon W, Baraitser M, Hayward R, Jones B, Malcolm S, Winter RM, Oldridge M, Slaney SF, *et al* (1995) Identical mutations in the FGFR2 gene cause both Pfeiffer and Crouzon syndrome phenotypes. Nat Genet 9: 173–176
238. Sailer HF, Haers PE, Zollikofer CP, Warnke T, Carls FR, Stucki P (1998) The value of stereolithographic models for preoperative diagnosis of craniofacial deformities and planning of surgical corrections. Int J Oral Maxillofac Surg 27: 327–333
239. Sailer HF, Kolb E (1995) Influence of craniofacial surgery on the social attitudes toward the malformed and their handling in different cultures and

at different times: a contribution to social world history. J Craniofac Surg 6: 314–326
240. Salyer KE, Hall CD (1989) Porous hydroxyapatite as an onlay bone-graft substitute for maxillofacial surgery. Plast Reconstr Surg 84: 236–244
241. Sculerati N, Gottlieb MD, Zimbler MS, Chibbaro PD, McCarthy JG (1998) Airway management in children with major craniofacial anomalies. Laryngoscope 108: 1806–1812
242. Sgouros S, Goldin JH, Hockley AD, Wake MJ, Natarajan K (1999a) Intracranial volume change in childhood. J Neurosurg 91: 610–616
243. Sgouros S, Hockley AD, Goldin JH, Wake MJ, Natarajan K (1999b) Intracranial volume change in craniosynostosis. J Neurosurg 91: 617–625
244. Shillito J, Jr, Matson DD (1968) Craniosynostosis: a review of 519 surgical patients. Pediatrics 41: 829–853
245. Shwyrkow MB, Shamsudinov AK (1989) Methods of simultaneous treatment of the mandible defects and the adjacent soft tissues. Acta Chir Plast 31: 226–235
246. Smith DW, Tondury G (1978) Origin of the calvaria and its sutures. Am J Dis Child 132: 662–666
247. Smith SW, Sachdeva RC, Cope JB (1999) Evaluation of the consolidation period during osteodistraction using computed tomography. Am J Orthod Dentofacial Orthop 116: 254–263
248. Snyder CC, Levine GA, Swanson HM, Browne EZ, Jr (1973) Mandibular lengthening by gradual distraction. Preliminary report. Plast Reconstr Surg 51: 506–508
249. Sömmerring ST (1800) Vom Baue des Menschlichen Körpers. Voss, Leipzig
250. Stavrou P, Sgouros S, Willshaw HE, Goldin JH, Hockley AD, Wake MJ (1997) Visual failure caused by raised intracranial pressure in craniosynostosis. Childs Nerv Syst 13: 64–67
251. Stelnicki EJ (2002) Endoscopic treatment of craniosynostosis. Atlas Oral Maxillofac Surg Clin North Am 10: 57–72
252. Stelnicki EJ, Hoffman W (1998a) Intracranial migration of microplates versus wires in neonatal pigs after frontal advancement. J Craniofac Surg 9: 60–64
253. Stelnicki EJ, Lee S, Hoffman W, Lopoo J, Foster R, Harrison MR, Longaker MT (1999) A long-term, controlled-outcome analysis of in utero versus neonatal cleft lip repair using an ovine model. Plast Reconstr Surg 104: 607–615
254. Stelnicki EJ, Vanderwall K, Harrison MR, Longaker MT, Kaban LB, Hoffman WY (1998b) The in utero correction of unilateral coronal craniosynostosis. Plast Reconstr Surg 101: 287–296
255. Stelnicki EJ, Vanderwall K, Hoffman WY, Harrison MR, Glowacki J, Longaker MT (1998c) A new in utero sheep model for unilateral coronal craniosynostosis. Plast Reconstr Surg 101: 278–286
256. Stewart KJ, Weyand B, Hof RJ, White SA, Lvoff GO, Maffulli N, Poole MD (1999) A quantitative analysis of the effect of insulin-like growth factor-1 infusion during mandibular distraction osteogenesis in rabbits. Br J Plast Surg 52: 343–350

257. Sugawara Y, Hirabayashi S, Sakurai A, Harii K (1998) Gradual cranial vault expansion for the treatment of craniofacial synostosis: a preliminary report. Ann Plast Surg 40: 554–565
258. Sweeney TM, Opperman LA, Persing JA, Ogle RC (1995) Repair of critical size rat calvarial defects using extracellular matrix protein gels. J Neurosurg 83: 710–715
259. Talisman R, Hemmy DC, Denny AD (1997) Frontofacial osteotomies, advancement, and remodeling by distraction: an extended application of the technique. J Craniofac Surg 8: 308–317
260. Tavakoli K, Yu Y, Shahidi S, Bonar F, Walsh WR, Poole MD (1999) Expression of growth factors in the mandibular distraction zone: a sheep study. Br J Plast Surg 52: 434–439
261. Tessier P (1967a) Dysostoses cranio-faciales (syndromes de Crouzon et d'Apert). Osteotomies totales de la face. In: Sanvenero-Rosselli G (ed) IV International Congress of Plastic and Reconstructive Surgery Excerpta Medica Foundation, Rome
262. Tessier P (1967b) Osteotomies totales de la face. Syndrome de Crouzon, syndrome d'Apert: oxycéphalies, scaphocéphalies, turricéphalies. Ann Chir Plast 12: 273–286
263. Tessier P (1971a) The definitive plastic surgical treatment of the severe facial deformities of craniofacial dysostosis. Crouzon's and Apert's diseases. Plast Reconstr Surg 48: 419–442
264. Tessier P (1971b) Relationship of craniostenoses to craniofacial dysostoses, and to faciostenoses: a study with therapeutic implications. Plast Reconstr Surg 48: 224–237
265. Tessier P (1979) Reflexions sur la chirurgie cranio-faciale d'aujourd'hui et son avenir chez l'enfant. Ann Chir Plast 24: 109–119
266. Tessier P, Guiot G, Rougerie J, Delbet JP, Pastoriza J (1967) Ostéotomies cranio-naso-orbito-faciales. Hypertélorisme. Ann Chir Plast 12: 103–118
267. Thaler ER, Kotapka M, Lanza DC, Kennedy DW (1999) Endoscopically assisted anterior cranial skull base resection of sinonasal tumors. Am J Rhinol 13: 303–310
268. Thompson DN, Malcolm GP, Jones BM, Harkness WJ, Hayward RD (1995) Intracranial pressure in single-suture craniosynostosis. Pediatr Neurosurg 22: 235–240
269. Tolarova MM, Harris JA, Ordway DE, Vargervik K (1997) Birth prevalence, mutation rate, sex ratio, parents' age, and ethnicity in Apert syndrome. Am J Med Genet 72: 394–398
270. Tong L, Buchman SR, Ignelzi MA, Jr, Rhee S, Goldstein SA (2003) Focal adhesion kinase expression during mandibular distraction osteogenesis: evidence for mechanotransduction. Plast Reconstr Surg 111: 211–222
271. Toth BA, Kim JW, Chin M, Cedars M (1998) Distraction osteogenesis and its application to the midface and bony orbit in craniosynostosis syndromes. J Craniofac Surg 9: 100–113; discussion 119–122
272. Tschakaloff A, Losken HW, Mooney MP, Siegel MI, Losken A, Swan J

(1994) Internal calvarial bone distraction in rabbits with experimental coronal suture immobilization. J Craniofac Surg 5: 318–326
273. Tutino M, Chico F, Goodrich JT, Monasterio FO (2000) Endoscopic intracranial craniofacial and monobloc osteotomies with the aid of a malleable high-speed pneumatic drill: a cadaveric and clinical study. Ann Plast Surg 44: 1–7
274. Tutino M, Chico F, Ortiz-Monasterio F (1998) Endoscopic dissection of dura and craniotomy with minimal trephines: a preliminary series. J Craniofac Surg 9: 154–161
275. Urist MR (1965) Bone: formation by autoinduction. Science 150: 893–899
276. Vannier MW, Marsh JL, Warren JO (1984) Three dimensional CT reconstruction images for craniofacial surgical planning and evaluation. Radiology 150: 179–184
277. Virchow R (1851) Über den Cretinismus, namentlich in Franken, und über pathologische Schädelformen. Verh Phys Med Gesellsch Wurzburg 2: 230–270
278. Warren SM, Mehrara BJ, Steinbrech DS, Paccione MF, Greenwald JA, Spector JA, Longaker MT (2001) Rat mandibular distraction osteogenesis: part III. Gradual distraction versus acute lengthening. Plast Reconstr Surg 107: 441–453
279. Weinzweig J, Panter KE, Pantaloni M, Spangenberger A, Harper JS, Lui F, James LF, Edstrom LE (1999) The fetal cleft palate: II. Scarless healing after in utero repair of a congenital model. Plast Reconstr Surg 104: 1356–1364
280. Whitaker LA, Schut L, Kerr LP (1977) Early surgery for isolated craniofacial dysostosis. Improvement and possible prevention of increasing deformity. Plast Reconstr Surg 60: 575–581
281. Wilkie AO, Slaney SF, Oldridge M, Poole MD, Ashworth GJ, Hockley AD, Hayward RD, David DJ, Pulleyn LJ, Rutland P, *et al* (1995a) Apert syndrome results from localized mutations of FGFR2 and is allelic with Crouzon syndrome. Nat Genet 9: 165–172
282. Wilkie AO, Slaney SF, Oldridge M, Poole MD, Ashworth GJ, Hockley AD, Hayward RD, David DJ, Pulleyn LJ, Rutland P *et al* (1995b) Apert syndrome results from localized mutations of FGFR2 and is allelic with Crouzon syndrome [see comments]. Nat Genet 9: 165–172
283. Wiltfang J, Merten HA, Schultze-Mosgau S, Schrell U, Wenzel D, Kessler P (2000) Biodegradable miniplates (LactoSorb): long-term results in infant minipigs and clinical results. J Craniofac Surg 11: 239–243; discussion 244–245
284. Wolfe SA (1999) Looking back, looking ahead: how we arrived at our current thinking about craniofacial and maxillofacial surgery. In: Posnick JC (ed) Craniofacial and maxillofacial surgery in children and young adults. W.B. Saunders, Philadelphia, pp 3–21
285. Wong L, Richtsmeier JT, Manson PN (1993) Craniofacial growth following rigid fixation: suture excision, miniplating, and microplating. J Craniofac Surg 4: 234–244; discussion 245–246
286. Wood-Smith D, Epstein F, Morello D (1976) Transcranial decompression of the optic nerve in the osseous canal in Crouzon's disease. Clin Plast Surg 3: 621–623

Technical Standards

Preoperative Clinical Evaluation, Outline of Surgical Technique and Outcome in Temporal Lobe Epilepsy

A. Immonen[1], L. Jutila[2], R. Kälviäinen[2], E. Mervaala[3], K. Partanen[4], J. Partanen[3], R. Vanninen[4], A. Ylinen[5], I. Alafuzoff[6], L. Paljärvi[6], H. Hurskainen[1], J. Rinne[1], M. Puranen[4], and M. Vapalahti[1]

[1] Department of Neurosurgery, Kuopio University Hospital, Kuopio, Finland
[2] Department of Neurology, Kuopio University Hospital, Kuopio, Finland
[3] Department of Clinical Neurophysiology, Kuopio University Hospital, Kuopio, Finland
[4] Department of Clinical Radiology, Kuopio University Hospital, Kuopio, Finland
[5] Department of Neuroscience and Neurology, University of Kuopio, Kuopio, Finland
[6] Department of Clinical Pathology, Kuopio University Hospital, Kuopio, Finland

With 12 Figures

Contents

Abstract

Temporal lobe epilepsy (TLE) is the most common type of refractory epilepsy. The mechanisms of epileptogenesis and seizure semiology of the mesial and neocortical temporal lobe epilepsy are discussed. The evaluation and selection of patients for TLE surgcry requires team work: the different clinical aspects of neuropsychological evaluation, magnetic resonance and functional imaging (positron emission tomography, single photon emission computed tomography and magnetoenephalography) are reviewed. In our programme of epilepsy surgery at Kuopio University Hospital, Finland, we have performed 230 temporal resections from 1988 until 2002. Preoperative diagnostic EEG-videotelemetry often required intracranial monitoring and it has proved to be safe and efficient. The indications and technique for tailored temporal lobe resection with amygdalohippocampectomy used in our institution, as well as the complications, are described. Our analysis of outcome after temporal lobe surgery included 140 consecutive adult patients between 1988 and 1999; one year after the operation in unilateral TLE the Engel I–II outcome was observed in 68% of the patients. Outcome of surgery improved significantly after introduction of the standardised MR imaging protocol from 1993; 74% of patients with unilateral TLE achieved Engel I–II outcome.

Keywords: Temporal lobe epilepsy; epilepsy surgery; postoperative outcome.

Introduction

Epilepsy surgery is defined as any neurosurgical intervention whose primary objective is to relieve medically intractable epilepsy (European Federation of Neurological Societies Task Force 2000). The aim of epilepsy surgery is to reduce the number and intensity of seizures, minimise neurological morbidity and antiepileptic drug (AED) toxicity, and improve quality of life. By definition, epilepsy surgery does not include normal surgical treatment of intracranial lesions where the primary goal is to diagnose and possibly remove the pathological target, often an advancing tumour. In these patients, epileptic seizures are only one symptom of the lesion and will be treated concomitantly as part of the procedure.

Temporal lobe epilepsy (TLE) is recognised as the most common type of refractory, focal epilepsy. In one third of all cases the neuronal systems responsible for the seizures that characterise this form of epilepsy fail to respond to currently available AEDs (Andermann F 2002). New imaging methods, especially magnetic resonance imaging (MRI), identify localising abnormalities in an increasing proportion of patients with intractable focal epilepsy. Consequently, the accuracy of the preoperative diagnostic procedures has been significantly improved during the last decade; and suitable candidates for surgery can be selected more reliably. Currently the main resources in most epilepsy surgery centres have been used to evaluate candidates for TLE surgery.

Evaluation of patients for epilepsy surgery requires a team approach, in which the neurosurgeon is an important member but becomes pivotal in the later phase of the evaluation. An experienced neurologist (epileptologist) together with a neurophysiologist makes the primary selection and excludes patients with inadequate or inappropriate medical treatment or with non-epileptic characteristics. Neuroradiology, neuropsychology and psychiatry are other important aspects of the early evaluation process. In our centre more than half of the patients referred for evaluation will never have surgery, because many can be helped without surgical intervention.

Epidemiology of TLE

Epilepsy is a chronic disorder characterised by recurrent spontaneous epileptic seizures. In Western Europe the prevalence of epilepsy varies from 0.3% to 1% of the population, and in 60% of all cases the onset of the seizures is focal. Of these focal epilepsies, in about 80% of the patients the seizures are initiated within the temporal lobes. In 30% of patients with focal epilepsy, the seizures continue in spite of adequate AED medication or the patient develops intolerable side effects (Cockerell OC *et al.* 1995). If we suppose that half of these patients should be evaluated for epilepsy surgery and that half of the evaluated patients would eventually benefit from epilepsy surgery, this means that about 4.5% of all patients with epilepsy (0.03% of the total population) could profit from epilepsy surgery (Engel J, Jr 1993).

Etiology and Pathology of TLE

The mechanisms of epileptogenesis in mesial TLE are presented in detail in another article of this series (Jutila L *et al.* 2002a). Currently, however, the molecular and network changes are best understood in subjects with symptomatic aetiology for TLE. Brain damage resulting in epileptogenesis can be induced by a variety of brain insults such as head trauma or prolonged febrile seizure. The mesial temporal structures, especially the hippocampus, but also the amygdala and the entorhinal cortex, are vulnerable to cell death caused by brain insults. Consequently, there is reorganisation of neuronal circuitries, including neuronal loss and axonal and dendritic plasticity of the surviving neurons, neurogenesis and changes in extracellular matrix. All these changes together form the basis for development of mesial temporal lobe epilepsy syndrome.

The hallmark pathology of mesial TLE is classical neuronal death and gliosis in the hippocampus, so-called mesial temporal or hippocampal sclerosis. Loss of neurons in the dentate gyrus and the hippocampus proper is found in 60–70% of the histologic specimens obtained from patients undergoing TLE surgery (Babb TL 1992). The etiology of neocortical TLE is basically the same as that of any cortical epilepsy and consists of various possible pathologies. Brain tumours, vascular malformations, cortical dysplasias and posttraumatic lesions are probably the most frequent courses of epileptogenesis in the temporal neocortex. These lesions may also occur within the mesial temporal lobe, and in such cases the epileptogenesis may be connected with the mechanism of mesial TLE. However, seizure onset in the temporal neocortex is often associated with spread of the seizure to mesial structures, which confuses the seizure semiology and may lead to secondary epileptogenesis. In these cases, dual pathology may be present, which includes hippocampal sclerosis in combination with another extrahippocampal structural lesion.

In our series, major larger pathological entities were identified from the surgical specimens: patients with hippocampal sclerosis or gliosis, patients with tumour or cystic lesion, and patients with cortical microdysgenesis (Jutila L *et al.* 2002b). Hippocampal sclerosis was confirmed in 41 of 46 symptomatic patients with hippocampal atrophy in preoperative MR imaging. Focal cortical dysplasia was observed in one patient, and benign or low-grade tumour in six patients evaluated preoperatively as probably having symptoms. Moreover, hippocampal sclerosis, hippocampal gliosis,

Table 1. *Pathological Examinations of Resected Tissue; 140 Consecutive Patients with TLE Operated on at Kuopio Epilepsy Surgery Centre During 1988–1999. (Jutila L et al. 2002b)*

	n
No diagnostic lesions	13
Sclerosis/gliosis or numerous corpora amylacea	69
Hippocampus	60
– hippocampal sclerosis	55
– hippocampal gliosis	5
Amygdala	4
Temporal lobe	5
Microdysgenesis	24
Hippocampus	7
Temporal lobe	17
Dysplasia	3
Temporal lobe	3
Tumour or cystic lesion	18
Oligodendroglioma (low grade)	4
Astrocytoma (low grade)	1
Ganglioglioma	3
Hamartoma	3
Dysplastic neuroepithelial tumour	3
Cavernous hemangioma	3
Cysta epidermoides	1
Perivascular hemosiderin ladded macrophages	1
Sample not availabe or not evaluable	12
Total	140

(Published with Permission from the BMJ Publishing Group, J Neurol Neurosurg Psychiatry; 2002, 73:486–494).

or hippocampal microdysgenesis were identified in 40% of the patients with preoperative probable symptomatic etiology of epilepsy (Jutila L *et al.* 2002b) (Table 1).

Temporal Lobe Seizure Symptomatology

TLE can be divided into two main subtypes: mesial temporal lobe epilepsy (MTLE) (Najm I *et al.* 2001, Wieser HG *et al.* 2001) and neocortical temporal lobe epilepsy (NTLE) (Kotagal P 2001, Pacia SV *et al.* 1996). The term lesional TLE is used to describe symptomatic TLE associated with specific localised epileptogenic lesions, such as tumours, developmental abnormalities or infectious processes. Probable symptomatic TLE refers to situations in which the etiology of TLE is presumed to be symptomatic but is, in fact, unknown.

Mesial TLE

MTLE often has onset in the latter half of the first decade of life and is associated with increased incidence of complex febrile seizures. Seizure semiology in MTLE includes epigastric, autonomic or emotional auras, as well as focal seizures with arrest, oroalimentary or complex automatisms and postictal memory deficit (Najm I *et al.* 2001, Wieser HG *et al.* 2001). Often some lateralising ictal or postictal symptoms can also be observed (Manford M *et al.* 1996). Unilateral ictal dystonia contralateral to the epileptogenic zone has been reported in one third of the patients with TLE. Ictal dystonia is often associated with ipsilateral hand automatisms (Kotagal P *et al.* 1989). Epigastric sensation followed by ictal nausea, on the other hand, is associated with seizures of the nondominant temporal lobe (Kramer RE *et al.* 1988).

Neocortical TLE

Mesial temporal and temporal neocortical seizures cannot be differentiated only by seizure semiology, but some general features can be distinguished. According to the analyses of Anand and coworkers (Anand I *et al.* 1997), typical clinical manifestations of temporal neocortical seizures are cephalic or auditory auras followed by behavioural arrest, complex gestures, anxiety or agitation, ictal speech, vocalisation, whole body movements, version and rapid secondary generalisation. Ictal speech lateralises the lesion to the nondominant hemisphere and postictal dysfasia to the dominant hemisphere. The diagnostic value of these clinical manifestations has led to a semiological classification of seizures, and localisation of the partial seizures is the topic of a second classification (Hamer HM *et al.* 2001, Luders H *et al.* 1998).

Selection of Candidates for Preoperative Assessment

Recent evidence suggests that patients with newly diagnosed focal seizures have a high risk of subsequent medical intractability if their seizures are not controlled by the first two AEDs taken (Kwan P *et al.* 2000, Schmidt D 2002). Consequently, many programs of epilepsy surgery for adults define medical intractability as persistent seizures despite two years and 2–3 maximally tolerated AED trials (European Federation of Neurological Societies Task Force 2000). However, the median duration of epilepsy in patients referred to presurgical evaluation is 15–20 years (Duncan JS 2001, Jutila L *et al.* 2002b).

Patients with medically intractable epilepsy are at considerably increased risk of a large number of adverse outcomes. Mortality and risk of accidents are higher compared to the general population and to patients with less severe epilepsy (Hauser WA *et al.* 2001). In some patients hippocampal damage may also progress as a function of repeated seizures (Kalviainen R *et al.* 2002). All these facts emphasize the importance of early surgical treatment in medically intractable TLE.

Neuropsychological Evaluation in TLE

The neuropsychological assessment comprises a comprehensive evaluation of cognitive functioning that includes intelligence, frontal "executive skills", memory, attention, visuospatial abilities, language, and motor skills (Jones-Gotman M *et al.* 2000). Neuropsychological assessment also includes determination of cerebral dominance for language, which in our centre is perfomed using the intracarotid amobarbital procedure (IAP, WADA). Some centres, including ours, recommend it for all surgery candidates, while others study only a selected subset of patients. In addition to determining language lateralisation, the WADA test is used to assist in determination and prediction of memory outcomes (Simkins-Bullock J 2000).

Neuropsychological assessment contributes to the presurgical evaluation in several ways. First, it enables assessment of the general cognitive functioning of the patient. Whereas normal intelligence is not a requisite for surgical consideration in TLE, a cognitive deficit raises the possibility of multifocal epileptogeneity and possible poor outcome for surgical intervention (Jones MW *et al.* 2000). Secondly, such assessment provides information about the localisation and lateralisation of the possible dysfunctional area of the brain. The relationships between the left (dominant) medial temporal lobe and verbal memory, and between the right (nondominant) medial temporal lobe and visual memory, have long been recognized. However, the relationship between the right temporal lobe and visual memory has been found to be much more variable than the relationship between the left temporal lobe and verbal memory, and the lateralized memory deficits in TLE have been detected more consistently in verbal memory (Hermann *et al.* 1994, Moore *et al.* 1996).

Even though there are questions about the inadequacies of the traditional memory tests, neuropsychological evaluation allows prediction of a patient's postsurgical risk for cognitive changes (Jones-Gotman M *et al.* 2000). Epilepsy surgery usually produces no general deterioration of cognitive functioning because most of the removed tissue is dysfunctional. It may even result in improved cognitive performance due to elimination of the adverse effects of seizures and the beneficial effects of reducing antiepileptic drug treatment. Cognitive functions may occasionally be impaired, depending the areas involved in the surgery. As most epilepsy surgery is directed to the temporal lobes, there may be a selective decrease in memory functioning.

Resections of the speech-dominant temporal lobe have been reliably associated with decrements in the capacity to learn and retain verbal material. Depending on the type of visual material, some investigators have found decrements in nonverbal memory following nondominant temporal lobe resections, while others have reported negative results. The greatest decrements in immediate and delayed episodic verbal memory have been found in patients without significant hippocampal pathology, measured either with histological analysis of neuronal cell densities or with MRI volumetrics. Patients with structurally and functionally "adequate" hippocampi demonstrate greater memory losses than those with hippocampal sclerosis. In addition, more frequent decrements in both verbal and visual memory have been seen among patients with good or high preoperative memory functions. The greater risk for memory decline after anterior temporal lobectomy has also been related to the larger extent of resection, earlier age at onset of epilepsy, older age at operation and male sex. Postoperative memory function also depends on possible dynamic changes in the hippocampal remnant (Baxendale SA *et al.* 2000).

WADA Test

In the classical WADA test, 125–175 mg sodium amobarbital is injected into each carotid artery. The barbiturate suppresses the ipsilateral functional capacity for a few minutes, enabling testing of speech and memory selectively in one hemisphere. Lateralisation of language functions with the WADA test is well-founded and validated (Wada J *et al.* 1960) Assessment of memory function in the WADA test is based on the hypothesis that pharmacologic inactivation of a single temporal lobe should not create global amnesia if the awake temporal lobe is healthy (Milner B 1997). Memory testing after ipsilateral injection measures whether the remaining hemisphere is capable of supporting memory functions independently ("reserve model"), and performance correlates with the risk of severe postoperative amnesia. Memory testing after contralateral injection measures the functional adequacy of the hemisphere to be operated ("adequacy model"), and the performance correlates with the material-specific memory disorder (Chelune GJ 1995, Martin RC *et al.* 2002).

Further research is needed to ascertain the risk factors for memory decline after epilepsy surgery. However, it is also important to evaluate the natural course of intractable/inoperable focal/temporal lobe epilepsy and the associated impairment of memory performance and to compare the long-term cognitive outcome of these patients with the outcome of patients who have undergone anterior temporal lobectomy. Only with long-term comparative studies can we evaluate whether temporal lobe surgery can be seen as an additional lesion that accelerates the memory decline caused by the epileptic process or as a treatment that delays memory decline, which otherwise would have been caused by the process (Chelune GJ *et al.* 2001, Lee TM *et al.* 2002, Rausch R 2002).

Magnetic Resonance Imaging (MRI) in TLE

MRI is a sensitive and specific method for detecting various abnormalities of the structures in the medial temporal lobe. It can detect most of the structural epileptogenic lesions that are associated with TLE and helps to identify the anatomic relationship between the surgical target and the functionally eloquent cortex (Duncan JS 1997).

Patients referred for MRI imaging because of TLE should undergo a specific imaging protocol that is tailored to evaluation of the pathologic conditions commonly seen in these patients, such as mesial temporal scleroses (MTS), malformations of cortical development, vascular malformations, tumours and acquired cortical damage. Standard axial proton density and T2-weighted fast spin echo imaging with 5 mm thick sections cover the whole brain and are used for efficient screening of any space-occupying lesion. Coronal T2-weighted fast spin echo images are necessary for precise anatomic localisation within the temporal lobe and are absolutely mandatory for adequate evaluation of MTS. A coronal T1-weighted gradient echo volume aquisition (3D) covers the whole brain with 1.5–2 mm partitions and in hippocampal volumetry allows reformatting in any orientation. The coronal sequences are best oriented at a right angle to the long axis of the hippocampus. Fluid attenuated inversion recovery sequence (FLAIR) is also included in the protocol. The studies are monitored by a neuroradiologist, and contrast material is administered for further characterisation only if the space-occupying lesion is identified.

MTS or hippocampal sclerosis is the most frequent finding in patients investigated and operated for drug-resistant TLE (Duncan JS 1997). The main imaging features of MTS are increased signal intensity in the hippocampus on T2-weighted images, reduced hippocampal size and distortion of the internal architecture of the hippocampus. Additional findings may be collateral atrophy of the white matter, enlarged temporal horn, diminished gray/white matter demarcation and smaller temporal lobe (Meiners LC *et al.* 1994) (Fig. 1). Visual assessment of MTS is very accurate and if concordant with scalp EEG findings, predicts an excellent surgical outcome (Jack CR, Jr 1993).

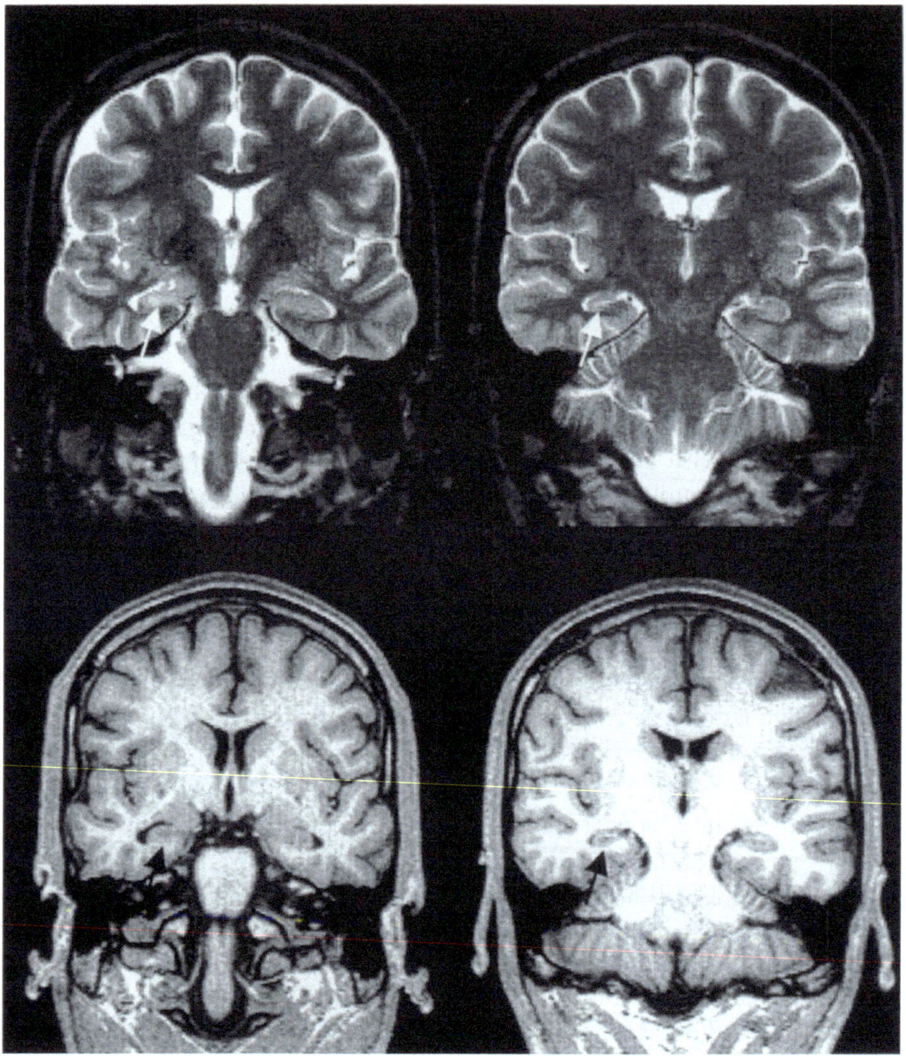

Fig. 1. Hippocampal sclerosis. T2-weighted (upper row) and T1-weighted (lower row) oblique coronal MR images of the hippocampal head (on the left) and body (on the right) of a 26-year-old male patient with TLE. Increased signal intensity on the T2 weighted images and the reduced size of the right hippocampus (arrows) are typical findings of hippocampal sclerosis with MRI

Volumetry

Quantitative methods such as MRI-based volumetry of the hippocampus (Fig. 2) (Jack CR, Jr *et al.* 1992, Kalviainen R *et al.* 1998), and recently also of the entorhinal cortex (Bernasconi N *et al.* 1999, Jutila L *et al.* 2001, Salmenpera T *et al.* 2000), are ideal tools for integrating research with

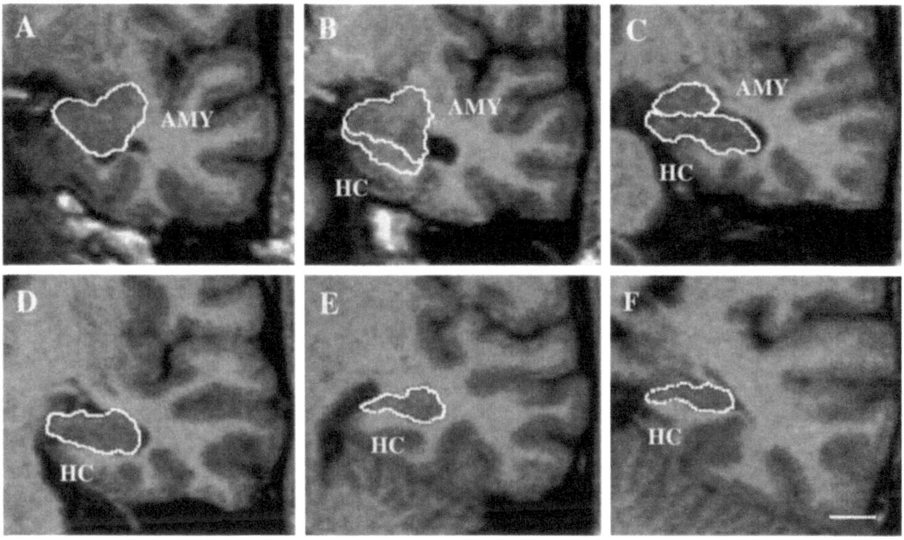

Fig. 2. MRI volumetry of hippocampus and amygdala. Manually traced boundaries of the amygdala and the hippocampus on coronal MR-images from six rostrocaudal levels of the medial temporal lobe. Panel A is the most rostral and panel F is the most caudal slice. *AMY* amygdala; *HC* hippocampus

clinical decision-making in epilepsy. Volumetry improves the diagnostic yield of MRI, as it may reveal subtle disease-related changes not apparent in routine clinical inspection and not associated with a visible lesion (Lencz T *et al.* 1992, Mackay CE *et al.* 2000). Although neuropathological examination rather than MRI is the reference standard for diagnosis of hippocampal sclerosis, measurement of unilateral hippocampal atrophy has proven to be the most accurate method for detecting hippocampal sclerosis (Kuzniecky RI *et al.* 1995) (Fig. 2). The measured reduction in hippocampal volume correlates with the severity of neuropathological cell loss (Cascino GD *et al.* 1991, Lee N *et al.* 1995). Many MRI studies have also found a positive correlation between reduced hippocampal volumes and EEG lateralization of the epileptogenic region in TLE (Jack CR, Jr *et al.* 1992, Lencz T *et al.* 1992).

NMR-Spectroscopy

Proton spectroscopy is an independent method for lateralisation of seizure onset in TLE. Both volume-of-interest single voxel and MRI spectroscopic imaging techniques have been used. A reduction in N acetyl aspartate, a neuronal marker in the epileptogenic temporal lobe, appears to be a reliable lateralising finding reported even in patients with normal MRI imaging (Fig. 3) (Connelly A *et al.* 1994, Connelly A *et al.* 1998, Vainio P *et al.* 1994). Other signals of interest are compounds containing creatine, phosphocreatine and cholin, where increased signals may indicate reactive astrocytosis. The predictive value of different metabolites for the outcome of temporal epilepsy surgery has been studied recently, and increasing NAA/Cr after surgery correlates with good seizure outcome (Serles W *et al.* 2001). A low NAA/Cr ratio before operation was an excellent predictor of postoperative seizure control as well, better than MRI volumetry; and even though the two methods are related statistically, they seem to be connected to different pathological processes (Kantarci K *et al.* 2002, Sawrie SM *et al.* 2001).

Lesions in MRI

MR reliably detects low grade tumours associated with focal epilepsies, such as ganglioglioma, dysembryoplastic neuroepithelial tumours (DNET), pilocytic astrocytomas, low-grade oligodendrogliomas and fibrillary astrocytomas. Typical features are small size, cortical or subcortical location, sharply defined borders and often little or no contrast enhancement. Additional features, such as cortical indentations in DNET and cysts in gangliogliomas, may be present. Calcifications, sometimes present in both tumours, may be more readily detectable in CT. The sensitivity of MR in identifying these tumours greatly exceeds the histological differential diagnostic capabilities of the method (Fig. 4 a–d). MRI is also highly accurate in finding vascular malformations and cavernous haemangiomas, which may be the underlining substrate in TLE.

The range of malformations of cortical development identified with MRI includes scizencephaly, subependymal heterotopia, bilateral subcortical laminar heterotopia, tuberous scleroses and focal cortical dysplasia (Raymond AA *et al.* 1995). For defining these pathologies, volumetric 3D acquisition with multiplanar reformations is essential in terms of both detection and differentiation. Focal cortical dysplasia is a challenge to MRI because of the often subtle findings and the need to differentiate it from low-grade tumours (Bronen RA *et al.* 1997). The classification of these malformations is in a state of evolution and is currently based largely on MRI features (Raymond AA *et al.* 1995).

Functional MRI

Functional MRI continues to be studied as a non-invasive alternative to the WADA test, which is currently used to assess both the language dominance and the memory competence of the hemisphere(s) in the surgical candidates (Rausch R 2002). The most important difference between

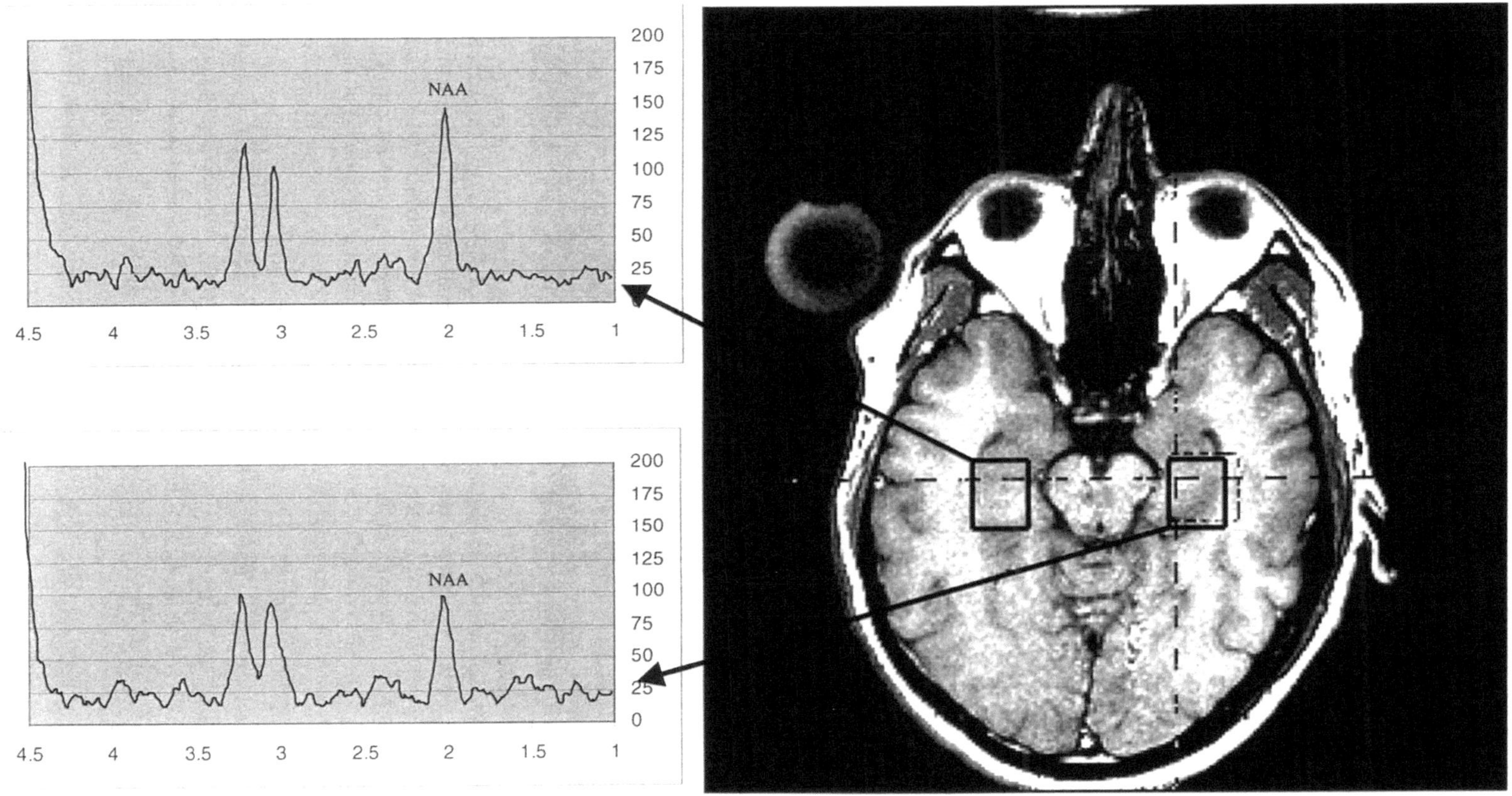

Fig. 3. NMR-spectroscopy of bilateral hippocampus in left TLE. A marked reduction in N acetyl aspartate (*NAA*) shown in the voxel of interest (*VOI*) in the hippocampus on the left side, reflecting reactive astrocytosis

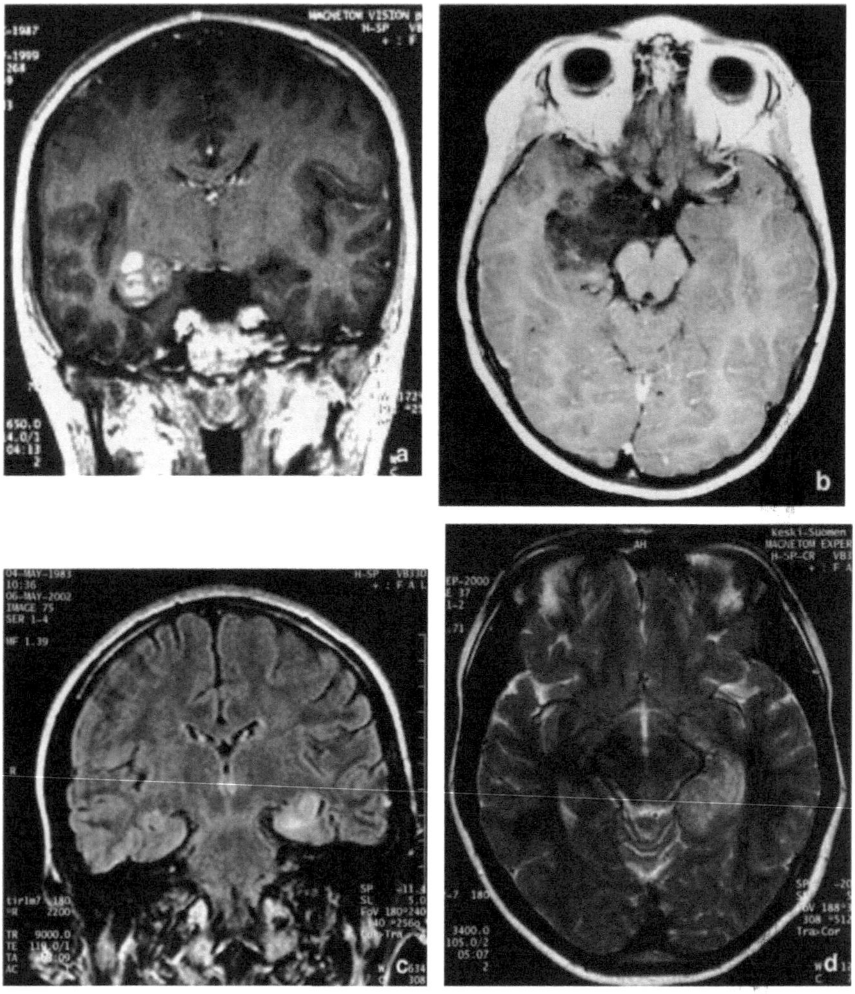

Fig. 4. (a–d) Mesial tumours with drug-resistant TLE. (a) 17-year-old woman with pilocytic astrocytoma on the right amygdala. (b) 6-year-old girl with oneyear history of mesial temporal seizures with aura. Lesion in the right amygdala and the hippocampus proved to be a dysembryoplastic neuroepithelial tumour (*DNET*). (c–d) 17-year-old woman with behavioural problems. Grade I–II oligodendroglioma in the left amygdala and the hippocampus

functional MRI and the WADA test is that functional MRI is a test of activation rather than inactivation. Functional MRI allows investigation of patients with no time limitations, and repeatedly if necessary. Functional MRI may provide a practical and reliable alternative in assessment of preoperative language lateralisation in epileptic patients (Adcock JE *et al.* 2003, Gaillard WD *et al.* 2002, Rutten GJ *et al.* 2002, Sabbah P *et al.* 2003). Additionally fMRI can also be used to detect asymmetries in memory activation in patients with TLE providing complementary information for presurgical evaluation (Detre JA *et al.* 1998, Golby AJ *et al.* 2002, Jokeit H *et al.* 2001, Killgore WD *et al.* 1999). Whether fMRI can be used to predict the risk of severe postoperative memory impairment, remains to be seen (Abou-Khalil B *et al.* 2002).

EEG-Videotelemetry in TLE

Interictal EEG

A seizure onset zone in TLE may be situated on the mesial area or neocortically in the temporal lobe (Pacia SV *et al.* 1999). Seizure discharge may also spread from the extratemporal primary focus in the ipsilateral frontal or occipital lobe. The seizure discharge may also spread rapidly from one mesial temporal lobe to the contralateral one, and especially in these cases, it is difficult to determine the side of seizure onset (King D *et al.* 1995). Interictal EEG gives evidence concerning the site of the irritative zone; but in many patients there may be several, bilateral, frontotemporal or poorly localizing interictal irritative abnormalities. Thus interictal EEG is, in practice, never sufficient for localisation of the epileptogenic zone for surgery; and ictal electro-clinical documentation of seizures is still considered to be the golden standard.

Diagnostic EEG-Videotelemetry

The first step is to determine the exact type of seizures of a given patient. This is done by diagnostic EEG-videotelemetry, usually with extracranial scalp and sphenoidal electrodes. Supraorbital, anterior temporal and mastoideus electrodes may also be used (Jayakar P *et al.* 1991). To increase the possibility to record habitual seizures, reduction of individual antiepileptic drugs reduction is used routinely. In about 80% of the adult patients with TLE, extracranial EEG-videotelemetry provides sufficient localization of seizure origin (Foldvary N *et al.* 2001) for making a decision about surgery. If the patient has mesial temporal lobe epilepsy (MTLE), the percentage is even higher (up to 90%). The extracranial ictal EEG shows focal/regional onset, which typically has a relatively slow frequency (i.e., in the alpha-theta range). A late onset extracranial EEG discharge in unitemporal electrodes occurring within 30 seconds *after* the onset of seizure is considered very highly localisatory in TLE. The ictal EEG onset is thus not necessarily seen at the time of clinical onset of a habitual seizure. Electro-clinical correlation relates EEG changes to ictal clinical manifestations temporarily. If an ictal lateralising EEG discharge is associated with a clinically habitual seizure and the MRI and neuropsychological and other studies agree with the side of the seizure onset, epilepsy surgery may be safely considered without intracranial EEG recordings.

In a case where it is not possible to localize seizure onset with extracranial EEG-videotelemetry, invasive EEG studies are usually required (Diehl B *et al.* 2000). At present we also have the possibility to visualize ictal intracranial EEG onset on a 3-D-MRI picture of the patient (Immonen A *et al.* 2003).

Intracranial EEG-Monitoring in TLE

Indications for Intracranial EEG

Today, due to the improvement in various diagnostic methods, only in about 20% of all TLE patients evaluated, precise delineation of the epileptogenic/symptomatogenic zone still requires semi-invasive (i.e., use of epidural and/or foramen ovale electrodes) or invasive EEG-monitoring (i.e., subdural and/or intracerebral depth electrodes). However, the diagnostic programs differ considerably between well-known and established centres.

In our series since 1990, 45% of those temporal lobe epilepsy patients who underwent resective surgery needed preoperative invasive intracranial recordings. The need for intracranial recordings may arise if ictal extracranial EEG and other pre-operative studies do not give sufficient evidence for the same localization, if their results are disconcordant, or if the suspected seizure onset is in or very near critical eloquent cortical areas (Hamer HM *et al.* 2001). Clinical examples include differential diagnosis of bitemporal or unilateral fronto-temporal onsets, extratemporal vs. temporal onset, or a remote extratemporal anatomical lesion (MRI) while the clinical seizure pattern suggests temporal lobe onset. In our own series, an indication for invasive EEG-investigations was suspected bitemporal (52%) or extratemporal (36%) onset of seizures or remote structural lesion (12%).

Intracranial EEG-Recording in Videotelemetry

The use of intracranial EEG necessitates careful and individual prejudgement of the hypotheses and goals for every planned recording (Rosenow F *et al.* 2001). Taking into account the crucial electrophysiological properties, discrete cortical regions are evaluated with superb temporal and spatial resolution, whereas adjacent areas just outside the recording surfaces remain totally invisible in EEG. In intracerebral depth electrodes the recording zone is even more restricted. On the other hand, minute epileptogenic phenomena missed in scalp recordings can be detected locally with invasive electrodes. Thus intracranial electrodes have a powerful ability to localise high frequency onset bursts of the primary epileptic focus if they are situated close to it.

In most of our TLE patients undergoing electrophysiolgical evaluation for epilepsy surgery, the most important strips for the analyses are the two temporal strips: one strip is positioned towards the inferomesial area, distal contacts close to amygdala and hippocampus, proximal contacts on the basal and lateral surface of TL. The other strip lies on the lateral neocortical surface of the TL, the distal contact located posteriorly. The other strips are either symmetric frontal, frontobasal or posterior. Depending on the pre-evaluation planning, with such bilateral implantation of symmetric strips, the onset and subsequent spread of seizure discharges can usually be observed. If one of the contacts is close to the epileptogenic/symptomatic zone, a high-frequency low-amplitude discharge is often seen, usually several seconds before the initial clinical symptoms appear. In such a case the possibility for successful resection is high. Another common type of initial EEG pattern consists of abrupt regional (i.e., within one strip) spike- and slow wave paroxysm, followed immediately by low-amplitude high frequency discharge at the same site. If a late-onset focal EEG discharge occurs after onset of the first clinical symptoms, the accuracy of seizure-onset localization decreases. The onset is then remote and the ictal EEG discharges spread to the symptomatogenic zone before they spread to the recording strip(s). A similar conclusion is reached when the EEG-onset is seen simultaneously in several strip contacts.

In selected patients we have also used subdural grid electrodes, most often with children, when there is a question of extratemporal onset vs. temporal spread. In these patients there has been the possibility of a temporo-occipital or temporo-parietal epileptogenic/symptomatogenic zone where the resection would be more restricted than in the anterior and mesial part of TL. Implantation of a grid electrode is always invasive and requires a craniotomy with its 32–64 contacts for the recording; and when indicated, we have performed an acute resection immediately after a few days of intracranial EEG-evaluation.

Previously we also applied foramen ovale electrodes. Since they may not sufficiently differentiate between mesial and neocortical origin of TLE seizure, however, this has not evolved to become a routine part of our protocol. The procedure of locating the electrode is quite familiar to neurosurgeons with experience in the treatment of trigeminal neuralgia with electrocoagulation.

Surgical Technique for Inserting the Electrodes

We have been using primarily 6- and 8-contact strip electrodes, with additional 4-contact electrodes if needed, and occasionally depth electrodes.

The electrodes are implanted under general anaesthesia and in TLE patients routinely through four burr holes, two of which are located bilaterally just above the zygomatic arch and anterior to the ear and two frontal burr holes anterior to the coronal suture in the midpupillary line. In order to avoid a postoperative leak of CSF, we have set an extra stitch percutaneously around the strip cable running in the subcutaneous tunnel. The localisations for electrode implantations, as well as the number of electrodes needed, is always individually tailored; but in a routine setting, 8×6-contact strips are implanted. They are implanted to record EEG bilaterally from subtemporal, temporal neocortical, frontal dorsolateral, and orbitofrontal cortices. Due to previous operations or arachnoid adhesions, in 18% of the patients we had problems sliding one or two strips into the subdural space. In that case we have chosen a shorter electrode or changed the direction of the strip slightly – there have been no haemorrhagic complications. Prophylactic antibiotics have been administered intravenously during the first postoperative day and then orally throughout the period of intensive recording, an average of 12 days.

In a few selected cases we have introduced intracerebral depth electrodes alone or in combination with subdural strips (Fig. 5). They are introduced stereotactically through the lateral temporal cortex and directed to the amygdala and hippocampus. The occipital longitudinal route has been used by many centres; and even if the risk of complications is thought to increase with invasiveness, with depth electrodes the rate of serious complications has been reported to be low (Behrens E *et al.* 1994, Blatt DR *et al.* 1997). It has been emphasized that no new type of seizure, related to this procedure itself, has developed afterwards (Espinosa J *et al.* 1994, Fernandez G *et al.* 1997).

In order to identify the precise anatomic location of the electrodes, we (as in many other centres) are using fusion of MRI image and postoperative CT-scan (Fig. 6) (So EL 2000, Winkler PA *et al.* 2000). The electrodes can be digitised, projected to the subdural space of the MRI and visualised. MRI alone may also localise the electrodes but it has been feared that the contacts in magnetic fields will heat up or that the device will move. There are reports that stainless steel or platinum electrodes scanned by MR have been used without documented complications (Davis LM *et al.* 1999, Meiners LC *et al.* 1996).

Complications of Invasive EEG-Monitoring

Chronic recording with subdural electrodes carries a low morbidity rate (Swartz BE *et al.* 1996, Wyler AR *et al.* 1991). Morbidity after implantation of depth electrodes is low as well, and in recent reports no permanent deficit has been found (Espinosa J *et al.* 1994, Fernandez G *et al.* 1997). In

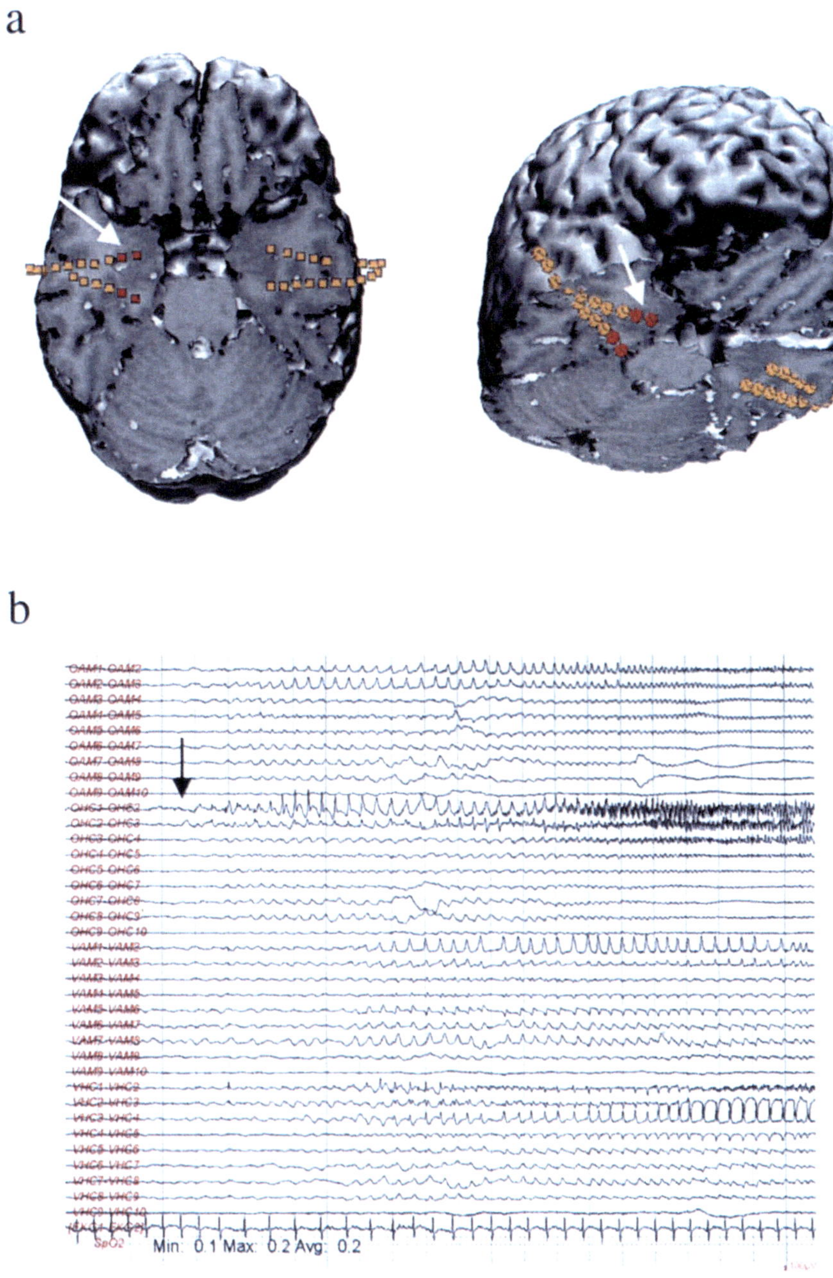

Fig. 5. (a–c) 3-D reconstructed MRI imaging of intracerebral electrodes. (a) Location of the seizure onset in 3D MR-images at the deepest contacts of the right temporomesial depth electrodes (arrow), (b) timing of onset in ictal invasive EEG (arrow), (c) overview of the monitoring

c

Fig. 5 (*continued*)

subdural electrodes the most common complication is infection, especially with subdural large-area grids. However, such electrodes are not used frequently in TLE surgery (Wiggins GC *et al.* 1999). In our own series the most common adverse effects were CSF-leakage, headache and low-grade fever. The amount of CSF leakage has diminished markedly when the tunnel has been closed with an extra stitch. No infection, permanent morbidity or neurological sequel has been noted (Table 2).

Functional Imaging in TLE

Most patients can be operated on without additional functional studies; however, there are many new noninvasive possibilities for lateralisation or localisation of an epileptic focus. Some of these are expensive and yet are considered to be developing research tools, which may improve the localisation of seizure onset. The clinical role and the possibilities to gain additional relevant data from functional MRI and NMR spectroscopy have been discussed previously.

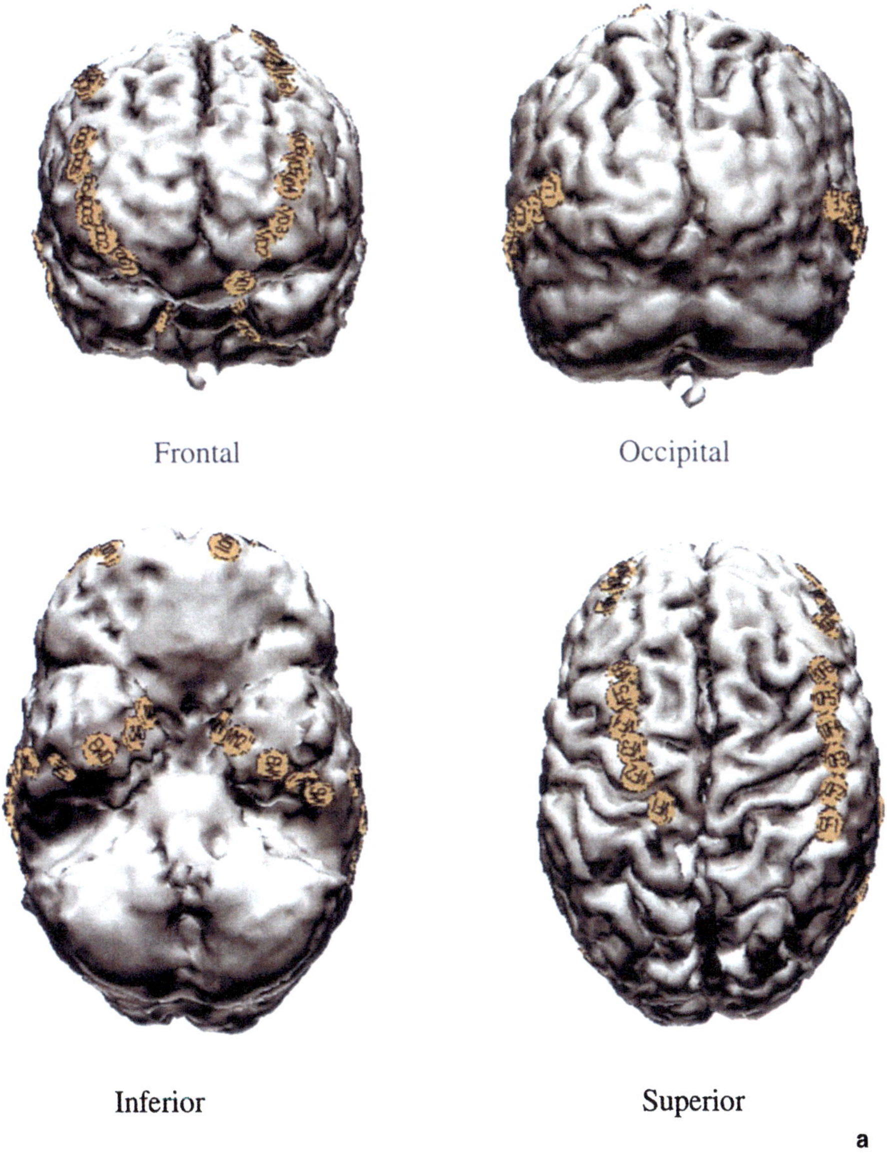

Fig. 6. Three dimensional (3-D) reconstructed MRI in localisation of subdural electrodes. Visualisation of the subdural electrodes in the 3-D reconstructed MR images after fusion of the digitised subdural electrodes from postoperative CT scan. Different views at (a and b)

Positron Emission Tomography, PET

Positron Emission Tomography (PET) images local blood flow, metabolism and brain transmitter systems in vivo, using short-lived radioisotopes as markers. Almost any molecule can be labelled by special ligands, but

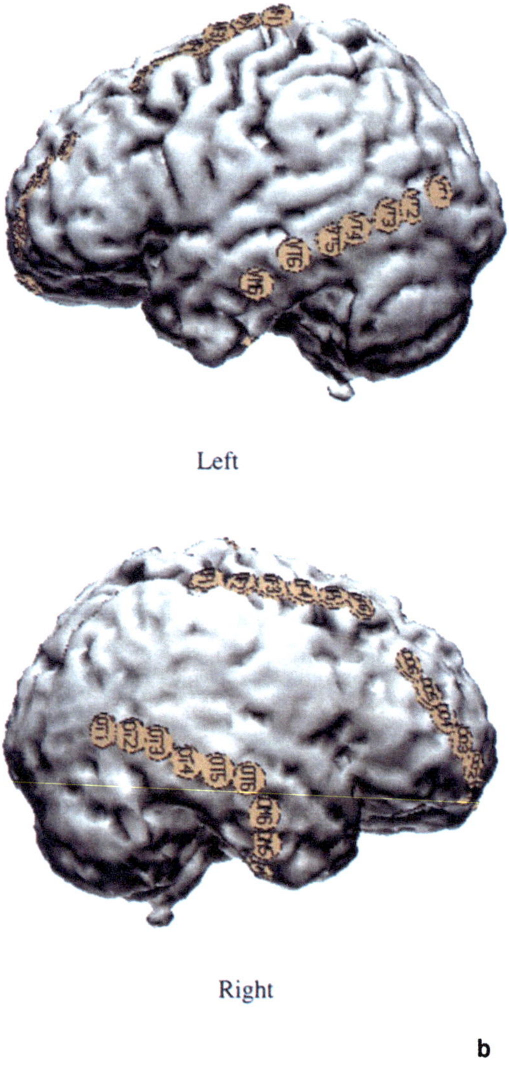

Fig. 6 (*continued*)

they have to be injected intravenously. Short-lived isotopes should be prepared just before injection; thus there must be a cyclotron on-site. PET is expensive and requires this local connection with the cyclotron. An epileptic focus is seen interictally as low glucose metabolism, e.g. [^{18}F]FDG-PET visualises a large hypometabolic area correlating with the focus in 80% of patients with focal TLE (Knowlton RC *et al.* 2001). Ictally there is increased glucose metabolism, but it is not advisable to try to obtain seizures during the PET study. For children, interictal [^{18}F]FDG-PET seems to be quite useful in focal epilepsy. PET can also image decreased GABA

Table 2. *Complications from Invasive Intracranial Monitoring; 80 Patients had 84 Invasive Diagnostic Procedures. Mean Recording Time at EEG-Videotelemetric Unit was 12 Days*

	N	*N%*
CSF leakage	12	14
Headache[1]	11	13
Low-grade fever	8	10
Nausea and vomiting	5	6
Infection[2]	3	4
Paresis of III nerve, transient	2	2
Subdural haematoma, asymptomatic[3]	1	1

[1] Need for daily repeated doses of painkillers; [2] suspected meningitis; no bacteriological verification, [3] patient with grid electrode.

metabolism in the mesial temporal area using [^{11}C]flumazenil as a tracer. An other possibility for imaging the epileptic focus is increased metabolism of the opiate receptors with 11C-carfentanil or MAO-enzymes with 11C-deprenyl. We studied the [^{11}C]flumazenil binding potential in 16 TLE patients with preoperative PET and MRI from the operative specimens of the hippocampus. In 12 patients the resected hippocampus was available for analysis of benzodiazepine binding. In all patients with pathological MRI volumetry, flumazenil benzodiazepine intake (BP) was reduced, but it was also reduced in 38% of the MRI-negative patients and in all patients who were PET positive but MRI negative (Lamusuo S *et al.* 2000). This indicates the sensitivity of [^{11}C]flumazenil PET in presurgical evaluation of TLE-epilepsy. However, we have used the PET technique mostly in diagnosing extratemporal focal epilepsy, especially in children with controversial findings. Multiple epileptic foci often mean difficulties in mapping the epileptogenic area, and all available non-invasive information is necessary (Fig. 7) (Choi JY *et al.* 2003).

Single Photon Emission Computed Tomography, SPECT

Single Photon Emission Computed Tomography (SPECT) is based on much cheaper technology: radioactive isotopes emitting gamma radiation with a much longer half time. The resolution of SPECT has been increasing continuously and it is already quite near that of PET. Information on GABA metabolism can be provided by [^{123}I]iomazenil-SPECT, and [^{99m}Tc]HMPAO-SPECT can be used for measurement of ictal cerebral

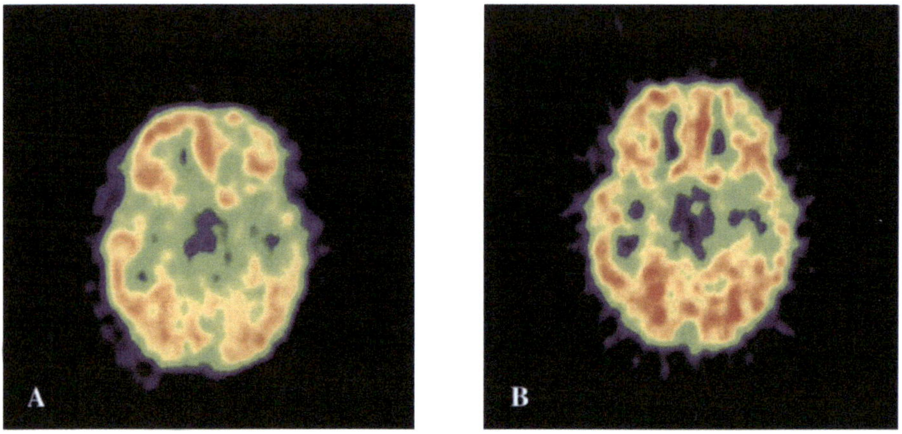

Fig. 7. [^{18}F]FDG-PET and [^{11}C]flumazenil-PET in localising TLE focus. (A) [^{18}F]FDG-PET shows an area of hypometabolism in the left temporal area. (B) [^{11}C]flumazenil-PET showed reduced binding also in the left temporal lobe but compared to the [^{18}F]FDG-PET the area was smaller and covered only the medial temporal lobe

blood flow (CBF) in the focal epileptic area. The same general principles used in PET also work for SPECT, but they are better for ictal studies. The ictal injection can be given during EEG-videotelemetry, confirming the precise timing of the tracer in concordance with the clinical seizure. We have found this to be extremely important, since ictal SPECT may also show focal hypoperfusion in the late ictal/immediate postictal phases (Kuikka JT *et al.* 1994).

In 18 patients with TLE we performed a prospective comparative PET-study, in which the localization from invasive subdural EEG-videotelemetry was used as a reference. Before surgery all patients were investigated with three radionuclides: interictal [^{18}F]FDG-PET and ictal [^{123}I]iomazenil-SPECT and [^{99m}Tc]HMPAO-SPECT. Fifteen patients had a localisatory subdural EEG; and of these patients, 13 had corresponding and concordant focal/regional findings in [^{18}F]FDG-PET studies. In 9 patients the [^{123}I]iomazenil-SPECT agreed, but with [^{99m}Tc]HMPAO-SPECT the yield of interictal findings was much less accurate (Lamusuo S *et al.* 1997).

Magnetoencephalography, MEG

Magnetoencephalography (MEG) was developed in the 1980's as a product of new technology in very-low-temperature physics. Interictal magnetic charges of the epileptic temporal neocortical area agree well with the epileptiform discharges seen in peroperative electrocorticography and acute peroperative depth electrodes during surgery (Tiihonen J *et al.* 1990). The second- and third-generation scalp scans with 122 and more channels already made MEG a potential alternative to more invasive studies, and further recent improvements plus over 200 detectors allow simultaneous measurement all over the scalp. MEG has been shown to be a potential tool in evaluating extratemporal epilepsy; but for evaluation of MEG as a presurgical tool for investigations in TLE, large clinical series are lacking (Wheless JW *et al.* 1999).

In a prospective study, eight TLE patients from our centre were evaluated with subdural EEG videotelemetry, which was used as a reference; and the surgical decision was based on this information. All patients also had presurgical PET and MEG studies. The information obtained from PET and MEG was correlated with the results of subdural EEG and the seizure outcome after possible surgery. In five out of six operated patients the PET and MEG were congruent with subdural EEG, and the outcome of the resection was good. In two patients one study showed bilateral and the other unilateral lesion/focal abnormalities; one had an operation with no change in seizures, the other has been treated conservatively (Lamusuo S *et al.* 1999). The value of preoperative MEG in surgery for TLE is at present still largely supportive, but needs further evaluation as a potential non-invasive and completing source of information (Fig. 8). The main disadvantage of applying MEG as a routine pre-operative method is the difficulty to obtain ictal recordings. The signal-to-noise ratio of MEG is low for deeply situated subcortical structures compared to cortical magnetic activity, which makes it a less ideal method for studying the mesial temporal lobe.

Indications for Surgery in TLE

Upon completion of the presurgical evaluation, the team should be able to decide whether surgery can be undertaken and define its risks and expected outcome (curative vs. palliative surgery). Firstly, a discrete structural abnormality should be identified, either directly with neuroimaging techniques or indirectly related to an associated functional deficit. Secondly, evidence of localised abnormal excitability of the actual or presumed structural lesion should be provided; and finally, the suspect region should not be situated within the eloquent cortex (Sperling MR *et al.* 1997). If these objectives are not controversial, there is a high likelihood of a satisfactory surgical result. However, no single test provides definitive information on which surgery can be based. The primary data for the decision comes from MRI and EEG recordings (including all interictal and ictal EEG:s); secondary data are obtained from the clinical seizure pattern,

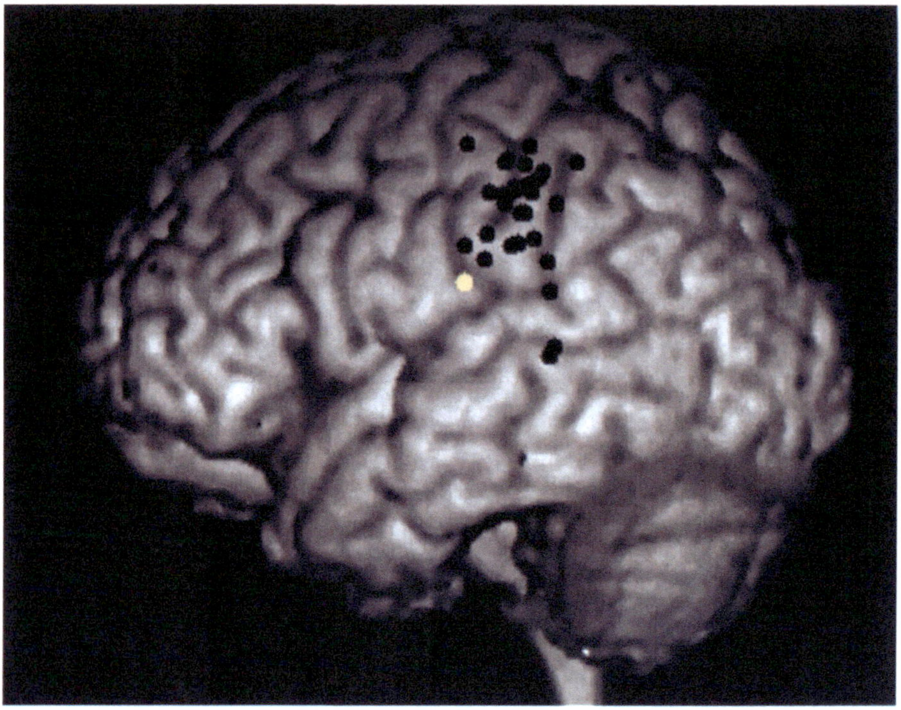

Fig. 8. MEG localising the focus. Sources of interictal spikes measured with whole-scalp MEG overlaid on the patient's MRI. The yellow dot shows the localisation of auditory cortex

neuropsychological evaluations and functional imaging (SPET, PET, fMRI). The basic rule for temporal lobe surgery might be positive findings in both primary investigations and supportive findings in two secondary investigations without conflicting data from other investigations. In the case of conflicting findings, intracranial recordings and/or other special methods (MEG) should be considered.

Surgery in Patients with Temporal Lobe Lesion

In normal clinical neurosurgical decision-making and when a specific epilepsy team approach is needed, the definition of a lesion is important in deciding which lesion should be taken care of. If the fresh epileptic symptoms are caused by an expansive malignant grade II–IV glioma, there is no question of the priority of the approach. The lesion needs to be diagnosed, resected and treated aggressively; and epilepsy is a secondary issue. With good quality MRI, however, almost all benign temporal tumours are now found soon after the first epileptic symptoms. It is not clear, whether a grade I astrocytoma, oligodendroglioma, hamartoma, ganglioglioma, dysembryoplastic neuroepithelioma or a benign cyst should be extirpated as soon as it is seen, in hope of preventing progressive development of mesial temporal changes (Fig. 9 a–g). In the case of simultaneous mesial and neocortical (double) pathology with refractory epilepsy, the surgery should remove both the tumour and the mesial structures. With a radiologically intact hippocampus, we follow the tradition of removing only the lesion (Lombardi D *et al.* 1997, Rossi GF *et al.* 1999). Acute perioperative ECoG with deep temporomesial electrodes may help in the decision to continue with resection, and a clearly irritative mesial area should be resected (Zentner J *et al.* 1997).

If the lesion is vascular and expressed as symptomatic epilepsy, we perform only an early lesionectomy, as do many centres with experience in vascular surgery (Maraire JN *et al.* 1995). The most frequent finding, cavernous haemangioma, will be removed microsurgically using neuronavigation; only cavernomas inside mesial structures should be removed together with the mesial TL (Fig. 10).

In the literature the seizure outcome of epilepsy surgery with pure lesionectomy is contradictory. Crandall and Mathern have compared their results with others and recommend always using lesionectomy with mesial resection, because of the slightly better seizure-free outcome in this group (Crandall PH *et al.* 2001). If the WADA test shows good verbal memory function and the MRI changes show no mesial sclerosis, we would suggest removing only the lesion especially if the history of epilepsy is short and the lesion is on the dominant side.

Surgery of temporal lobe epilepsy in children and adults is not technically an age-dependent issue, but the pathology is very different. Because the silent period in the progression of epilepsy often occurs late in childhood, it may not yet be possible to diagnose the mesial changes with MRI. On the other hand, developmental lesions and typical benign paediatric tumours are well demarcated MRI-findings. In our series of 27 children with temporal lobe lesion, hippocampectomy was included in only 11 resections.

Technique of Tailored Temporal Lobe Resection with Amygdalo-Hippocampectomy

There are several approaches for resection of neocortical and mesial epileptogenic areas in temporal lobe epilepsy, beginning from large temporal lobectomies to more tailored neocortical resections combined with mesial procedures and finally, selective amygdalo-hippocampectomies from different directions, as presented by M Sindou (Sindou M *et al.* 2002). Very few centres have a possibility for the use of intraoperative MRI that may

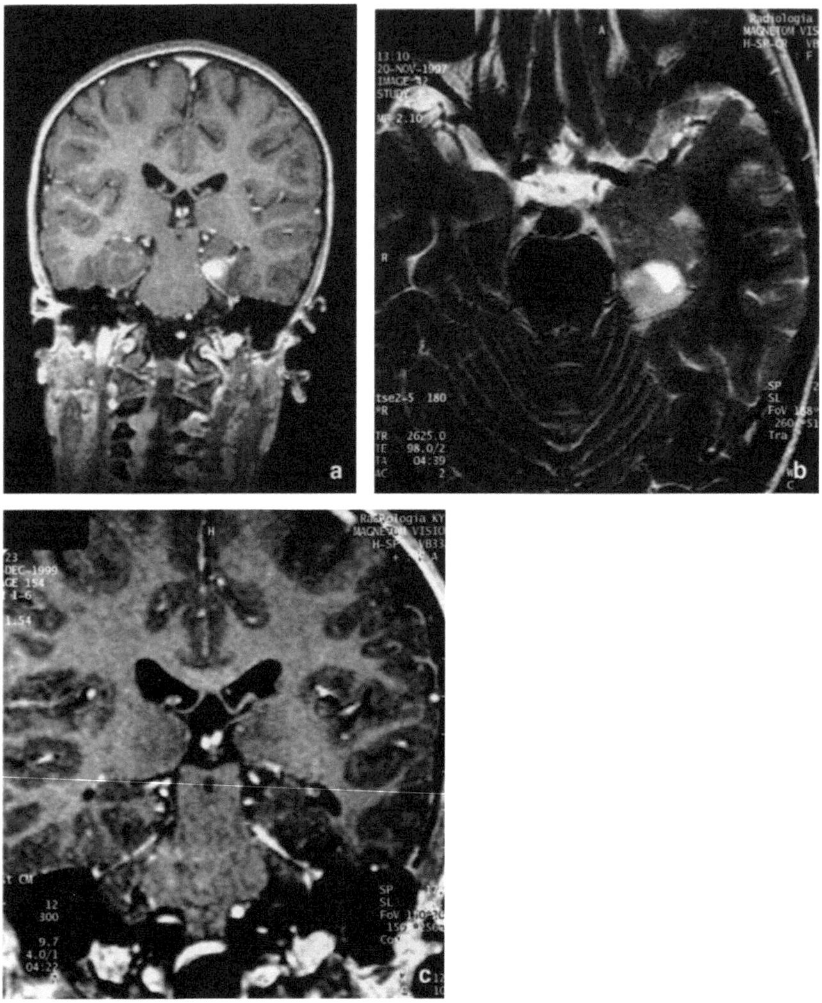

Fig. 9. (a–g) Surgical resection in temporal lobe lesions. (a–c) 6-year-old boy with two-year history of cognitive difficulties and absence seizures. Enhancing tumour in left posterior parahippocampus (a and b). Lesionectomy with neuronavigation confirmed ganglioglioma. Four years later no seizures, and postoperative MRI revealed a cyst (c). (d–e) 7-year-old boy with delayed speech and dyspraxia combined with complex partial seizures. MRI imaging shows a large right neocortical tumour extending to the hippocampus. Mesial temporal lobectomy with lesionectomy was performed. PAD: dysembryoplastic neuroepithelial tumour (*DNET*). Seizure-free at 6-year follow-up. (f–g) 30-year-old woman with a history of seizures since childhood. T2-weighted MRI shows amygdala signalintensity on the right side (f). Inferior temporal lobe together with gliotic amygdala and hippocampus were removed (g). Patient is seizure-free; the final PAD was ganglioglioma

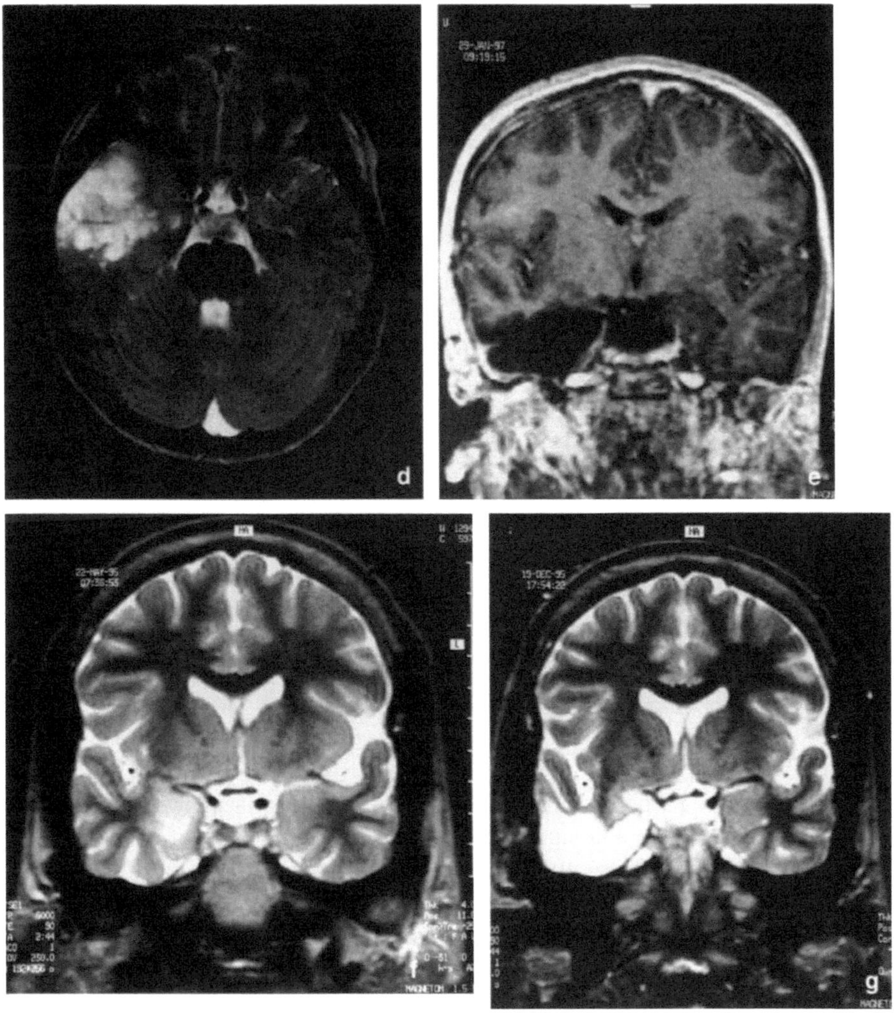

Fig. 9 (*continued*)

increase the control of resection and localisation of the lesion (Buchfelder M *et al.* 2002). The role of non-surgical therapeutic procedures, such as Gamma-Knife radiosurgery in MTLE, is still not established and more prolonged follow-up studies are necessary (Regis J *et al.* 2002).

In our programme of epilepsy surgery we have performed 230 temporal resections from 1988 until 2002. Of these 230 procedures, 140 consecutive tailored lobectomies with amygdalo-hippocampectomy were performed for adults between 1988–1999 (Jutila L *et al.* 2002b). Patients with only temporal lesionectomies were excluded, as were patients in whom extratemporal excisions were carried out in addition to the temporal

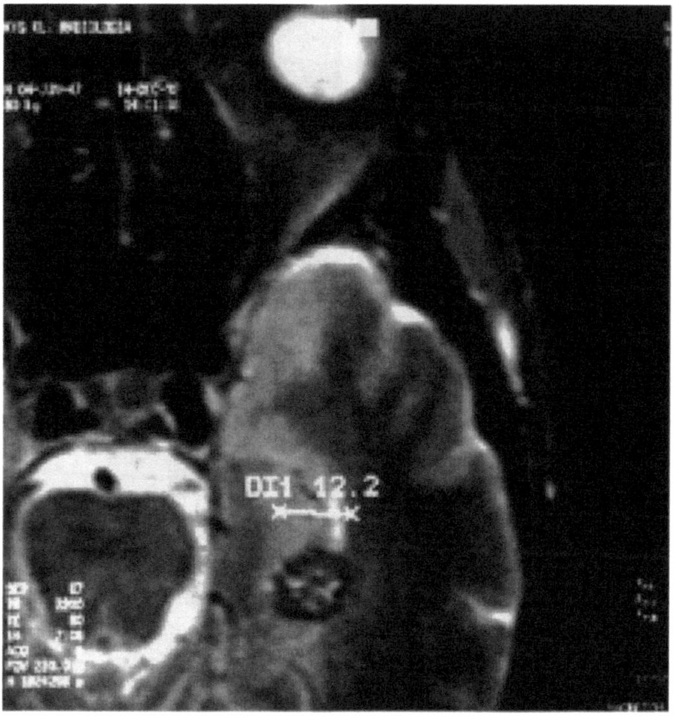

Fig. 10. Hippocampal cavernous haemangioma. 45-year-old construction worker with a 4-year history of partial epilepsy. MRI shows cavernous haemangioma in the left hippocampus. The lesion and hippocampus were resected; the patient was seizure-free at the ten-year follow-up

resection. We have used the lateral approach during all these years. In order to decide the extent of neocortical resection and assess the electric activity of mesial temporal lobe, we have used peroperative neocortical electrocorticography (EcoG) with two acute mesial intracerebral electrodes.

For our epilepsy surgery patients we do not use awake anaesthesia. We prefer controlled respiration with relaxants and low concentration isoflurane/NO2-oxygen with normal sedatives as premedication and barbiturates during intubation. The patients are under light anaesthesia after the craniotomy before the EcoG; and after the electrophysiological studies, the level of anaesthesia is again deepened during the resection and closing of the craniotomy. The neuronavigation instrument is calibrated and checked before the operative area is prepared. The anterior temporal pole is exposed from a frontotemporal skin flap, which is turned forward together with the temporal muscle. The bone is removed as a lambeau, and the craniotomy is further enlarged basally so that the inferior temporal gyrus can be seen well when the dura is opened and lifted up.

Peroperative Electrocorticography

With neuronavigation the position of the temporal horn is checked, as is the location of the hippocampus and amygdala. For electrocorticography, two 4-contact intracerebral electrodes are applied in both structures together with 16 surface electrodes placed neocortically. Both neocortical and intracerebral spontaneous EEG is registered for 5–10 minutes, and spontaneous spikes are carefully analysed by a neurophysiologist. After that, intravenous methohexital (or recently propofol) is used to enhance the occurrence of further focal epileptic discharge. Depending on the local irritation and earlier video-EEG information, the extent of neocortical resection is decided; and if the neocortex is silent, only the hippocampus and amygdala with accompanying parahippocampal gyrus is removed selectively.

If there is neocortical focal irritation independent of mesial discharge, that area will first be removed *en block*. Even if there is no functional hindrance, we do not go more than 3.5–4 cm posterior from the temporal pole.

Selective hippocampectomy without neocortical resection is performed with an initial 20 mm cortical resection through the inferior medial temporal gyrus with the help of neuronavigator to localise the temporal horn. CSF drainage from the ventricle helps give more space, and the hippocampus with its fimbria is seen. The ventricle and plexus will be the upper and mesial border of the resection, and is usually lightly retracted medially and upwards so that the plexus is left under the sponge and retractor. Hippocampectomy can be performed from that approach but is more atraumatic when the cortex is further resected by cutting the inferior temporal gyrus at the level of 1.5–2 cm from the temporal pole using the basal arachnoid as the limit (Fig. 11 a). The hippocampus and amygdala are separated by continuing this resection down to the mesial arachnoid. The hippocampus is then cut about 3–3.5 cm behind the anterior resection in the temporal horn and is separated from the temporal lobe at the level of the arachnoid posteriorly, laterally and anteriorly with an ultrasound aspirator and bipolar forceps. Mesially, light ultrasound can also be used; the vascular stilus comes from veins and the artery of the choroidea anterior, which should be identified, separated, coagulated and cut. The hippocampus is lifted out in one piece for further pathological investigations. Again using the anterior and basal arachnoid as the border, the amygdala is separated from the lateral arachnoid as a 15–20 mm thick layer and then removed in one piece together with the uncus. The entorhinal cortex is removed by ultrasound during resectioning of the parahippocampal gyrus. The typical postoperative MRI is shown in Fig. 11 b–c.

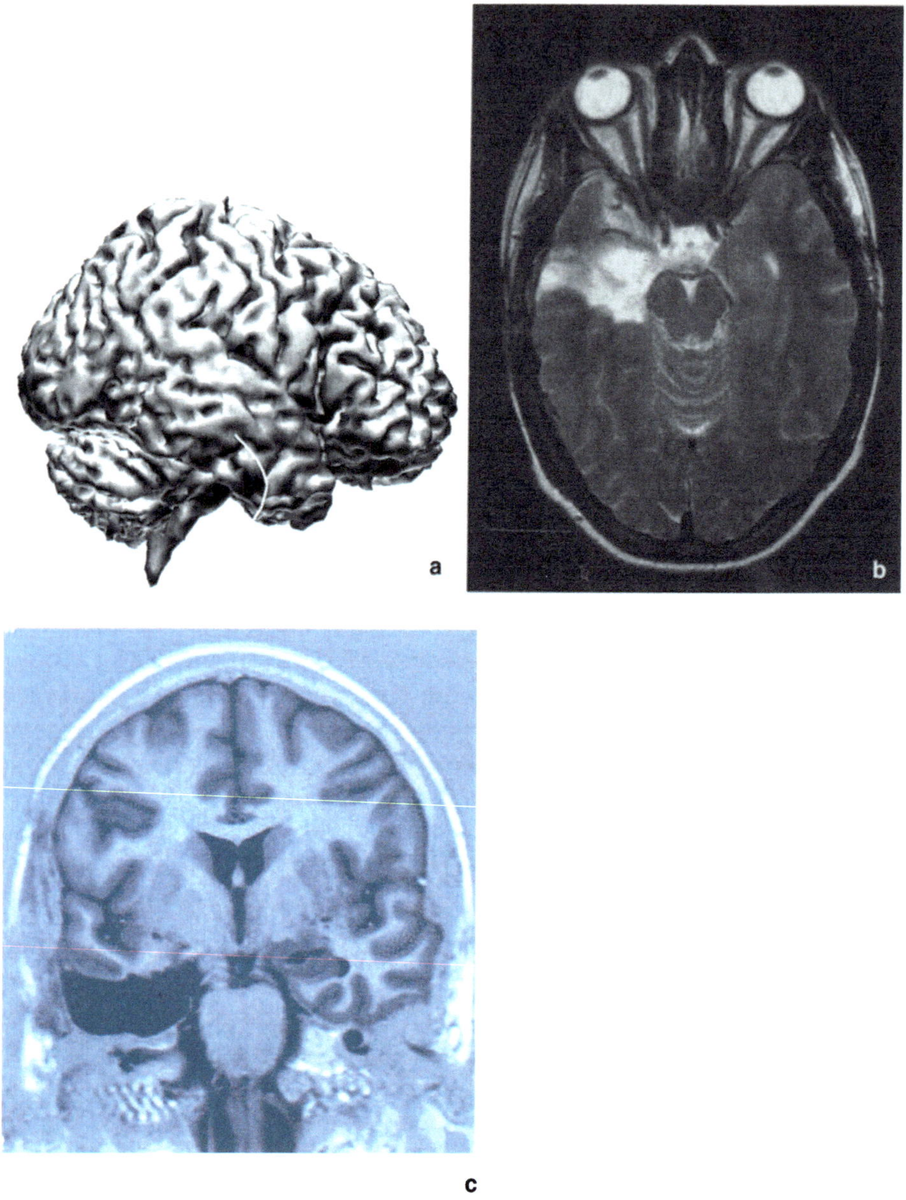

Fig. 11. (a–c) Right lateral approach to the temporomesial structures and postoperative MR-image after selective amygdalohippocampectomy. (a) Cortical incision of the inferior medial temporal gyrus. Postoperative (b) axial and (c) coronal views in MR-images

For many years we performed control EcoG after resection but noticed that this information did not actually change our procedure, so we have given up this step. The closing of the wound follows the normal steps of intracerebral surgery. Bleeding is usually minimal, the dura is closed and lifted up, the bone is fixed with two platinum plates and the soft tissues are closed in layers. The patients receive their own AED orally in the morning of the operation day and again from the first day postoperatively. Fosphenytoin is given intravenously for the first three days and prophylactic antibiotics during the operation. In general, the patient leaves the hospital on the 5th postoperative day. Routine follow-up by the team is 3 months, one year, 3 years and 10 years later with neurological, psychiatric and neuropsychological evaluations, including neurophysiological and MRI-investigations.

Outcome of Surgery with Respect to Seizures

Most of the data on outcome of TLE surgery have been derived from a series of individual centres. Only a single randomised controlled study is available (Wiebe S *et al.* 2001). In addition, there are significant differences in the preoperative evaluation strategies, in the extent of the resection, in the definition of surgical complications, in the length of postoperative follow-up and in the inclusion of different measures of outcome.

The most common way of assessing the postoperative surgical outcome is outcome in terms of seizures. Until recently the original classification adapted from Engel has been used almost exclusively (Table 3) (Engel JJ 1987). Already in his original publication, Engel differentiated between auras and seizures that impair consciousness; and patients with postoperative auras only were considered "seizure free". This definition has subsequently been adopted in most studies, including our own. However, it is certainly true that typical temporal lobe auras are focal seizures originating from the temporal lobe structures. A new classification by The International League Against Epilepsy (ILAE) now separates seizure-free patients and patients with postoperative auras only, considering them to be two distinct entities (Wieser HG *et al.* 2001). The proposed classification also counts seizure days rather than total number of postoperative seizures.

A detailed meta-analyses of 126 studies published between 1900 and 2000 showed that 70% (median) of patients become seizure free (range 33%–93%) (McIntosh AM *et al.* 2001). We have summarized some recent follow-up studies of temporal lobe epilepsy in Table 4. Our own consecutive series of patients includes 140 adults with TLE operated on between 1988 and 1999. In addition to children, patients with selective lesionectomy were excluded as were those patients for whom any extratemporal

Table 3. *Seizure Outcome Classification by Engel J. (Adapted from Engel J et al. 1987)*

Class I: seizure-free	
A	Completely seizure free since surgery
B	Auras only since surgery
C	Some seizures after surgery but seizure free for at least 2 years
D	Atypical generalized convulsions with antiepileptic drug withdrawal only
Class II: rare seizures (fewer than 3 per year)	
A	Initially seizure free but has rare seizures now
B	Rare seizures since surgery
C	More than rare seizures since surgery but has rare seizures now (for at least 2 years)
Class III: worthwhile improvement	
A	Creater than 80% seizure reduction and improved lifestyle; or lesser than 80% seizure reduction but resolution of generalized tonic clonic seizures
Class IV: no worthwhile improvement	
A	Significant (at least 50%) seizure reduction but no lifestyle benefit
B	No change from preoperative seizure frequency and severity of seizures
C	Seizures more frequent or severe than preoperatively

resection was part of the procedure. The aim of surgery was determined preoperatively as "curative" for 74% ($n = 103$) of patients with unilateral TLE. Surgery was defined as "palliative" for 26% ($n = 37$) of the patients. The palliative category included, for example, patients with bitemporal seizure onset (49%), multifocal epilepsy, dual pathology, or patients in whom the epileptic focus could not be removed completely. Outcome of surgery was assessed independently in unilateral TLE and in palliative patients due to the significant difference in expected outcome (Fig. 12) (Jutila L *et al.* 2002b). In unilateral TLE, 45% of the patients were completely seizure-free and 12% had only postoperative auras. Engel I–II outcome was observed in 68% of the patients. In the palliative group a minimum of 80% seizure reduction (Engel I–III) was achieved by 62% of patients (27% were completely seizure free and 3% had only postoperative auras). Outcome of surgery improved significantly after introduction of the standardised MR imaging protocol: 74% of patients with unilateral TLE achieved Engel I–II outcome in the one-year follow-up (61% were free of seizures, 11% had only postoperative auras, and 2% had rare seizures). In the long-term follow-up seizure reduction of at least 80% was achieved by 71% of the palliative patients. Our study demonstrated that patients without restricted unilateral seizure focus can also benefit from surgery.

Table 4. *Outcome with Respect to Seizures after Temporal Lobe Epilepsy Surgery*

Authors	SF	Length of follow-up
Elwes *et al.* 1991	56%	one year
	67% (in previous year)	five years
Rougier *et al.* 1992	66%	one year
	62% (from surgery)	five years
Sperling *et al.* 1996	55% (from surgery)	five years
	70% (in previous year)	five years
So *et al.* 1997	74%	one year
	69% (in previous year)	five years
Wyllie *et al.* 1998[a]	78%	mean 3.6 years
Salanova *et al.* 1999	66%	one year
	60% (in previous year)	five years
Foldvary *et al.* 2000	63%	one year
	52% (from surgery)	five years
Wiebe *et al.* 2001	58%	one year
Boling *et al.* 2001[b]	61%	mean 64 months
Jutila *et al.* 2002	57%	one year
	56%	mean 5.4 years
Clusmann *et al.* 2002	71%	mean 38 months
Wieser *et al.* 2003[c]	67%	Median 7.2 years

SF Percentage of seizure-free patients.
[a] Children and adolescents.
[b] Patients 50 years of age or older.
[c] Selective amygdalohippocampectomy.

Outcome at one year or at two years postoperatively predicts the long-term outcome in operated patients with TLE (Foldvary N *et al.* 2000, Rougier A *et al.* 1992). In our study 86% of all seizure relapses occurred within one year after surgery. Late seizure relapses (>2 years after surgery) were observed in only 5% of all patients, and these were often preceded by a specific explanatory factor such as withdrawal of antiepileptic medication (Jutila L *et al.* 2002b). In multivariate analyses, unitemporal MR imaging abnormalities, early onset of epilepsy and the predominance of seizure type (focal seizures with impaired consciousness and focal ictal EEG) were identified as predictors for successful postoperative outcome.

Complications

Like any major neurosurgical procedure, surgical treatment of TLE poses a risk of postoperative complications (Table 5). However, most of these are transient, and both morbidity and mortality are low. In a recent na-

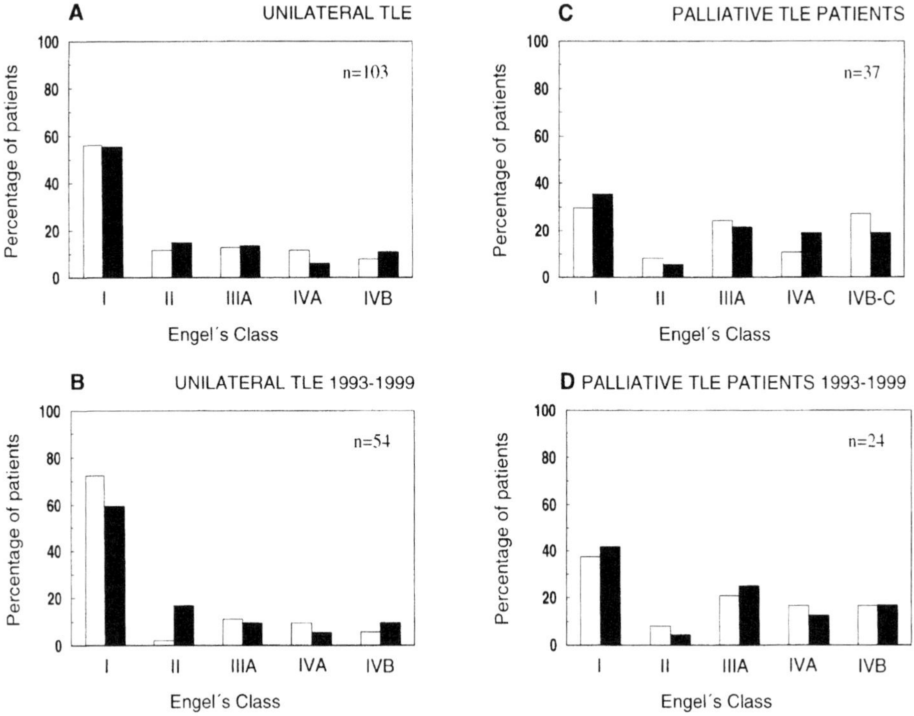

Fig. 12. Outcome of surgery with respect to seizures in 140 concomitant adult patients with TLE, Epilepsy Surgery Centre, Kuopio 1988–1999. Outcome of surgery with respect to seizures as percentage of patients in different Engel's classes: children, patients with pure lesionectomy or extended extratemporal resections excluded. (A) All patients with unilateral TLE (mean follow-up 5.4 years); (B) Patients with unilateral TLE imaged with a standardised MR imaging protocol (1993–1999, median follow-up 3.8 years); (C) All palliatively-operated TLE patients (mean follow-up 4.4 years); and (D) Palliatively-operated TLE patients imaged with a standardised MR imaging protocol (1993–1999, median follow-up 3.1 years). White columns refer to one year and black columns to long-term follow-up data. Abbreviations: *I* seizure-free patients (including patients having only auras); *II* patients with rare seizures (fewer than three seizures per year); *III* patients with a seizure reduction of at least 80%; *IVA* patients with a seizure reduction of at least 50%, *IVB* patients with no change in seizure frequency; *IVC* patients with more frequent seizures. (Jutila L *et al.* 2002b). (Published and modified with permission from the BMJ Publishing Group, J Neurol Neurosurg Psychiatry; 2002, 73:486–494)

tional multicentre survey by Rydenhag and Silander (Rydenhag B *et al.* 2001) the risk of postoperative major complicatios was 3.1% and that of minor complications 8.9%. A complication was defined as major if it included any significant neurological deficit or affected activities of daily living. Minor complications resolved within three months. In younger patients the risk of complications was related to age (under 35 years). The number of complications was comparable to an earlier German series,

Table 5. *Complications in 140 Adult Patients Operated on for TLE. Patients with Lesionectomies Only or with any Added Extratemporal Excisions were Excluded*

Complication	Minor	Major
Surgical		
Haematoma or haemorrhage	4[a]	1
Hydrocephalus	1	
Aseptic meningitis	5[b]	
Infection	1	
Deep vein thrombosis	1[c]	
Neurological		
Homonyme hemianopia		1
Dysphasia	3[b,c]	1
Total number of complications	15	3
Total number of patients with complications	13	3

[a] Three chronic subdural haematomas and one subdural effusion.
[b] One patients with transient dysphasia had also aseptic meningitis.
[c] One patient with transient dysphasia had also deep venous trombosis.
(Published with permission from the BMJ Publishing Group, J Neurol Neurosurg Psychiatry; 2002, 73: 486–494).

describing minor complications in 7.5% and major complications in 4.7% of operated patients (Behrens E *et al.* 1997). However, it is still somewhat uncertain to what extent reduced seizure frequency and improved quality of life in operated patients with TLE reduces mortality. Mortality is reduced in seizure-free patients compared to patients with recurrent postoperative seizures (Sperling MR *et al.* 1999). TLE surgery does not, however, normalise the overall mortality associated with chronic epilepsy (Hennessy MJ *et al.* 1999).

The presence of psychiatric symptoms is relatively common during the early months after surgery for TLE. However, the majority of patients with these symptoms also display a preoperative history of psychiatric disorders. In a prospective study by Ring *et al.* (Ring HA *et al.* 1998) half of the patients with normal preoperative psychiatric status had developed symptoms of anxiety and depression six weeks after the operation. At three months the depressive states tended to persist. A Danish series showed de novo depression in 8% of the patients after temporal lobectomy (Naylor AS *et al.* 1994).

Conclusions

In drug-refractory temporal lobe epilepsy, surgery is superior to prolonged medical treatment and the postoperative outcome is good in a majority of the patients with 70% (median) of patients becoming seizure free. In selected patients without restricted unitemporal seizure focus, palliative treatment is also beneficial. Many patients with TLE can be evaluated at relatively moderate expense and without invasive monitoring; however, new methods of investigation including functional MRI, are still needed. In future the emphasis should be placed on early prediction of medically intractable TLE in children and adolescents, since early surgery probably improves the overall outcome, especially in younger patients. Further evaluation for long-term cognitive outcome is also warranted.

Acknowledgements

This study was supported by the Academy of Finland, the North-Savo Regional Fund of the Finnish Cultural Foundation, the Kuopio University Hospital Research Fund (EVO Fund 577 27 19), the University of Kuopio, and the Vaajasalo Foundation.

References

1. Abou-Khalil B, Schlaggar BL (2002) Is it time to replace the Wada test? Neurology 59:160–161
2. Adcock JE, Wise RG, Oxbury JM, Oxbury SM, Matthews PM (2003) Quantitative fMRI assessment of the differences in lateralization of language-related brain activation in patients with temporal lobe epilepsy. Neuroimage 18:423–438
3. Anand I, Kotagal P, Hammel J (1997) Seizure semiology of lateral vs mesial temporal lobe epilepsy using statistical analyses. Neurology 48 [Suppl] 2:240–241
4. Andermann F (2002) Temporal pole and mesiotemporal epilepsy. Epileptic Disord 4 [Suppl] 1:7–8
5. Babb TL (1992) Research on the anatomy and pathology of epileptic tissue. In: Luders HO (ed) Epilepsy surgery. Raven Press, New york, pp 719–727
6. Baxendale SA, Thompson PJ, Kitchen ND (2000) Postoperative hippocampal remnant shrinkage and memory decline: a dynamic process. Neurology 55:243–249
7. Behrens E, Schramm J, Zentner J, Konig R (1997) Surgical and neurological complications in a series of 708 epilepsy surgery procedures. Neurosurgery 41:1–9; discussion 9–10
8. Behrens E, Zentner J, van Roost D, Hufnagel A, Elger CE, Schramm J (1994) Subdural and depth electrodes in the presurgical evaluation of epilepsy. Acta Neurochir (Wien) 128:84–87
9. Bernasconi N, Bernasconi A, Andermann F, Dubeau F, Feindel W, Reutens DC (1999) Entorhinal cortex in temporal lobe epilepsy: a quantitative MRI study. Neurology 52:1870–1876

10. Blatt DR, Roper SN, Friedman WA (1997) Invasive monitoring of limbic epilepsy using stereotactic depth and subdural strip electrodes: surgical technique. Surg Neurol 48:74–79
11. Boling W, Andermann F, Reutens D, Dubeau F, Caporicci L, Olivier A (2001) Surgery for temporal lobe epilepsy in older patients. J Neurosurg 95:242–248
12. Bronen RA, Vives KP, Kim JH, Fulbright RK, Spencer SS, Spencer DD (1997) Focal cortical dysplasia of Taylor, balloon cell subtype: MR differentiation from low-grade tumors. AJNR Am J Neuroradiol 18:1141–1151
13. Buchfelder M, Fahlbusch R, Ganslandt O, Stefan H, Nimsky C (2002) Use of intraoperative magnetic resonance imaging in tailored temporal lobe surgeries for epilepsy. Epilepsia 43:864–873
14. Cascino GD, Jack CR, Jr, Parisi JE, Sharbrough FW, Hirschorn KA, Meyer FB, Marsh WR, O'Brien PC (1991) Magnetic resonance imaging-based volume studies in temporal lobe epilepsy: pathological correlations. Ann Neurol 30:31–36
15. Chelune GJ (1995) Hippocampal adequacy versus functional reserve: predicting memory functions following temporal lobectomy. Arch Clin Neuropsychol 10:413–425
16. Chelune GJ, Najm I (2001) Risk factors associated with postsurgical decrements in memory. In: Luders HO, Comair YG (eds) Epilepsy surgery. Lippincott Williams and Wilkins, Philadelphia, pp 497–504
17. Choi JY, Kim SJ, Hong SB, Seo DW, Hong SC, Kim BT, Kim SE (2003) Extratemporal hypometabolism on FDG PET in temporal lobe epilepsy as a predictor of seizure outcome after temporal lobectomy. Eur J Nucl Med Mol Imaging 30:581–587
18. Clusmann H, Schramm J, Kral T, Helmstaedter C, Ostertun B, Fimmers R, Haun D, Elger CE (2002) Prognostic factors and outcome after different types of resection for temporal lobe epilepsy. J Neurosurg 97:1131–1141
19. Cockerell OC, Sander JW, Shorvon SD (1995) Remission of epilepsy. The NGPS. National General Practice Study of Epilepsy. Lancet 346:1228
20. Connelly A, Jackson GD, Duncan JS, King MD, Gadian DG (1994) Magnetic resonance spectroscopy in temporal lobe epilepsy. Neurology 44:1411–1417
21. Connelly A, Van Paesschen W, Porter DA, Johnson CL, Duncan JS, Gadian DG (1998) Proton magnetic resonance spectroscopy in MRI-negative temporal lobe epilepsy. Neurology 51:61–66
22. Crandall PH, Mathern GW (2001) Surgery for lesional temporal lobe epilepsy. In: Luders HO, Comair YG (eds) Epilepsy surgery. Lippincott Williams & Wilkins, Philadelphia, pp 653–665
23. Davis LM, Spencer DD, Spencer SS, Bronen RA (1999) MR imaging of implanted depth and subdural electrodes: is it safe? Epilepsy Res 35:95–98
24. Detre JA, Maccotta L, King D, Alsop DC, Glosser G, D'Esposito M, Zarahn E, Aguirre GK, French JA (1998) Functional MRI lateralization of memory in temporal lobe epilepsy. Neurology 50:926–932

25. Diehl B, Luders HO (2000) Temporal lobe epilepsy: when are invasive recordings needed? Epilepsia 41:S61–S74
26. Duncan JS (1997) Imaging and epilepsy. Brain 120:339–377
27. Duncan JS (2001) The outcome of epilepsy surgery. J Neurol Neurosurg Psychiatry 70:432
28. Elwes RD, Dunn G, Binnie CD, Polkey CE (1991) Outcome following resective surgery for temporal lobe epilepsy: a prospective follow up study of 102 consecutive cases. J Neurol Neurosurg Psychiatry 54:949–952
29. Engel J, Jr (1993) Update on surgical treatment of the epilepsies. Summary of the Second International Palm Desert Conference on the Surgical Treatment of the Epilepsies (1992). Neurology 43:1612–1617
30. Engel JJ (1987) Outcome with respect to epileptic seizures. In: Engel JJ (ed) Surgical treatment of the epilepsies. Raven Press, New York, pp 553–571
31. Espinosa J, Olivier A, Andermann F, Quesney F, Dubeau F, Savard G (1994) Morbidity of chronic recording with intracranial depth electrodes in 170 patients. Stereotact Funct Neurosurg 63:63–65
32. European Federation of Neurological Societies Task Force (2000) Presurgical evaluation for epilepsy surgery – European standards. European Federation of Neurological Societies Task Force. Eur J Neurol 7:119–122
33. Fernandez G, Hufnagel A, Van Roost D, Helmstaedter C, Wolf HK, Zentner J, Schramm J, Elger CE (1997) Safety of intrahippocampal depth electrodes for presurgical evaluation of patients with intractable epilepsy. Epilepsia 38:922–929
34. Foldvary N, Klem G, Hammel J, Bingaman W, Najm I, Luders H (2001) The localizing value of ictal EEG in focal epilepsy. Neurology 57:2022–2028
35. Foldvary N, Nashold B, Mascha E, Thompson EA, Lee N, McNamara JO, Lewis DV, Luther JS, Friedman AH, Radtke RA (2000) Seizure outcome after temporal lobectomy for temporal lobe epilepsy: a Kaplan-Meier survival analysis. Neurology 54:630–634
36. Gaillard WD, Balsamo L, Xu B, Grandin CB, Braniecki SH, Papero PH, Weinstein S, Conry J, Pearl PL, Sachs B, Sato S, Jabbari B, Vezina LG, Frattali C, Theodore WH (2002) Language dominance in partial epilepsy patients identified with an fMRI reading task. Neurology 59:256–265
37. Golby AJ, Poldrack RA, Illes J, Chen D, Desmond JE, Gabrieli JD (2002) Memory lateralization in medial temporal lobe epilepsy assessed by functional MRI. Epilepsia 43:855–863
38. Hamer HM, Luders HO (2001) A new approach for classification of epileptic syndromes and epileptic seizures. In: Luders HO, Comair YG (eds) Epilepsy surgery. Lippincott Williams & Wilkins, Philadelphia, pp 79–94
39. Hamer HM, Morris HH (2001) Indications for invasive video-electroencephalographic monitoring. In: Luders HO, Comair YG (eds) Epilepsy surgery. Lippincott Williams & Wilkins, Philadelphia, pp 559–566
40. Hauser WA, Hesdorffer DC (2001) Epidemiology of intractable epilepsy. In: Luders HO, Comair YG (eds) Epilepsy surgery. Lippincott Williams & Wilkins, Philadelphia, pp 55–61

41. Hennessy MJ, Langan Y, Elwes RD, Binnie CD, Polkey CE, Nashef L (1999) A study of mortality after temporal lobe epilepsy surgery. Neurology 53:1276–1283
42. Immonen A, Jutila L, Kononen M, Mervaala E, Partanen J, Puranen M, Rinne J, Ylinen A, Vapalahti M (2003) 3-D reconstructed magnetic resonance imaging in localization of subdural EEG electrodes. Case illustration. Epilepsy Res 54:59–62
43. Jack CR, Jr (1993) Epilepsy: surgery and imaging. Radiology 189:635–646
44. Jack CR, Jr, Sharbrough FW, Cascino GD, Hirschorn KA, O'Brien PC, Marsh WR (1992) Magnetic resonance image-based hippocampal volumetry: correlation with outcome after temporal lobectomy. Ann Neurol 31:138–146
45. Jayakar P, Duchowny M, Resnick TJ, Alvarez LA (1991) Localization of seizure foci: pitfalls and caveats. J Clin Neurophysiol 8:414–431
46. Jokeit H, Okujava M, Woermann FG (2001) Memory fMRI lateralizes temporal lobe epilepsy. Neurology 57:1786–1793
47. Jones MW, Andermann F (2000) Temporal lobe epilepsy surgery: definition of candidacy. Can J Neurol Sci 27:20–21
48. Jones-Gotman M, Harnadek MC, Kubu CS (2000) Neuropsychological assessment for temporal lobe epilepsy surgery. Can J Neurol Sci 27:50–52
49. Jutila L, Immonen A, Mervaala E, Partanen J, Partanen K, Puranen M, Kalviainen R, Alafuzoff I, Hurskainen H, Vapalahti M, Ylinen A (2002b) Long term outcome of temporal lobe epilepsy surgery: analyses of 140 consecutive patients. J Neurol Neurosurg Psychiatry 73:486–494
50. Jutila L, Immonen A, Partanen K, Partanen J, Mervaala E, Ylinen A, Alafuzoff I, Paljarvi L, Karkola K, Vapalahti M, Pitkanen A (2002a) Neurobiology of epileptogenesis in the temporal lobe. Adv Tech Stand Neurosurg 27:5–22
51. Jutila L, Ylinen A, Partanen K, Alafuzoff I, Mervaala E, Partanen J, Vapalahti M, Vainio P, Pitkanen A (2001) MR volumetry of the entorhinal, perirhinal, and temporopolar cortices in drug-refractory temporal lobe epilepsy. AJNR Am J Neuroradiol 22:1490–1501
52. Kalviainen R, Salmenpera T (2002) Do recurrent seizures cause neuronal damage? A series of studies with MRI volumetry in adults with partial epilepsy. Prog Brain Res 135:279–295
53. Kalviainen R, Salmenpera T, Partanen K, Vainio P, Riekkinen P, Pitkanen A (1998) Recurrent seizures may cause hippocampal damage in temporal lobe epilepsy. Neurology 50:1377–1382
54. Kantarci K, Shin C, Britton JW, So EL, Cascino GD, Jack CR, Jr (2002) Comparative diagnostic utility of 1H MRS and DWI in evaluation of temporal lobe epilepsy. Neurology 58:1745–1753
55. Killgore WD, Glosser G, Casasanto DJ, French JA, Alsop DC, Detre JA (1999) Functional MRI and the Wada test provide complementary information for predicting post-operative seizure control. Seizure 8:450–455
56. King D, Spencer S (1995) Invasive electroencephalography in mesial temporal lobe epilepsy. J Clin Neurophysiol 12:32–45

57. Knowlton RC, Laxer KD, Klein G, Sawrie S, Ende G, Hawkins RA, Aassar OS, Soohoo K, Wong S, Barbaro N (2001) In vivo hippocampal glucose metabolism in mesial temporal lobe epilepsy. Neurology 57:1184–1190
58. Kotagal P (2001) Neocortical temporal lobe epilepsy. In: Luders HO, Comair YG (eds) Epilepsy surgery. Lippincott Williams & Wilkins, Philadelphia, pp 105–109
59. Kotagal P, Luders H, Morris HH, Dinner DS, Wyllie E, Godoy J, Rothner AD (1989) Dystonic posturing in complex partial seizures of temporal lobe onset: a new lateralizing sign. Neurology 39:196–201
60. Kramer RE, Luders H, Goldstick LP, Dinner DS, Morris HH, Lesser RP, Wyllie E (1988) Ictus emeticus: an electroclinical analysis. Neurology 38:1048–1052
61. Kuikka JT, Mervaala E, Vanninen E, Kalviainen R (1994) Does technetium-99m bicisate image local brain metabolism in late ictal temporal lobe epilepsy? Eur J Nucl Med 21:1247–1251
62. Kuzniecky RI, Jackson GD (1995) Temporal lobe epilepsy. In: Kuzniecky RI, Jackson GD (eds) Magnetic resonance in epilepsy. Raven Press, New York, pp 107–182
63. Kwan P, Brodie MJ (2000) Early identification of refractory epilepsy. N Engl J Med 342:314–319
64. Lamusuo S, Forss N, Ruottinen HM, Bergman J, Makela JP, Mervaala E, Solin O, Rinne JK, Ruotsalainen U, Ylinen A, Vapalahti M, Hari R, Rinne JO (1999) [^{18}F]FDG-PET and whole-scalp MEG localization of epileptogenic cortex. Epilepsia 40:921–930
65. Lamusuo S, Pitkanen A, Jutila L, Ylinen A, Partanen K, Kalviainen R, Ruottinen HM, Oikonen V, Nagren K, Lehikoinen P, Vapalahti M, Vainio P, Rinne JO (2000) [^{11}C]Flumazenil binding in the medial temporal lobe in patients with temporal lobe epilepsy: correlation with hippocampal MR volumetry, T2 relaxometry, and neuropathology. Neurology 54:2252–2260
66. Lamusuo S, Ruottinen HM, Knuuti J, Harkonen R, Ruotsalainen U, Bergman J, Haaparanta M, Solin O, Mervaala E, Nousiainen U, Jaaskelainen S, Ylinen A, Kalviainen R, Rinne JK, Vapalahti M, Rinne JO (1997) Comparison of [^{18}F]FDG-PET, [^{99m}Tc]-HMPAO-SPECT, and [^{123}I]-iomazenil-SPECT in localising the epileptogenic cortex. J Neurol Neurosurg Psychiatry 63:743–748
67. Lee N, Tien RD, Lewis DV, Friedman AH, Felsberg GJ, Crain B, Hulette C, Osumi AK, Smith JS, VanLandingham KE, *et al* (1995) Fast spin-echo, magnetic resonance imaging-measured hippocampal volume: correlation with neuronal density in anterior temporal lobectomy patients. Epilepsia 36: 899–904
68. Lee TM, Yip JT, Jones-Gotman M (2002) Memory deficits after resection from left or right anterior temporal lobe in humans: a meta-analytic review. Epilepsia 43:283–291
69. Lencz T, McCarthy G, Bronen RA, Scott TM, Inserni JA, Sass KJ, Novelly RA, Kim JH, Spencer DD (1992) Quantitative magnetic resonance imaging

in temporal lobe epilepsy: relationship to neuropathology and neuropsychological function. Ann Neurol 31:629–637
70. Lombardi D, Marsh R, de Tribolet N (1997) Low grade glioma in intractable epilepsy: lesionectomy versus epilepsy surgery. Acta Neurochir (Wien) [Suppl] 68:70–74
71. Luders H, Acharya J, Baumgartner C, Benbadis S, Bleasel A, Burgess R, Dinner DS, Ebner A, Foldvary N, Geller E, Hamer H, Holthausen H, Kotagal P, Morris H, Meencke HJ, Noachtar S, Rosenow F, Sakamoto A, Steinhoff BJ, Tuxhorn I, Wyllie E (1998) Semiological seizure classification. Epilepsia 39:1006–1013
72. Mackay CE, Webb JA, Eldridge PR, Chadwick DW, Whitehouse GH, Roberts N (2000) Quantitative magnetic resonance imaging in consecutive patients evaluated for surgical treatment of temporal lobe epilepsy. Magn Reson Imaging 18:1187–1199
73. Manford M, Fish DR, Shorvon SD (1996) An analysis of clinical seizure patterns and their localizing value in frontal and temporal lobe epilepsies. Brain 119:17–40
74. Maraire JN, Awad IA (1995) Intracranial cavernous malformations: lesion behavior and management strategies. Neurosurgery 37:591–605
75. Martin RC, Grote CL (2002) Controversies in epilepsy and behaviour. Does the Wada test predict memory decline following epilepsy surgery. Epilepsy & Behaviour 3:4–15
76. McIntosh AM, Wilson SJ, Berkovic SF (2001) Seizure outcome after temporal lobectomy: current research practice and findings. Epilepsia 42:1288–1307
77. Meiners LC, Bakker CJ, van Rijen PC, van Veelen CW, van Huffelen AC, van Dieren A, Jansen GH, Mali WP (1996) Fast spin-echo MR of contact points on implanted intracerebral stainless steel multicontact electrodes. AJNR Am J Neuroradiol 17:1815–1819
78. Meiners LC, van Gils A, Jansen GH, de Kort G, Witkamp TD, Ramos LM, Valk J, Debets RM, van Huffelen AC, van Veelen CW, *et al* (1994) Temporal lobe epilepsy: the various MR appearances of histologically proven mesial temporal sclerosis. AJNR Am J Neuroradiol 15:1547–1555
79. Milner B (1997) Amobarbital memory testing: some personal reflections. Brain Cogn 33:14–17
80. Najm I, Babb TL, Mohamed A, Diehl B, Ng TC, Bingaman WE, Luders H (2001) Mesial temporal lobe sclerosis. In: Luders HO, Comair YG (eds) Epilepsy surgery. Lippincott Williams & Wilkins, Philadelphia, pp 95–103
81. Naylor AS, Rogvi-Hansen B, Kessing L, Kruse-Larsen C (1994) Psychiatric morbidity after surgery for epilepsy: short-term follow up of patients undergoing amygdalohippocampectomy. J Neurol Neurosurg Psychiatry 57:1375–1381
82. Pacia SV, Devinsky O, Perrine K, Ravdin L, Luciano D, Vazquez B, Doyle WK (1996) Clinical features of neocortical temporal lobe epilepsy. Ann Neurol 40:724–730

83. Pacia SV, Ebersole JS (1999) Intracranial EEG in temporal lobe epilepsy. J Clin Neurophysiol 16:399–407
84. Rausch R (2002) Epilepsy surgery within the temporal lobe and its short-term and long-term effects on memory. Curr Opin Neurol 15:185–189
85. Raymond AA, Fish DR, Sisodiya SM, Alsanjari N, Stevens JM, Shorvon SD (1995) Abnormalities of gyration, heterotopias, tuberous sclerosis, focal cortical dysplasia, microdysgenesis, dysembryoplastic neuroepithelial tumour and dysgenesis of the archicortex in epilepsy. Clinical, EEG and neuroimaging features in 100 adult patients. Brain 118:629–660
86. Regis J, Bartolomei F, Hayashi M, Chauvel P (2002) What role for radiosurgery in mesial temporal lobe epilepsy. Zentralbl Neurochir 63:101–105
87. Ring HA, Moriarty J, Trimble MR (1998) A prospective study of the early postsurgical psychiatric associations of epilepsy surgery. J Neurol Neurosurg Psychiatry 64:601–604
88. Rosenow F, Luders H (2001) Presurgical evaluation of epilepsy. Brain 124: 1683–1700
89. Rossi GF, Pompucci A, Colicchio G, Scerrati M (1999) Factors of surgical outcome in tumoural epilepsy. Acta Neurochir (Wien) 141:819–824
90. Rougier A, Dartigues JF, Commenges D, Claverie B, Loiseau P, Cohadon F (1992) A longitudinal assessment of seizure outcome and overall benefit from 100 cortectomies for epilepsy. J Neurol Neurosurg Psychiatry 55:762–767
91. Rutten GJ, Ramsey NF, van Rijen PC, Alpherts WC, van Veelen CW (2002) FMRI-determined language lateralization in patients with unilateral or mixed language dominance according to the Wada test. Neuroimage 17:447–460
92. Rydenhag B, Silander HC (2001) Complications of epilepsy surgery after 654 procedures in Sweden 1990–1995: a multicenter study based on the Swedish National Epilepsy Surgery Register. Neurosurgery 49:51–56; discussion 56–57
93. Sabbah P, Chassoux F, Leveque C, Landre E, Baudoin-Chial S, Devaux B, Mann M, Godon-Hardy S, Nioche C, Ait-Ameur A, Sarrazin JL, Chodkiewicz JP, Cordoliani YS (2003) Functional MR imaging in assessment of language dominance in epileptic patients. Neuroimage 18:460–467
94. Salanova V, Markand O, Worth R (1999) Longitudinal follow-up in 145 patients with medically refractory temporal lobe epilepsy treated surgically between 1984 and 1995. Epilepsia 40:1417–1423
95. Salmenpera T, Kalviainen R, Partanen K, Pitkanen A (2000) Quantitative MRI volumetry of the entorhinal cortex in temporal lobe epilepsy. Seizure 9:208–215
96. Sawrie SM, Martin RC, Knowlton R, Faught E, Gilliam F, Kuzniecky R (2001) Relationships among hippocampal volumetry, proton magnetic resonance spectroscopy, and verbal memory in temporal lobe epilepsy. Epilepsia 42:1403–1407
97. Schmidt D (2002) The clinical impact of new antiepileptic drugs after a decade of use in epilepsy. Epilepsy Res 50:21–32

98. Serles W, Li LM, Antel SB, Cendes F, Gotman J, Olivier A, Andermann F, Dubeau F, Arnold DL (2001) Time course of postoperative recovery of N-acetyl-aspartate in temporal lobe epilepsy. Epilepsia 42:190–197
99. Simkins-Bullock J (2000) Beyond speech lateralization: a review of the variability, reliability, and validity of the intracarotid amobarbital procedure and its nonlanguage uses in epilepsy surgery candidates. Neuropsychol Rev 10:41–74
100. Sindou M, Guenot M (2002) Surgical anatomy of the temporal lobe for epilepsy surgery. Adv Tech Stand Neurosurg 28
101. So EL (2000) Integration of EEG, MRI, and SPECT in localizing the seizure focus for epilepsy surgery. Epilepsia 41:S48–S54
102. So EL, Radhakrishnan K, Silbert PL, Cascino GD, Sharbrough FW, O'Brien PC (1997) Assessing changes over time in temporal lobectomy: outcome by scoring seizure frequency. Epilepsy Res 27:119–125
103. Sperling MR, Feldman H, Kinman J, Liporace JD, O'Connor MJ (1999) Seizure control and mortality in epilepsy. Ann Neurol 46:45–50
104. Sperling MR, O'Connor MJ, Saykin AJ, Plummer C (1996) Temporal lobectomy for refractory epilepsy. Jama 276:470–475
105. Sperling MR, Shewmon DA (1997) General principles for presurgical evaluation. In: Engel J (ed) Epilepsy: A Comprehensive textbook. Lippincott, Williams and Wilkins, Philadelphia
106. Swartz BE, Rich JR, Dwan PS, DeSalles A, Kaufman MH, Walsh GO, Delgado-Escueta AV (1996) The safety and efficacy of chronically implanted subdural electrodes: a prospective study. Surg Neurol 46:87–93
107. Tiihonen J, Hari R, Kajola M, Nousiainen U, Vapalahti M (1990) Localization of epileptic foci using a large-area magnetometer and functional brain anatomy. Ann Neurol 27:283–290
108. Wada J, Rasmussen T (1960) Intracarotid injection of sodium Amytal for the lateralization of cerebral speech dominance: Experimental and clinical observations. J Neurosurg 17:226–282
109. Vainio P, Usenius JP, Vapalahti M, Partanen K, Kalviainen R, Rinne J, Kauppinen RA (1994) Reduced N-acetylaspartate concentration in temporal lobe epilepsy by quantitative 1H MRS in vivo. Neuroreport 5:1733–1736
110. Wheless JW, Willmore LJ, Breier JI, Kataki M, Smith JR, King DW, Meador KJ, Park YD, Loring DW, Clifton GL, Baumgartner J, Thomas AB, Constantinou JE, Papanicolaou AC (1999) A comparison of magnetoencephalography, MRI, and V-EEG in patients evaluated for epilepsy surgery. Epilepsia 40:931–941
111. Wiebe S, Blume WT, Girvin JP, Eliasziw M (2001) A randomized, controlled trial of surgery for temporal-lobe epilepsy. N Engl J Med 345:311–318
112. Wieser HG, Blume WT, Fish D, Goldensohn E, Hufnagel A, King D, Sperling MR, Luders H, Pedley TA (2001) ILAE Commission Report. Proposal for a new classification of outcome with respect to epileptic seizures following epilepsy surgery. Epilepsia 42:282–286
113. Wieser HG, Ortega M, Friedman A, Yonekawa Y (2003) Long-term seizure outcomes following amygdalohippocampectomy. J Neurosurg 98:751–763

114. Wiggins GC, Elisevich K, Smith BJ (1999) Morbidity and infection in combined subdural grid and strip electrode investigation for intractable epilepsy. Epilepsy Res 37:73–80
115. Winkler PA, Vollmar C, Krishnan KG, Pfluger T, Bruckmann H, Noachtar S (2000) Usefulness of 3-D reconstructed images of the human cerebral cortex for localization of subdural electrodes in epilepsy surgery. Epilepsy Res 41:169–178
116. Wyler AR, Walker G, Somes G (1991) The morbidity of long-term seizure monitoring using subdural strip electrodes [see comments]. J Neurosurg 74:734–737
117. Wyllie E, Comair YG, Kotagal P, Bulacio J, Bingaman W, Ruggieri P (1998) Seizure outcome after epilepsy surgery in children and adolescents. Ann Neurol 44:740–748
118. Zentner J, Hufnagel A, Wolf HK, Ostertun B, Behrens E, Campos MG, Elger CE, Wiestler OD, Schramm J (1997) Surgical treatment of neoplasms associated with medically intractable epilepsy. Neurosurgery 41:378–386; discussion 386–387

Motor Evoked Potential Monitoring for Spinal Cord and Brain Stem Surgery

F. SALA, P. LANTERI, and A. BRICOLO

Department of Neurological Sciences and Vision, Section of Neurosurgery, University of Verona

With 17 Figures

Contents

Abstract

Intraoperative Neurophysiology (ION) has established itself as one of the means by which modern neurosurgery can improve surgical results while minimizing morbidity. The advent of motor evoked potential (MEP) monitoring represents a landmark in this recent progress.

ION consists of *monitoring* (the continuous "on-line" assessment of the functional integrity of neural pathways) and *mapping* (the functional identification and preservation of anatomically ambiguous nervous tissue) techniques. In this chapter we have attempted to critically review the evolution of MEP use during *monitoring* and *mapping* techniques for neurosurgical procedures in the brainstem and the spinal cord, providing the neurophysiological theoretical background and practical aspects of clinical applications. According to the experience from our and other groups involved in ION, we suggest the following:

1) ION is mandatory whenever neurological complications are expected as predicted by a known pathophysiological mechanism. It is therefore advisable to perform ION when dealing with brain stem and intramedullary spinal cord lesions.
2) MEP *monitoring* after transcranial electrical stimulation is today a feasible and reliable technique for use under general anesthesia. MEP *monitoring* is the most appropriate technique to assess the functional integrity of descending motor pathways in the brainstem and, foremost, in the spinal cord.
3) *Mapping* of the corticospinal tract at the level of the cerebral peduncle as well as mapping of the VII, IX–X and XII cranial nerve motor nuclei on the floor of the fourth ventricle is of great value with which to identify "safe entry zones" into the brainstem.
4) Other techniques, although safe and feasible, still lack rigorous validation in terms of prognostic value and correlation with the postoperative neurological outcome. These techniques include *mapping* of the corticospinal tract within the spinal cord and *monitoring* of the corticobulbar tracts. These techniques, however, are expected to open new perspectives in the near future.

Keywords: Motor evoked potentials; intraoperative neurophysiology; spinal cord monitoring; brain stem surgery; spinal cord surgery.

Introduction

The last decade has seen the introduction of a number of innovative techniques throughout different fields in the Neurosciences. Intraoperative Neurophysiology (ION) has established itself as one of the means by which modern neurosurgery can improve surgical results while minimizing morbidity. The advent of motor evoked potential (MEP) use represents a landmark in this recent progress.

Nowadays, it is feasible to monitor MEPs under general anesthesia and provide invaluable feedback to the surgeon with regards to the functional integrity of motor pathways. This method has proven to be reliable and to correlate highly with neurological motor outcome, particularly in the field of spinal cord surgery. Brainstem surgery has also benefited from the introduction of MEPs: continuous monitoring of the corticobulbar pathways, direct mapping of the corticospinal tracts (CTs) and identification of the cranial nerve motor nuclei on the floor of the fourth ventricle are all currently employed to assist the surgeon during different surgical approaches.

In this chapter we will begin by clarifying some misleading concepts that have historically affected the credibility of ION. After reviewing some modern concepts of ION of the motor system, a more clinical approach will be presented with two aims: 1) to identify warning signs and corrective measures aimed at preventing irreversible injuries to the brainstem and spinal cord motor pathways and 2) to predict the long-term motor outcome of monitored patients on the basis of ION data. Illustrative cases will be presented to discuss the role of ION during different spinal cord and brainstem procedures.

Part I: Intraoperative Neurophysiological Techniques

The Past

Historical Background of MEP Monitoring

Although the pioneering work of Wilder Penfield on the intraoperative electrical stimulation of the motor cortex was published in 1937 [42], it was only late in the past century that investigators focused again on the neurophysiology of the motor system and its intraoperative applications. Efforts to search for reliable methods to assess the functional integrity of the motor pathways were initially started because of the problematic nature of using somatosensory evoked potentials (SEPs) to do so. In the late 1970s, SEPs were routinely used to monitor the functional integrity of the somatosensory system during surgery for scoliosis [17]; a few years later they were also introduced as a neurophysiological monitoring tool during the embolization of spinal cord vascular lesions [1]. Although SEPs have played an important role in decreasing the incidence of neurological deficits during orthopedic, neuroradiological and neurosurgical procedures in the spine and the spinal cord [1, 25, 38], attempts to extrapolate information about the functional integrity of the motor system from SEPs disappointingly resulted in so-called false negative results (e.g. patient being paraplegic in spite of unchanged intraoperative SEPs) [18, 29, 33, 62]. In addition, SEP monitoring suffers from two other major drawbacks: first they need long averaging times (at least 10 to 40 sec.) which inhibits the vital flow of quick and "on line" feedback to the surgeon; second, in the case of spinal cord surgery, they are often lost during the initial myelotomy so that ION is no longer possible [8].

Unfortunately, these results and limitations induced a long-lasting lack of trust in the reliability of ION amongst the neurosurgical community.

Although SEPs were the only monitoring technique originally available for monitoring the long tracts of the spinal cord, today there is no justification for their use when the goal of monitoring is to assess the functional integrity of the motor pathways. SEPs retain a specific role for the assessment of the dorsal column pathways, but they are not designed to provide specific and sensitive information about the CT integrity (for which MEPs should be used).

As illustrated in Fig. 1, the anterior spinal artery (ASA), through its perforating sulcocommissural arteries, supplies the anterior two thirds to four fifths of the spinal cord, including the anterior gray matter and the anterior and lateral CTs. This arterial axis therefore vascularizes the neural structures involved in the propagation of MEPs through the cord, namely the CTs and, partially, the propriospinal system. The two posterior spinal arteries (PSAs), conversely, are assumed to provide the vascular supply to the posterior horns of the gray matter and to the dorsal columns, which are involved in the propagation of SEPs [6]. Data from neurophysiological monitoring during endovascular procedures suggests that such a simplistic pattern has to be taken "cum grano salis" [48]. Exceptions to the rule exist because of the extreme variability of spinal cord vascularization and the existence of anastomotic patterns between anterior and posterior cord circulation. Nevertheless, a neurophysiological approach to spinal cord vascularization supports the observation of preserved SEPs in patients who are paraplegic because of an anterior spinal artery syndrome [62].

The methodology for eliciting MEPs has its roots in the mid 1950s when Patton and Amassian [40] recorded descending volleys from the pyramidal tract exposed at the level of the medulla, after direct electrical stimulation of the motor cortex in monkeys. Three decades later Merton and Morton for the first time elicited MEPs in awake volunteers using a transcranial electrical stimulation (TES) [32]. However, TES causes discomfort and was not suitable for extensive clinical use in awake patients. Subsequently, the use of transcranial magnetic stimulation (TMS) to activate the CTs in the brain was much more easily tolerated by awake patients and virtually replaced TES in clinical practice. In the operative setting, however, both single stimulus TES and TMS techniques turned

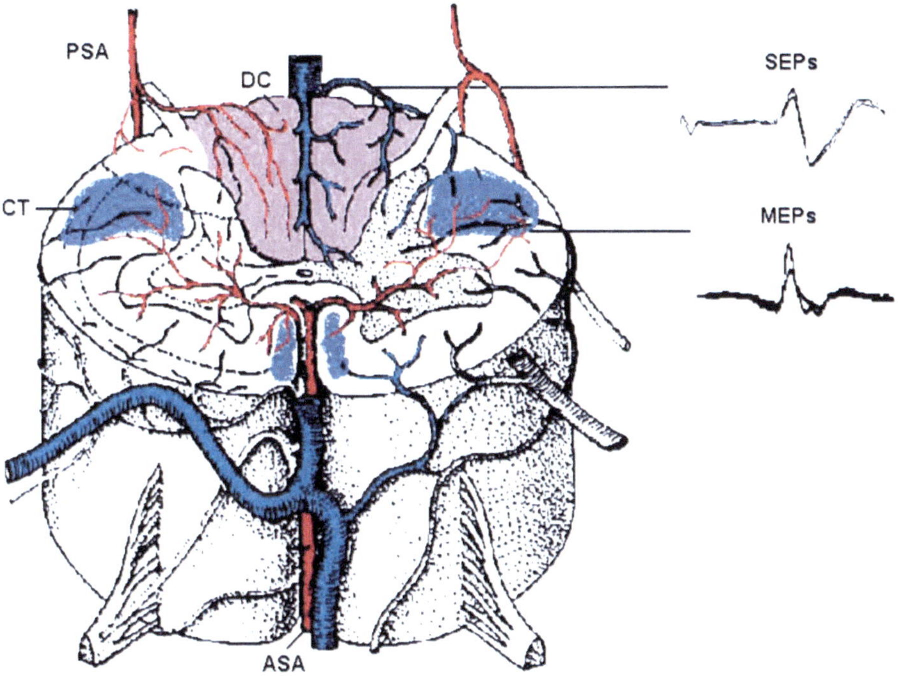

Fig. 1. Schematic representation of spinal cord vascular anatomy and its relationship with long tracts involved in the generation of somatosensory and motor evoked potentials. *ASA* Anterior spinal artery, *PSA* posterior spinal artery, *DC* dorsal column, *CT* corticospinal tract, *SEPs* somatosensory evoked potentials, *MEPs* motor evoked potentials. (Modified from Nieuwenhuys R, Voogd J, van Huijzen C (1988) The human central nervous system: a synopsis and atlas, rev 3rd edn. Springer, Berlin Heidelberg New York Tokyo

out to be unsatisfactory given their inability to consistently elicit muscle MEPs (mMEPs) in patients under general anesthesia. This limitation has been overcome in the past fifteen years due to the advent of the spinal epidural recording of the motor evoked response (D-wave) [3, 4, 7, 23] and the use of a multipulse (i.e., train of stimuli) technique to peripherally record the motor response from the limb muscles [22, 41, 56]. These techniques will be described in details further in the text.

Misconceptions about MEP Monitoring

Before describing the techniques we currently use to monitor MEPs, a short review of other monitoring modalities is presented.

The "spinal cord to spinal cord" monitoring technique was introduced in 1986 [55] providing a non-selective stimulation of the spinal cord with non-selective recordings from the distant segments of the spinal cord. Its major limitation was its inability to provide specific information on the dorsal columns and the CTs since only an overall evaluation of the preservation of long tracts within the cord was possible. "Spinal cord to peripheral nerve" [39] or "spinal cord to muscle" [30] techniques have also been introduced. However, α-motoneurons are activated not only by the CTs but also by any descending tract within the cord and/or antidromically by segmental branches of the dorsal columns that mediate the H-reflex [7]. Therefore, selective recordings from the peripheral nerves or muscles generated by electrical stimulation of the spinal cord most likely do not arise purely from the CT [57]. Recently, Minahan *et al.* described two patients with postoperative paraplegia in spite of preservation of a spinally elicited peripheral nerve response, also known as "a neurogenic motor evoked response" [9, 33]. Moreover, injuries to the motor system rostral to the upper cervical spinal cord cannot be detected using these techniques due to methodological limitations.

This being said, it should be recognized that both "spinal cord to spinal cord" and "spinal cord to peripheral nerve/muscle" technique can grossly assess the functional integrity of long tracts within the spinal cord. Consequently, this limitation may be acceptable for orthopedic surgery where damage to the spinal cord is most likely diffuse so that multiple pathways are simultaneously affected (with the exception of anterior spinal artery syndrome). Surgery for spinal cord lesions, however, requires a more selective approach since sensory and motor pathways (or even different motor pathways i.e. corticospinal, propriospinal, rubrospinal, etc.), can be independently affected [8].

The Present

ION consists of two main categories of techniques: monitoring and mapping techniques. Semantically, the term *monitoring* refers to the continuous assessment of the functional integrity of neural pathways. The principal goal of monitoring is to expeditiously identify the source of surgically induced neurophysiological changes in order to allow for prompt correction of the cause before neurological impairment occurs. *Mapping* techniques are those which allow the functional identification and preservation of anatomically ambiguous nervous tissue. An example of neurophysiological mapping is the identification of displaced motor cranial nerve nuclei during surgery for brain stem tumors.

MEP Monitoring Techniques for Spinal Cord and Brainstem Procedures

Current techniques to intraoperatively monitor MEPs after TES have their origin in the work of Merton and Morton [32]. Since then, two methodologies for intraoperatively monitoring the motor pathways have been developed and are currently used at our institution.

Transcranial Electrical Stimulation of the Motor Cortex and Muscle Recordings (Multipulse Technique). We performed TES using corkscrew-like electrodes inserted in the scalp, since they are secure and provide low impedance. In children, where the fontanelle is still open, or in patients with subcutaneous shunts, cup electrodes maybe preferred. Short trains of 5 to 7 square-wave stimuli of 500 μs duration and interstimulus interval (ISI) of 4 ms are applied at a repetition rate up to 2 Hz through electrodes placed at C1 and C2 scalp sites, according to the International 10/20 EEG system. A C1/C2 montage preferentially elicits mMEPs in the right limb muscles, while C2/C1 favors recordings from the left limb muscles. For monitoring of lower extremities muscles, a Cz-C6cm montage is usually preferred, where Cz is placed 1 cm behind the typical Cz point. The stimulation intensity usually does not exceed 200 mA, but in healthy subjects upper limb mMEPs are sometimes recordable after stimulation intensities as low as 40–50 mA. Muscle responses are recorded via needle electrodes inserted into upper and lower extremity muscles. We usually monitor the abductor pollicis brevis and the extensor digitorum communis for the arm and the tibialis anterior and the abductor hallucis for the leg. We use the Axon Sentinel-4 evoked potential system with modified software (AXON Systems, Inc., Hauppage, NY) for stimulation and recording.

A schematic representation of this technique is presented in Fig. 2.

The main advantage of the multipulse TES is the ability to overcome the effects of anaesthetics on a multisynaptic pathway and record mMEPs under general anesthesia [22, 41, 56]. Furthermore, the presence of mMEPs indicates that the functional integrity of not only the motor cortex and the CT but also the α-motor neuron, the peripheral nerve and the neuromuscular junction has been maintained. One direct consequence of this technique is the ability to assess which extremity is going to be affected (see illustrative cases n. 3 and n. 4). One can, for example, intraoperatively observe the disappearance of the left tibialis anterior mMEP while the response from the left abductor pollicis brevis and from right side muscles is still present.

Unlike SEPs, mMEPs need no averaging and, at a stimulation rate of 1 to 2 Hz, they provide rapid "on-line" feedback. Since mMEPs are generated through a polysynaptic pathway, however, they are very sensitive to the effect of anesthesia and muscle relaxant so that wide variation in mMEPs amplitude and latency can be observed [12, 22, 44, 58]. This variability explains the lack of a linear correlation between intraoperative changes in mMEPs amplitude and/or latency and the motor outcome.

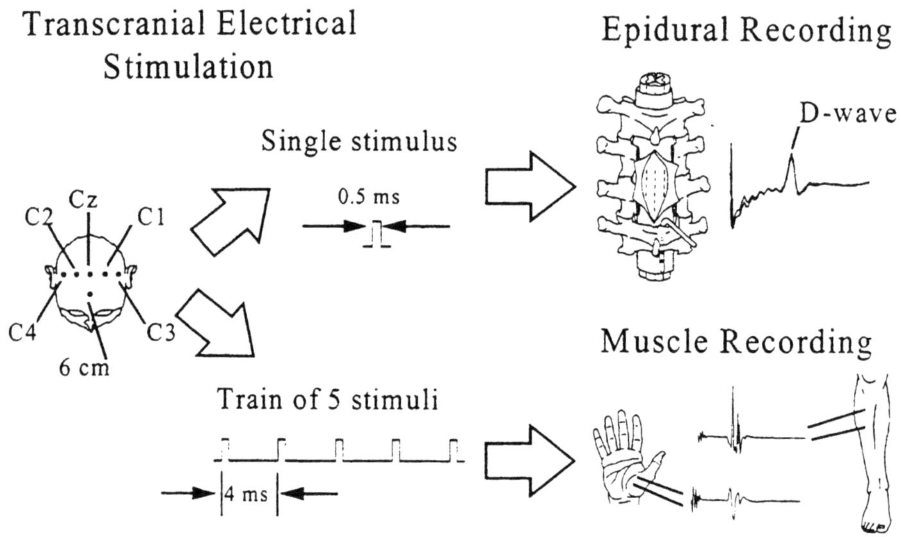

Fig. 2. *Motor evoked potentials for spinal cord surgery.* (Left) schematic illustration of electrode positions for transcranial electrical stimulation of the motor cortex according to the International 10–20 EEG system. The site labelled "6 cm" is 6 cm anterior to CZ. (Top right) schematic diagram of the position of the epidural catheter electrode placed caudal to the lesion to monitor the incoming signal (D-wave) passing through the site of surgery. A single stimulus of 0.5 ms duration is used. (Bottom right) recording of mMEPs from the abductor pollicis brevis and tibialis anterior muscles after eliciting them with a short train of electrical stimuli (multipulse technique), 4 ms apart. Modified from Kothbauer K, Deletis V, Epstein FJ (2000) Intraoperative neurophysiological monitoring. In: Crockard A, Hayward R, Hoff JT (eds) Neurosurgery: the scientific basis of clinical practice, 3rd edn, p 1042

Also, based also on reports where the only consistent finding was a correlation between intraoperative mMEP loss and postoperative motor deterioration [22, 61] we prefer to use a "yes/no" criterion for mMEPs during spinal cord surgery [27]. It has to be stressed, however, that this "yes/no criterion" are valid only for Intramedullary Spinal Cord Tumor (ISCT) surgery, not for supratentorial procedures [37].

Transcranial Electrical Stimulation of the Motor Cortex and Epidural Recordings (Single Pulse Technique). A single electrical stimulus applied transcranially or directly to the exposed motor cortex elicits a so-called direct (D) wave that can be recorded by a catheter electrode placed epi- or subdurally adjacent to the spinal cord (Fig. 2, top right). This wave form is a highly reliable parameter for monitoring the functional integrity of the CTs intraoperatively because it represents the direct activation of a population of synchronized fast conducting fibers of the CT [5, 23, 40]. The single stimulus technique is advantageous because it produces no muscle twitches and the D-wave is very robust under general anesthesia [51] because no synapses are involved in its generation. A decrease in amplitude has been observed only at high doses of volatile anaesthetics but the interpretation of this phenomenon remains debated [10, 20]. Temperature changes can affect the D-wave latencies [7, 10]. The D-wave latency can also be shortened by a stronger stimulation intensity since the CT fibers are activated deeper in the white matter [5].

The D-wave, or epidural MEP (eMEP), is usually recorded in a "single-stimulus-single-response" fashion. The stimulation rate of 0.5 to 2 Hz, however, provides a fast feedback even when a few averages are needed. Signals are amplified 10.000 times and the filter bandpass is set from 1.5 to 1700 Hz. As the D-wave provides a semi-quantitative assessment of the amount of preserved CT fibers (fast neuron of the CT fibers), a decrease in the peak-to-peak amplitude mirrors a reduction in the number of preserved fibers. Fortunately, from a clinical perspective, the D-wave amplitude deteriorates in a stepwise incremental pattern. Thus, warning signs can be observed and corrective measures can be taken before irreversible damage to the spinal cord occurs [13, 27]. Today, during surgery for ISCTs, we use a 50% drop of the baseline amplitude as a warning sign to transiently stop surgery or abandon the removal (if the signals do not recover). Baseline D-waves are recorded after exposing the dura and the epidural catheter is initially placed. This criterion is based on clinical experience more than on a strong neurophysiological background. In fact, there is still no information available on the percentage of CT fibers necessary to support locomotion. Experimental data suggest that in cats this number can be as low as 10% [2], but in humans this proportion might be different. Furthermore, there currently does not exist a test to quantify the number of preoperatively damaged CT fibers in a paretic patient undergoing surgery, although the collision technique (described later in this chapter) is opening new perspectives. Conversely, data from over 100 monitored ISCT surgeries strongly suggest that the arbitrary cut off of 50% works appropriately to predict the long-term motor outcome [27].

There are, however, some drawbacks with this technique. In about one third of the patients [26], the D-wave is not recordable. The D-wave cannot be recorded below the T10 bony level because of an insufficient number of CT fibers (i.e., insufficient D-wave amplitude). A control electrode, which is usually placed above the level of the lesion to detect non-surgically induced changes in the D-wave, cannot be placed cranial to the lesion in the case of high cervical lesions. The presence of scar tissue and/or massive dural adhesions may prevent the placement of the epidural catheter, especially in patients who underwent previous surgery or spinal irradiation. Positioning the catheter electrode subdurally, although feasible, is sometimes viewed unfavourably by neurosurgeons because of the risk of venous plexus haemorrhages behind the level of the laminotomy. Finally, in patients where extensive tumors, previous surgery and/or radiation to the cord have impaired the conduction properties of the CTs, the conduction velocities of individual axons may vary greatly and the D-wave can be so dispersed that is not recordable. This phenomenon has been described as a "desynchronization" of the D-wave [10].

The combined use of the single and multipulse techniques utilizes beneficial features of both while compensating for their disadvantages, and allows predictions on short and long-term neurological outcome, as will be described in Table 1 of this chapter. Both these techniques are routinely used at our institution during surgery for ISCT. During brainstem surgery we usually record only mMEPs after multipulse TES. However, the recording of eMEPs during brainstem surgery is feasible anytime an epidural electrode can be inserted from the exposed calamus scriptorius downward to the upper cervical levels of the cord. As an alternative, for brainstem procedures where the floor of the fourth ventricle is not exposed, a percutaneous placement of the epidural catheter at the cervical level can be done using a Tuohy needle, although we suggest this manoeuvre should be performed by skilled anaesthesiologists.

Mapping of the Corticospinal Tract During Brainstem Procedures

This is a recently described technique used to map the CT within the brainstem at the level of the cerebral peduncle [15]. To identify the CT we use a hand-held monopolar stimulator (tip diameter 0.75 mm) as cathode, with a needle electrode inserted in nearby muscle as anode. If the response is recorded from an epidural electrode, a single stimulus is used. Conversely, if the response is recorded as compound muscle action potential from one or more muscle of controlateral limbs, a train of stimuli should be used (Fig. 3).

We usually increase the stimulation intensity up to 2 mA. When a motor response is recorded, the probe is then moved in small increments of 1 mm in order to find the lowest threshold to elicit that response. This technique is particularly useful for midbrain tumors that have displaced the CT from its original position (see illustrative case n. 1). Usually, the so-called midbrain lateral vein described by Rhoton [43] represent a useful anatomical landmarks since it allows an indirect identification of the CT, located anteriorly to the vein. However, when an expansive lesion distorts anatomy, only neurophysiological mapping allows the identification of the CT and, consequently, a safe entry zone to the lateral midbrain. In the case of cystic midbrain lesion (i.e. a pilocytic astrocytoma, common in this

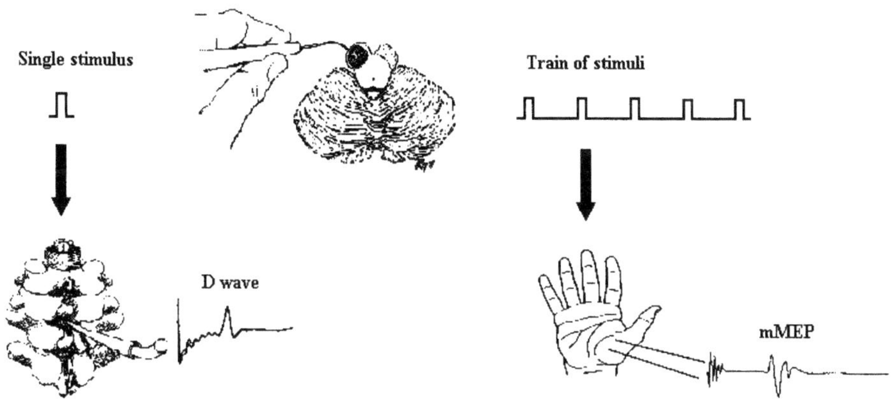

Fig. 3. Mapping of the CT on the cerebral peduncle. (Middle) the cerebral peduncle is being mapped by a hand-held monopolar probe. (Left) epidural recording of the D-wave (= eMEP) after single stimulus. (Right) muscle recording (mMEP) from the right abductor pollicis brevis after train of stimuli. Modified from [15]

location), sometimes mapping of the CT is negative at the beginning of the procedure but a positive response could be recorded when mapping from within the cystic cavity towards the antero-lateral cystic wall.

The "Future"

Mapping of the Corticospinal Tract During Spinal Cord Procedures: the Collision Technique

This is a new technique recently proposed by Deletis *et al.* to precisely and specifically localize the CT within the exposed spinal cord [11]. The neurophysiological rationale is based on the fact that a small stimulus (max. 2 mA) delivered to the spinal cord through a hand held bipolar electrode is antidromically carried through the CT and can collide with a descending D-wave simultaneously elicited using TES (Fig. 4). If the probe is in close proximity to the CT, collision will occur and the amplitude of the descending D-wave, recorded from an epidural catheter placed cranial to the stimulating point, will diminish. Conversely, if the spinal cord is stimulated far from the CT, collision will not occur and the D-wave amplitude remains unchanged. This technique, which is still under investigation, is opening new perspectives because it will allow the functional identification of the CT within the spinal cord—analogous to what is already feasible at the level of the cerebral peduncle—with obvious implications for surgery of intramedullary lesions.

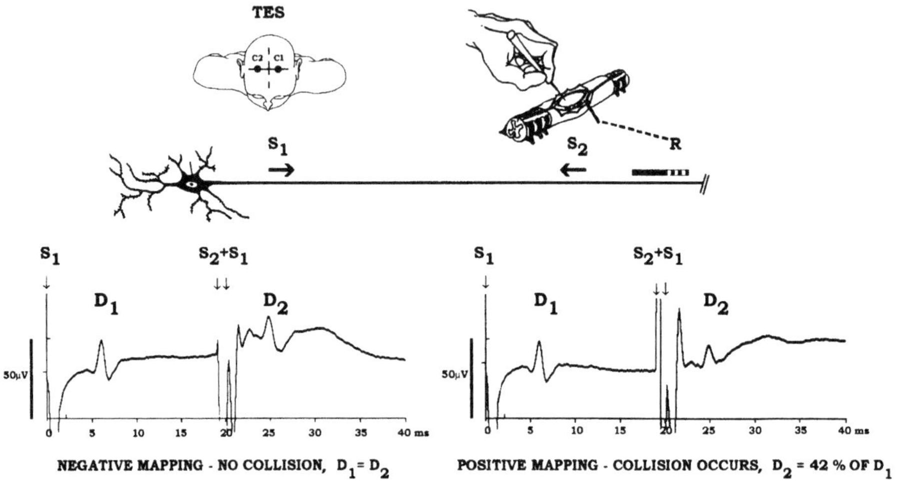

Fig. 4. Mapping of the CT by D-wave collision technique. *S1* Transcranial electrical stimulation (*TES*). *S2* Spinal cord electrical stimulation (*SpES*). *D1* Control D-wave (TES only). *D2* D-wave after combined stimulation of the brain and spinal cord. *R* D-wave recording electrode in the spinal epidural space. (Below left) negative mapping results (*D1* = *D2*). (Below right) positive mapping results (D-wave amplitude significantly diminished after collision). Reprinted from [11]

Monitoring of the Corticobulbar Tracts

Brainstem surgery has undoubtedly benefited from the introduction of mapping techniques to identify the motor nuclei of the cranial nerves. These techniques have become standard procedures for helping the surgeon to identify a "safe" entry zone to the brainstem whenever pure anatomical landmarks cannot be recognized [24, 35, 52–54]. The motor nuclei of cranial nerves III through XII can today be identified using direct stimulation of the dorsal brainstem surface and recording from needle electrodes inserted in the muscles innervated by these nerves (Fig. 5). It must be noted, however, that this mapping technique allows the functional identification of these anatomical neural structures without a continuous "on-line" assessment of their functional integrity. Also, brainstem mapping cannot detect injury to the supranuclear tracts originating in the motor cortex and ending on the cranial nerve motor nuclei. Consequently, a supranuclear paralysis would not be detected although lower motor neuron integrity has been preserved.

Standard techniques for continuously assessing the functional integrity of motor cranial nerves are today limited to the evaluation of spontaneous electromyography (EMG) activity in the muscles innervated by motor cranial nerves [16, 19, 45, 50]. Although several criteria have been proposed with which to identify EMG activity patterns suspicious of nerve

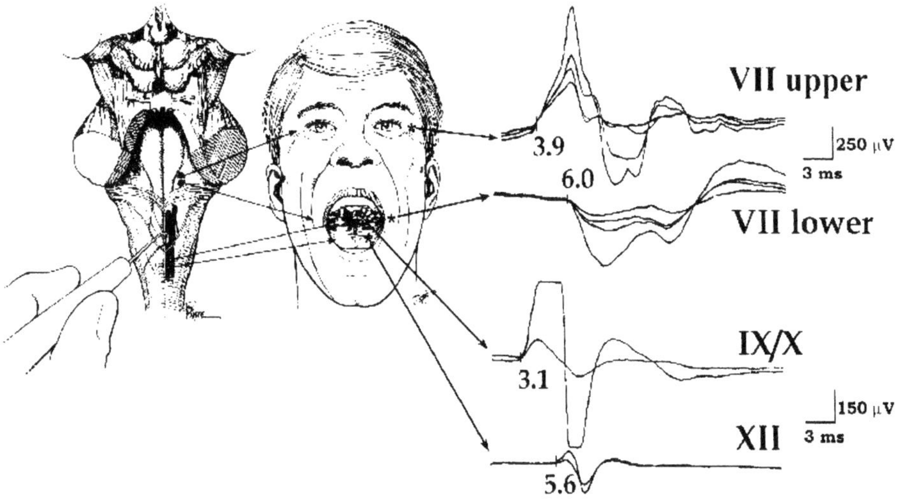

Fig. 5. Mapping of the brainstem cranial nerve motor nuclei. (Left) Schematic drawing of the exposed floor of the fourth ventricle with the surgeon's hand-held stimulator in view. The motor nuclei of the VIIth, IX/Xth and XIIth cranial nerved are illustrated. (Middle) sites of insertion of wire hook electrodes for recording the muscle responses. (Right) compound muscle action potentials recorded from the orbicularis oculi and oris muscles after stimulation of the upper and lower facial nuclei (upper two panels) and compound action potentials recorded from the pharyngeal wall and tongue muscles after stimulation of the motor nuclei of cranial nerves IX/X and XII (lower two panels). Modified from [15]

injury, the terminology remains somewhat confusing and convincing data regarding a clinical correlation between EMG activity and clinical outcome are still lacking [19, 50].

The possibility of extending the principles of CT monitoring to the corticobulbar tracts is currently under investigation in different centers in Japan, U.S. and Italy. We use TES with a train of 4 stimuli at a rate of 1–2 Hz and intensity between 60 and 100 mA. The electrode montage is usually C3/Cz for right side muscles and C4/Cz for left side muscles. For recording mMEPs, EMG wire electrodes are inserted in the orbicularis oris and orbicularis oculi muscles for the VIIth nerve, in the posterior pharyngeal wall for the IXth and Xth motor cranial nerves, and in the tongue muscles for the hypoglosseal nerve. The recording electrodes are Teflon-coated wire electrodes with 2 mm bare hooked tips encased in a 27-gauge "guide" needle that can be manipulated with a long needle holder; this facilitates insertion of the electrode in the appropriate pharyngeal muscles, under direct laryngoscopy. A reproducible mMEP can be continuously recorded from the facial, pharyngeal and tongue muscles while the brainstem is surgically manipulated (Fig. 6). The main advantage of

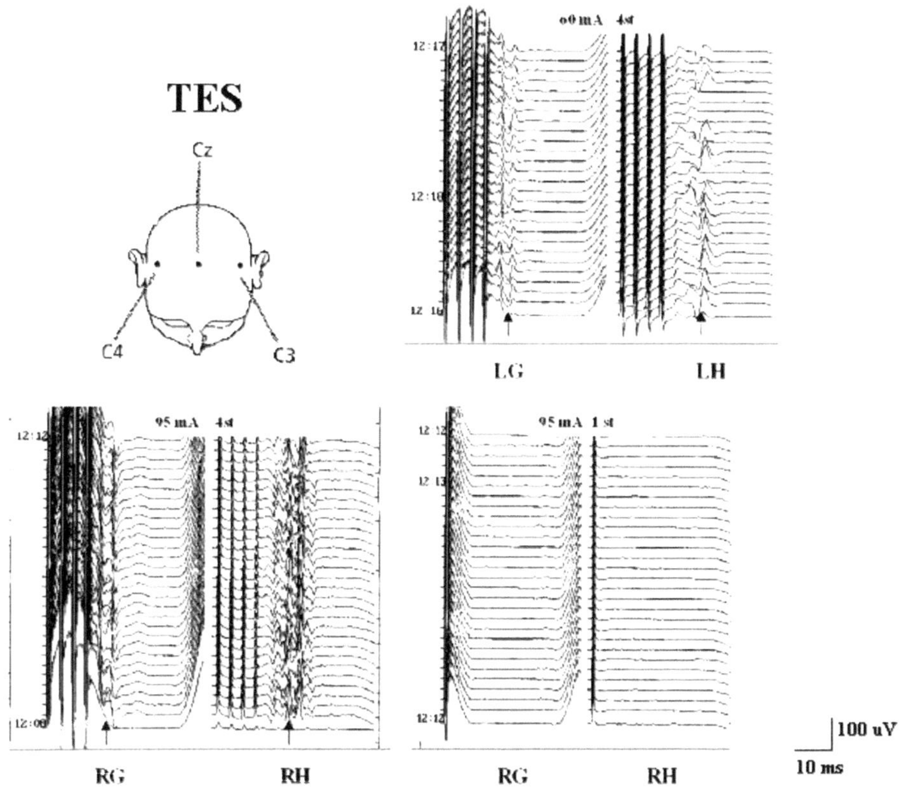

Fig. 6. Continuous monitoring of the corticobulbar pathways. (Upper left) Stimulation: TES with a train of 4 stimuli at a rate of 1–2 Hz is used. Electrode montage is usually C3/Cz for right side muscles and C4/Cz for left side muscles. Stimulation intensity usually ranges between 60 and 100 mA. (Upper right) Recording: compound muscle action potentials are recorded "on line" from needle electrodes inserted in the left pharyngeal wall *(LG)* and tongue *(LH)* muscles. (Lower panels) In order to exclude the possibility of a direct activation of the peripheral cranial nerve, anytime a corticobulbar mMEP is recorded using multipulse TES (left), we repeat the TES with the stimulation reduced to a single stimulus and all other stimulation parameters remaining the same (right). If no mMEPs responses are recorded after a single TES (right), this stimulation intensity is considered specific for a proximal activation of the corticobulbar pathway at the level of the motor cortex. *RG* right pharyngeal wall muscles, *RH* right tongue muscles

this technique is that, unlike direct brainstem mapping, the functional integrity of the corticobulbar pathways is assessed continuously. Moreover, the entire pathway, from the motor cortex down to the neuromuscular junction, is covered so that a supranuclear injury can be detected.

There are, nevertheless, some theoretical and practical drawbacks that have so far limited the widespread use of this technique. First, from a neurophysiological perspective, using a lateral montage with C3 or C4 as anodal stimulating electrode increases the risk that a strong TES may activate the corticobulbar pathways deep in the brain or even at the level of brainstem/foramen magnum [10]. The possibility of a direct activation of the peripheral cranial nerve, especially the facial, can also not be excluded. Accordingly, an injury to the corticobulbar pathways rostral to the point of activation may be masked by a misleading preservation of the mMEP. Although we have not experienced false negative results using this technique, this possibility should be taken into account and the stimulation intensity should be kept as low as possible. As discussed previously in the text, a single pulse TES does not allow recording of mMEPs under general anesthesia. Therefore, anytime a corticobulbar mMEP is recorded using multipulse TES, we repeat the TES with the stimulation reduced to a single stimulus and all other stimulation parameters remaining the same. If a muscle response is still present (taking into account the shortening of the latency due to the smaller number of stimuli) we interpret this response as direct activation of the cranial nerve. Therefore this response is not reliable for monitoring. Vice versa, if no mMEPs responses are recorded after a single TES (Fig. 6 bottom right), we assume that the stimulation intensity threshold is specific for a proximal activation of the corticobulbar pathway at the level of the motor cortex. Given the continuous fluctuations in the threshold to elicit mMEPs intraoperatively—because of room temperature, anesthesiological regimen, physiological variability in mMEP threshold, etc.—it is recommended that the appropriate threshold for monitoring corticobulbar pathways be rechecked throughout the surgical procedure.

A second limitation of this technique is that spontaneous EMG activity, which is rather common during manipulation of the brainstem and motor cranial nerves, can hinder the recording of reliable mMEPs from the same muscles. In our experience, this spontaneous activity appears to be more common in the pharyngeal muscles as compared to the facial and tongue muscles. Finally, due to the limited experience with this technique, robust data about the prognostic role of these mMEPs with regard to the postoperative facial nerve paresis, dysphagia and tongue paralysis are still lacking and warrant further investigation. In particular, with regards to the preservation of the swallowing and coughing reflexes, available intraoperative techniques are insufficient to provide reliable prognostic data since only the efferent arc of the reflex can be tested. Recently, however, Sakuma *et al.* have succeeded in monitoring glossopharyngeal nerve compound action potentials after stimulation of the tongue in dogs, opening the possibility of a new field of investigation in humans [46].

It is expected that the continuous monitoring of corticospinal and corticobulbar tracts will remarkably improve the impact of ION during brainstem surgery since in the past, using only classical methods such as

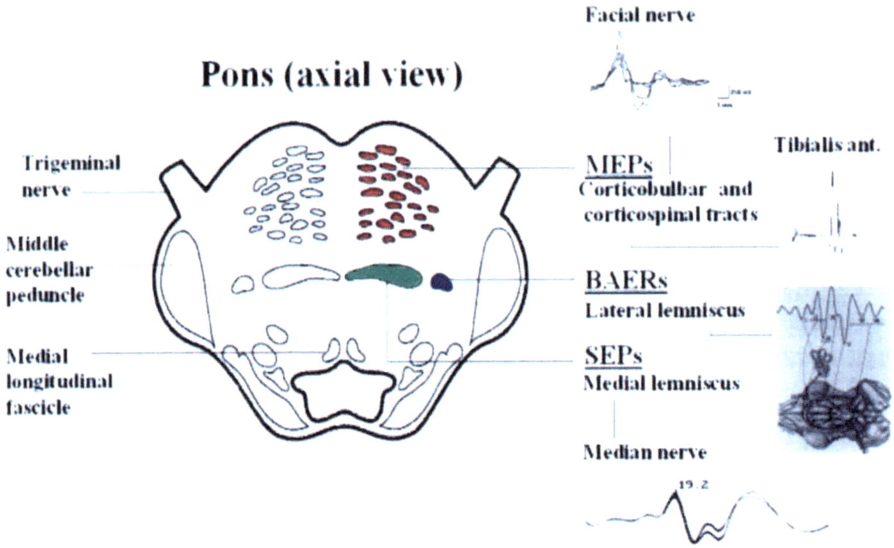

Fig. 7. Schematic illustration of the brainstem. Axial view at the level of the pons. Anatomical landmarks are indicated on the left. On the right side, neural pathways that can be monitored through brainstem intraoperative neurophysiological methods are illustrated: MEPs—including corticospinal and corticobulbar pathways-, BAERs and SEPs. Modified from: Kothbauer K, Deletis V, Epstein FJ (2000) Intraoperative neurophysiological monitoring. In: Crockard A, Hayward R, Hoff JT (eds) Neurosurgery: the scientific basis of clinical practice, 3rd edn, p 1042

brainstem auditory evoked responses (BAERs) and SEPs monitoring, less than 20% of the brainstem area could be evaluated (Fig. 7).

Part II: Clinical Application

Correlation Between Intraoperative MEP Recordings and Post-Operative Motor Outcome

The prognostic reliability of MEPs has progressively advanced over the past 5 years. In a preliminary report from 1997, Morota *et al.* [34] concluded that the eMEP monitorability appeared to be a better predictor of functional outcome than the patient's preoperative motor status. Through the years, this predictive value has been further reinforced by the introduction of mMEPs [13, 14, 26, 27].

It is our opinion that the best available monitoring technique for ISCT removal is a combination of muscle and epidural recordings. A retrospective analysis of over 100 ISCT surgeries monitored with both these techniques has created the predictive criteria summarized in Table 1.

Table 1. *Correlations Between Intraoperative eMEP (D-Wave) and mMEP with Postoperative Motor Outcome. Reprinted from [8]*

D-Wave	Muscle MEP*	Motor outcome
Unchanged or 30–50% decrease	preserved	unchanged
Unchanged or 30–50% decrease	lost uni- or bilaterally	transient motor deficit
>50% decrease	lost bilaterally	long term motor deficit

* In the tibial anterior muscle(s).

Whenever D-waves are preserved up to 50% of baseline values and muscle MEPs are preserved at the end of surgery, we expect no additional motor deficits post-operatively. If mMEP are lost intraoperatively but the D-wave is decreased to less than 50%, the patient will suffer a so-called "transient paraplegia", but will ultimately recover [13, 26, 27]. Finally, mMEP loss and a D-wave amplitude drop of more than 50% of baseline values will result in a long-term or even permanent postoperative motor deficit.

The complementary role played by epidural and muscle MEP recordings is due to some clinical evidence that, to some extent, these monitoring modalities assess the functional integrity of different component of the descending motor system. The phenomenon of "transient paraplegia", where D-waves are preserved up to 50% but mMEP have disappeared at the end of surgery, is likely due to the fact that the supportive system of the spinal cord (meaning non-corticospinal descending tracts and the propriospinal system) is predominantly affected by surgery while fast-conducting corticospinal fibers are mostly preserved. In other words, the spinal cord retains the potentiality to induce voluntary motor control, which will be possible as soon as the supportive system recovers. Typically, patients showing transient paraplegia or monoplegia recover within a few days or weeks after surgery. It can be speculated that during this recovery time other descending systems compensate for the lack of propriospinal tonic influence on α-motoneurons [10].

Applying the criteria presented in Table 1, the combination of eMEPs and mMEPs monitoring showed a sensitivity of 100% and a specificity of 91% in a series of 100 consecutive procedures for ISCT [27].

How to Avoid Irreversible MEP Loss

A corollary of the observation that intraoperative MEPs retain a better predictive value that the patient's pre-operative motor status during surgery for ISCT [34] is that ION should become a critical variable in the decision-making process whether or not to attempt a total tumor removal. The role played by ION is especially critical in the controversial field of low-grade astrocytomas for which the benefit of an aggressive surgical removal is still debatd. Based on a recent review of pediatric ISCTs, a consensus has emerged indicating gross total removal as the standard recommendation for ependymomas, and a conservative approach with tumor debulking and adjuvant therapy as the wisest approach for malignant gliomas. More controversial appears to be the role of surgery for low grade astrocytomas where preservation of the function must be the priority, while a gross total removal can only be considered an option as determined by the surgical plane [36].

Warning Signals

It should be stressed that the criteria presented in Table 1, besides their prognostic value, are invaluable during the intraoperative decision-making process of determining the most safe and rational surgical strategy to adopt. In fact, a sudden disappearance of both the D-wave and mMEPs is very rare and almost exclusively related to an acute anterior spinal artery syndrome that can occur if bipolar cauterization is used on the ventral cleavage plane between the spinal cord and the tumor. In the majority of cases, however, warning signals exist and should be recognized by the neurophysiologist performing ION. Before a complete loss of mMEPs occur, we have often observed an increase in the threshold necessary to elicit a consistent response or, keeping the stimulation intensity stable, the response begins to fluctuate and is recorded in an "on and off" fashion. At a given stimulation intensity, a decrease in amplitude is usually more common than latency shifts. Disappearance of mMEPs usually precedes changes in the D-wave although the D-wave may remain stable or drop insignificantly in spite of complete mMEP loss. Very rarely, the D-wave amplitude decreases without significant changes in the mMEPs. In any case, D-wave deterioration occurs gradually so that, if this event is recognized, there is usually time to take corrective measures.

It is currently our policy to encourage the surgeon to proceed with surgical removal as long as both eMEPs and mMEPs are stable. No significant changes in the post-operative motor status should be expected under these circumstances. If mMEPs are lost, unilaterally or bilaterally, tumor removal can progress if the D-wave remains above 50% of baseline values. At this point, neurophysiological parameters indicate that a so-called "transient" paraplegia phenomenon may occur. Because great care should be taken by the surgeon and the neurophysiologist not to compromise the preservation of the D-wave, this situation is probably the most challenging part of ISCT surgery and ION.

It has to be stressed that if the "point of no return" is reached (meaning loss of mMEPs and D-wave amplitude deterioration > 50%—see further in the text-) surgery should be immediately arrested and corrective measures should be taken to favor the recovery of MEPs. If these do not reappear, surgery should be stopped since the patient is at high risk of a long-term severe or complete motor deficit.

Another particularly challenging situation occurs where baseline mMEPs are present while the D-wave is absent. This occur in about 30% of the cases [27] and requires special considerations since a mMEP loss during tumor removal would not allow for differentiation between the possibility of permanent and transient post-operative motor deficit. In this case, therefore, if the surgeon decides to continue surgery despite mMEP disappearance, he should be aware that the patient is exposed to a significant risk of complete and permanent motor deficit. This example stresses the extreme importance of communication between the neurosurgeon and the neurophysiologist such that each understands the others job. It has to be very clear to both that D-wave is the gold standard for ION during spinal cord surgery and that no reliable prognostic evaluation is possible if the D-wave is not recordable at the beginning of surgery.

Although mMEP loss usually indicates a postoperative impairment of voluntary movement, about 10% of cases exhibit false positive results (i.e., patients who intraoperatively lost mMEPs but did not show any significant motor deficit after surgery) [27]. One possible explanation for this discrepancy, which essentially represents an exception to the "transient paraplegia" phenomenon, is the possibility that preserved fast conducting CT fibers can immediately compensate for an injury that is most likely limited to the supportive motor system. Thus no clinical motor deficits are present post-operatively. This has almost exclusively happened for ISCT at the thoracic level, and has very rarely occurred during cervical ISCT surgeries.

Finally, it should taken into consideration that the specificity of mMEP monitoring increases with the number of muscles that are monitored. If, for example, only the tibialis anterior muscle is monitored during surgery at the thoracic level, and this mMEP is lost during tumor removal while the D-wave is preserved, then a post-operative transient paraplegia would be predicted. However, if we also monitor the abductor hallucis and or the quadriceps femoris, we may see that a response from these muscles is still present. Therefore, any motor deficit expected post-operatively would be limited to the function of the tibialis anterior muscle and not to the entire limb.

Corrective Measures for the Management of Deteriorating Signals

By correlating surgical maneuvers with ION data, we have identified some critical steps during surgeries for ISCTs. It is a common observation that MEPs are often lost when the surgeon attempts to detach the tumor from the spinal cord using both the bipolar coagulator and/or the CUSA in close proximity to the spinal cord white matter. These maneuvers could result in a serious lesion to the CT. During surgery for astrocytomas, which lack a true cleavage plane from the surrounding spinal cord, mMEP deterioration can occur at any time. Most often, however, deterioration occurs when the surgical manipulation involves the antero-lateral borders of the tumor because of CT are often displaced anteriorly or laterally [21]. During surgery for ependymomas, hazard to the cord more often occurs near the end of surgery while attempting to remove the "last piece of tumor" which is usually adherent to the anterior spinal artery axis [31].

Again, whenever mMEPs disappear and the D-wave drops more than 50% of baseline values, corrective measures must be taken. If these do not promote MEP recovery, surgery should be abandoned. Factors we have found useful for promoting the recovery of lost MEPs or deteriorated SEPs during ION of ISCTs can be easily recalled using the acronym T.I.P.: Time, Irrigation, Papaverine/Pressure (blood pressure).

The variable "Time" is likely to be one of the key factors, which can make the difference in the motor outcome of a patient. We have repeatedly observed that if surgery is transiently stopped immediately after mMEP have disappeared or D-wave has significantly deteriorated, these potentials often spontaneously recover. At this point the spinal cord is again able to sustain the further manipulation necessary to remove the remaining tumor. Conversely, to ignore these events and continue or, even worse, speed up the use of cavitron ultrasound aspirator (CUSA) or any other cord manipulation would likely transform a reversible injury into an irreversible one.

The mechanism behind the beneficial effect of warm irrigation of the surgical field has not yet been explained. Nevertheless, it is a common observation that this application accelerates evoked potential recovery. A possible explanation is that irrigation facilitates the washout of extracellular potassium, which has accumulated during the disruption of cell membranes and as a result of depolarization [59, 60].

Local infusion of papaverine and increasing the systolic blood pressure are both methods to improve local perfusion in an effort to counteract an incipient ischemia. Sometimes, MEPs are dramatically correlated with blood pressure values and a sustained hypotension may affect MEPs and unfavorably affect the outcome [49].

Does MEP Monitoring Really Make a Difference?

ION during spinal cord surgery is the only available tool to continuously evaluate the functional integrity of CTs. However, to propose such monitoring as a "standard of care" would require validation of its advantages through a randomised prospective control study comparing monitored versus unmonitored surgeries. So far support for MEP monitoring as an essential component of spinal cord surgery has been based more on the lack of bad results than on documented outcomes of patients who have benefited from ION when compared to an unmonitored control group. Such a trial is not feasible due to different levels of surgical and ION experience at different centers, and to attempt such a study at a single institution where ION is routinely performed would be ethically arguable.

We have recently started a case/"historical control" study which suggests that, on a very early outcome at discharge from hospital, patient operated on with the assistance of ION present a trend towards better rate of unchanged or improved motor status as compared to those operated without ION. Moreover, in the monitored group, only one patient (2.6%) deteriorated to grade 4 of the McCormick Scale (severe neurological deficit, wheelchair dependent) after surgery, as compared to 6 patients (15.7%) in the control group [47]. The 6-month outcome on a larger series of patients is currently under investigation. A further consolidation of the invaluable role of MEPs during ISCT surgery will also rely on Magnetic Resonance Imaging (MRI) studies. Today, the question may rise whether or not the use of ION results in a more conservative approach because the deterioration of the neurophysiological signals may stop surgery before a radical removal is reached. Based on the subjective opinion of those neurosurgeons operating on ISCT with the assistance of ION, the opposite seems to be true. Nevertheless, opinions need to be supported by postoperative MR imaging data demonstrating that the volume of residual tumor (especially for astrocytomas) is equal or even smaller in monitored patients.

Illustrative Cases

Illustrative Case n. 1: Cerebral Peduncle Mapping

A 4-year old child was admitted to our Department following a 2-year history of strabismus that had progressively worsened over time; recently this was associated with a left hemiparesis and an intentional tremor of the left hand. A MRI study documented an intra-axial midbrain lesion on the right side, which enhanced dishomogenously and extended to the lower diencephalon (Fig. 8). The neurological exam on admission revealed a paralysis of the right oculomotor nerve and a mild left hemiparesis. At

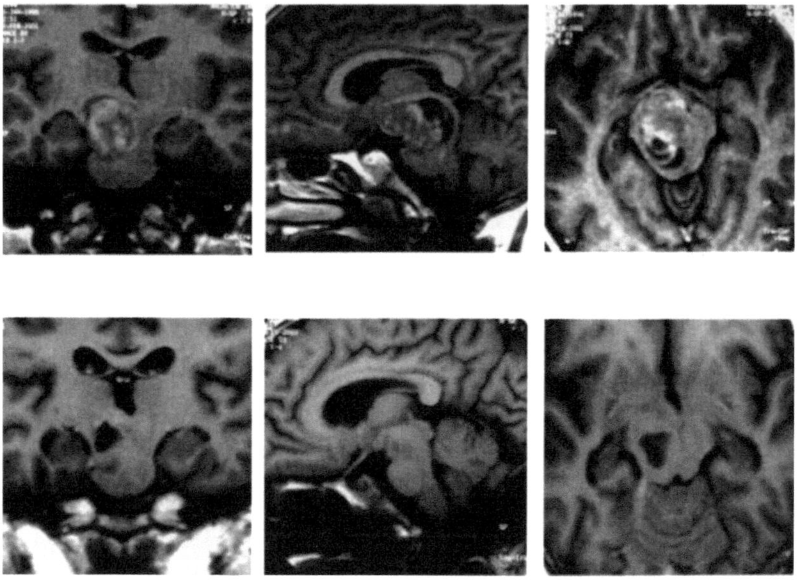

Fig. 8. (Upper panels) Pre-operative contrast enhanced T1-weighted MR images of a right midbrain pylocitic astrocytoma in a 4-year-old child (Lower panels) Total removal of the tumor is documented at the 4-month follow-up MRI study. The tumor was removed through a suboccipital craniectomy and a right supracerebellar infratentorial approach (see illustrative case n. 1)

surgery, the tumor was approached through a suboccipital craniectomy and a right supracerebellar infratentorial route. Once the right dorsolateral aspect of the midbrain was exposed, the bulging tumor prevented the identification of the lateral mesencephalic sulcus. At this point the cerebral peduncle was mapped to avoid injury to the CT (Fig. 9). A consistent low threshold (1.5 mA) response from the left abductor pollicis brevis (LA) was recorded. Then a short longitudinal myelotomy was performed a few millimetres posteriorly to the stimulated point and to the lateral mesencephalic vein. The tumor was then grossly totally removed under continuous mMEP monitoring. At closure, all mMEPs from upper and lower extremities were present. Post-operatively the child experienced only a slight transient worsening of the pre-existing left hemiparesis that subsided by the time of discharge. A 4-month follow-up control MRI documented the removal of the tumor (Fig. 8).

Illustrative Case n. 2: Brainstem Mapping and Corticobulbar Monitoring

A 12-year-old female underwent resection of a giant astrocytoma of the pons on the left side (Fig. 10). Preoperatively she presented a right hemiparesis which subsided after therapy with steroids.

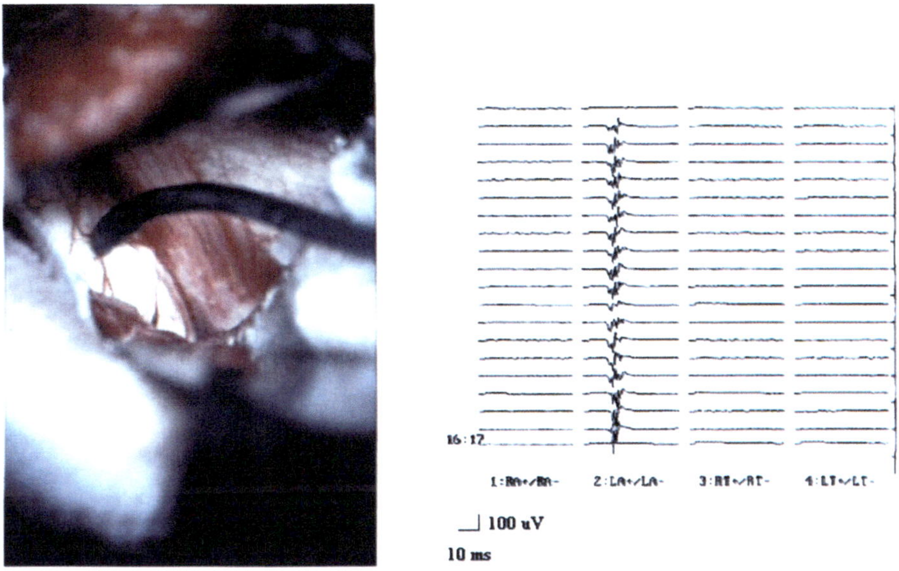

Fig. 9. (Left) Intraoperative view of the direct mapping of the CT on the exposed right lateral surface of the midbrain using a hand-held monopolar probe. A short train of 5 stimuli with stimulation intensity of 1.5 mA at 1 Hz was used. (Right) A consistent response from the left abductor pollicis brevis (*LA*) was recorded. No response was recorded in the right abductor pollicis brevis (*RA*) and in the tibialis anterior muscles (*RT and LT*)

The tumor was approached through a median suboccipital craniectomy. When the floor of the fourth ventricle was exposed, the median sulcus appeared dislocated to the right and the left median eminence was expanded.

EMG wire electrodes were inserted bilaterally in the orbicularis ori (LL and RL) and orbicularis oculi (LU and RU) muscles for mapping and monitoring of the VIIth nerve, and in the left and right abductor pollicis brevis for continuous monitoring of the CT integrity.

Because of the distorted anatomy, direct mapping of the floor of the fourth ventricle was performed to localize the motor nuclei of the facial nerve and, consequently, the safest entry zone to the pons (Fig. 11). We initially stimulated on the left side, about 1.5 cm rostral to the striae medullares, where the motor nuclei were expected to be according to normal brainstem functional anatomy. A response was obtained from the left orbicularis oculi (LU) at a stimulation intensity of 1.5 mA (Fig. 11A). By moving the stimulating probe caudally and to the right, a consistent response from the left orbicularis oris (LL) was recorded at a stimulation intensity of 0.5 mA (Fig. 11B). At this point we moved the stimulation probe more laterally to the right side, about 1 cm above the striae medullares, and a clear response was recorded from the right orbicularis oris

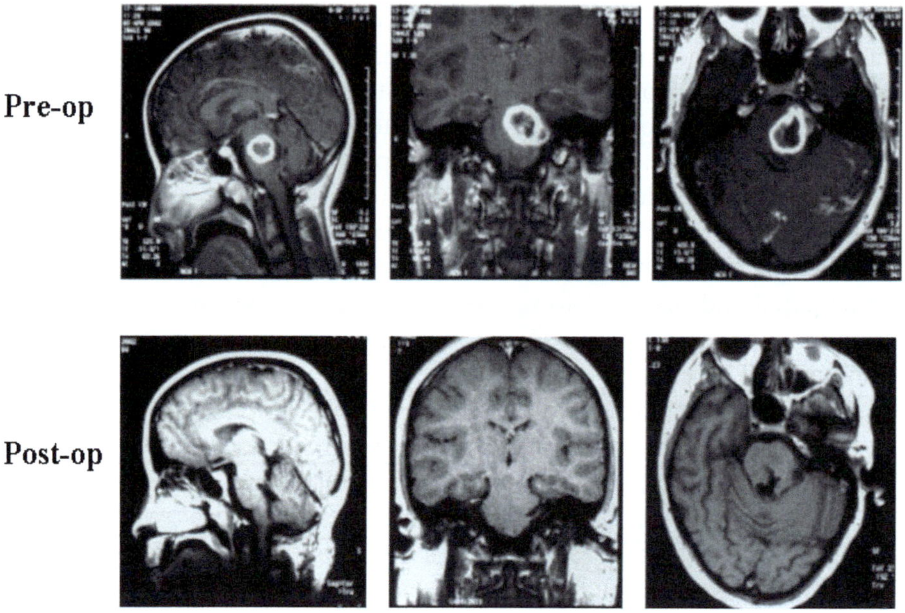

Fig. 10. Pre-operative (upper panels) contrast enhanced T1-weighted MR images of a left pontine astrocytoma: sagittal (left), coronal (middle) and axial (right) view (see illustrative case n. 2). (Lower panels) total removal of the tumor is documented at the follow-up MRI study

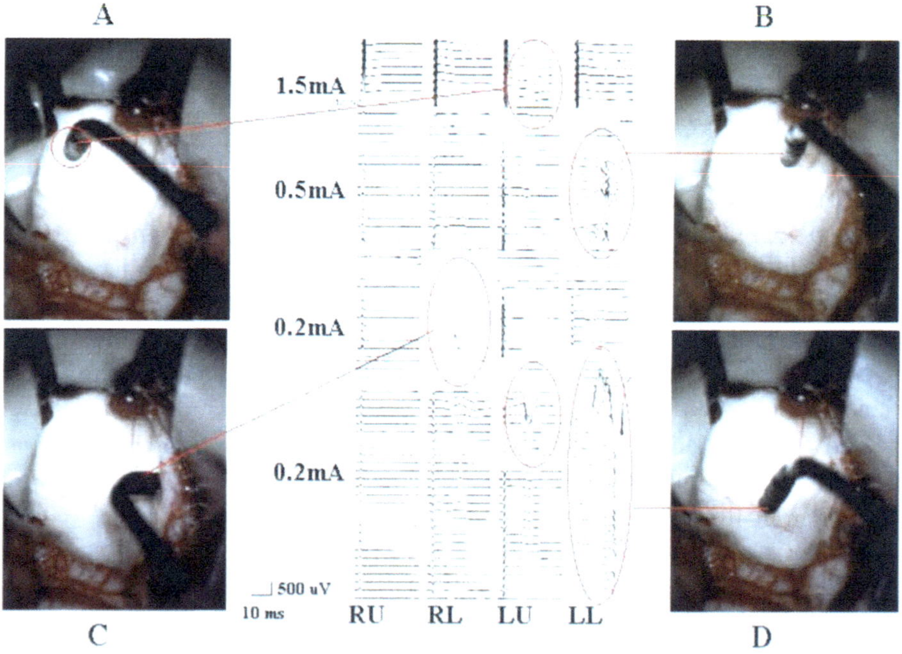

Fig. 11. Mapping of the facial nerve motor nuclei on the floor of the fourth ventricle in the same patient as in Fig. 10 (see text for details)

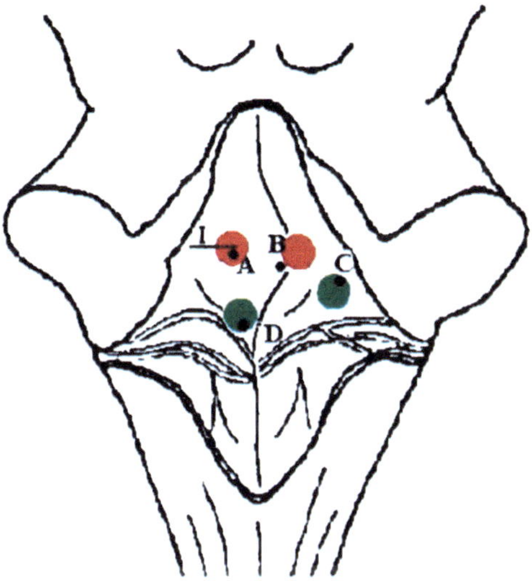

Fig. 12. Schematic summary of mapping results in the same patient as in Fig. 11. In red is represented original position of left and right facial colliculi, as expected according to brainstem anatomy. *A*, *B*, *C*, *D* correspond on the stimulating point illustrated in Fig. 11. *C and D* corresponded also to the lower threshold to elicit consistent response from respectively right and left muscles innervated by the facial nerve. The conclusion was made that real location of facial nerve motor nuclei (in green) was more caudal than expected, especially on the left side, due to the tumor mass effect. Accordingly, initial incision (*I*) was carried on transversally in correspondence of stimulating point A

(RL) at the lowest threshold intensity of 0.2 mA (Fig. 11C). Finally, by moving the stimulating probe paramedially to the left side, a few mm above the striae medullares, a consistent response was recorded from both the left orbicularis oris (LL) and orbicularis oculi (LL), using the same low threshold (0.2 mA) (Fig. 11D). The conclusion was drawn that the tumor displaced the facial nerve motor nuclei caudally especially on the left side (Fig. 12).

Based on these mapping results, the surgeon decided to enter the brainstem on the left side in correspondence to the high threshold stimulating point (Fig. 12). During tumor removal we continuously monitored mMEPs from the left facial (LL) and abductor pollicis brevis (LA) muscles (Fig. 13). These potentials remained unchanged throughout the procedure and the patient suffered no further postoperative motor deficits.

Illustrative Case n. 3: Spinal Cord Monitoring

A 40-year-old man underwent resection of a T3-T6 ependymoma with associated syringomyelia (Fig. 14). Preoperatively he had no motor deficit

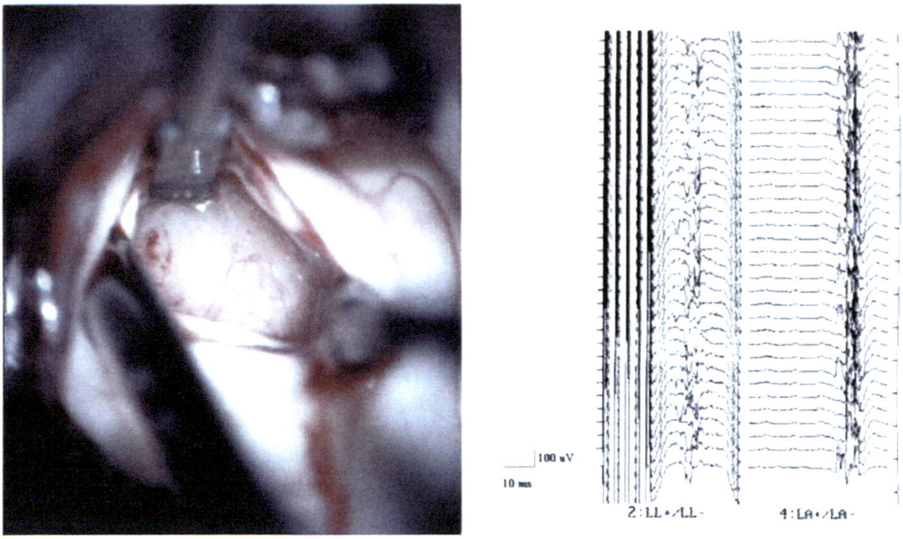

Fig. 13. (Left) Intraoperative view during tumor removal in the same patient as in Fig. 10. (Right) Continuous neurophysiological monitoring of mMEPs during tumor removal. Electrodes were inserted in the left orbicularis ori (*LL*) and abductor pollicis brevis (*LA*) muscles for continuous monitoring of respectively the corticobulbar and CT integrity, after TES (Electrode montage C4/Cz; short train of 4 stimuli; intensity 50 mA)

but only dysaesthesiae in the right arm and right thoracic region. During tumor removal, bilateral loss of the tibialis anterior mMEPs and a progressive decrement of the D-wave amplitude of more than 50% of initial values occurred (Fig. 15a). At this point the decision was made to tran-

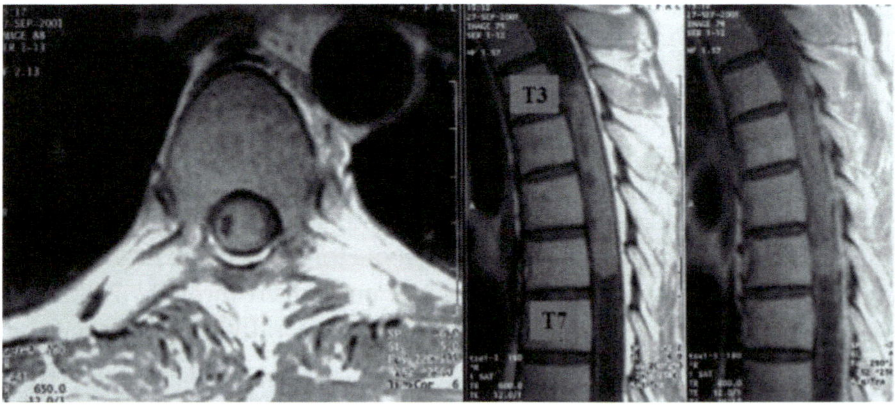

Fig. 14. Axial (left) and sagittal (right) contrast enhanced T1-weighted MR images of a T3-T6 spinal cord ependymoma with associated syringomyelia (see illustrative case n. 3)

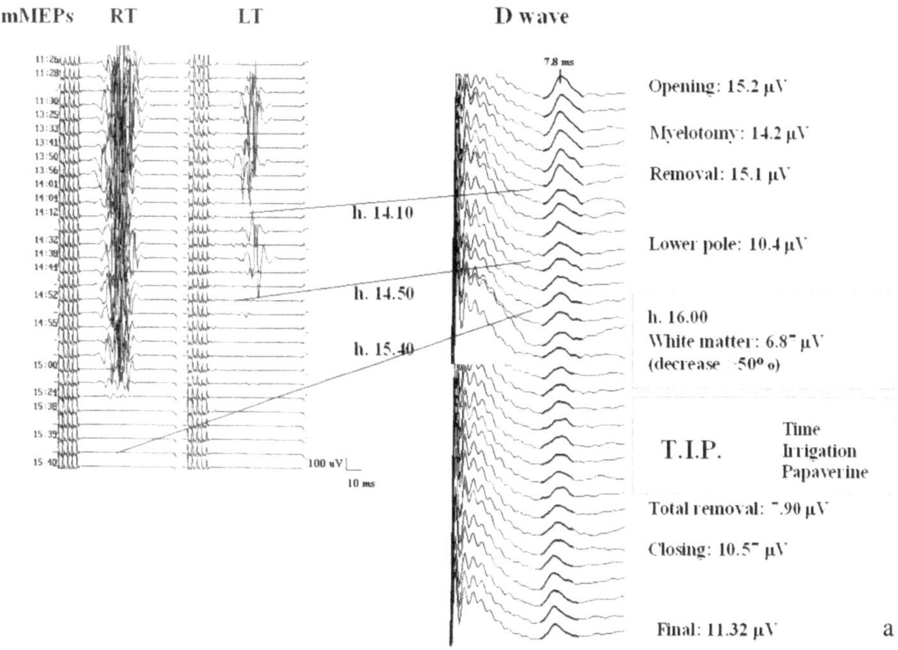

Fig. 15. (a) Motor evoked potential monitoring during surgery for the spinal cord ependymoma illustrated in Fig. 14. (Left panel) mMEPs recordings from the right (*RT*) and left (*LT*) tibialis anterior muscles, after TES. (Right panel) continuous recording of the D wave from an epidural catheter inserted below the level of the lesion; same TES stimulation parameters as for mMEPs are used. At 14.50 the LT mMEP is lost and D-wave has decreased to about 65% of initial values (10.4 vs 15.2 µV). At 15.40, also the RT mMEP disappeared, but D-wave amplitude was still above the 50% drop threshold. Therefore, the decision was made to attempt the removal of the last piece of the tumor, which was adherent to the anterolateral spinal cord white matter. Shortly thereafter, the D-wave amplitude dropped to less than 50% of initial values (6.87 vs 15.2 µV). Surgery was therefore stopped and corrective measures were taken. About 20 minutes later, after warm irrigation and local infusion of papaverine, the D-wave recovered to more than 50% of initial values and it was possible to carefully proceed to a complete tumor removal. (b) D-wave (left panel) and mMEPs from the bilateral abductor pollicis brevis (*RA*, *LA*) and tibialis anterior (*RT*, *LT*) muscles (right panel), before (*OP*) and after (*CL*) tumor removal. The D-wave amplitude declined throughout the critical part of the procedure but at the end recovered to about 70% of initial values. Muscle MEPs were preserved at the end of the procedure. The patient had no significant postoperative motor deficits

siently stop surgery. The surgical field was irrigated with warm saline and the blood pressure was pharmacologically increased. In about 20 minutes the D-wave recovered up to 65–70% of initial values although mMEPs in the tibialis anterior muscles were still absent. Based on the recovery of the D-wave above the 50% threshold, we encouraged the surgeon to com-

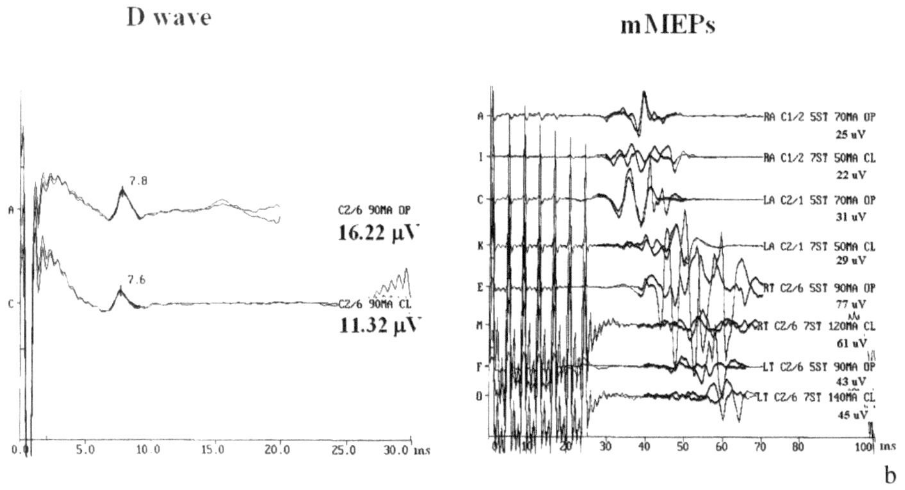

Fig. 15 (*continued*)

plete the tumor removal. At the end of the procedure, D-wave amplitude was 70% of the baseline value and muscle MEPs from the tibialis anterior muscles reappeared (Fig. 15b). The patient had no postoperative motor deficits.

Illustrative Case n. 4: Spinal Cord Monitoring

A 24-year-old female underwent resection of a C6-T1 ependymoma with a syringomyelic cavity extending down to T10 (Fig. 16). Preoperatively she had no motor deficit. At the dura opening (time 12.00) both D-wave and mMEPs from the right and left tibialis anterior (RT and LT) and right and left abductor hallucis (AR and AL) were present (Fig. 17). During microsurgical gross total resection of the lesion, unilateral loss of mMEPs from RT (h 13.00), AR (h 13.45), LT (time 15.30) and AL (time 16.00) occurred. Contemporary, D-wave amplitude dropped by almost 80% of initial values (h 15.30). Surgery was therefore stopped and corrective measures such as warm irrigation and infusion of papaverine were taken. About 30 minutes later, D-wave amplitude recovered to about 50% of initial values and reappearance of mMEPs was documented in LT and AL. At the end of the procedure D-wave amplitude was decreased to 65% of the initial values, and all mMEPs were present except the RT mMEP. Postoperatively the patient experienced a mild transitory paresis of the right leg that recovered by the time of discharge from hospital, one week after surgery.

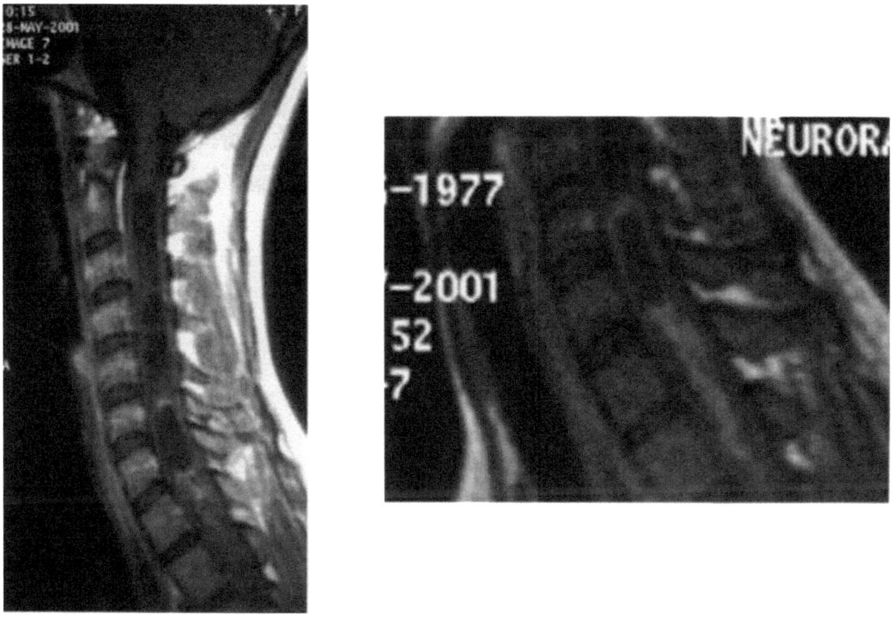

Fig. 16. Sagittal contrast enhanced T1-weighted MR images of a C7-T5 ependymoma associated with a syringomielia in a 24 year old female (see illustrative case n. 4)

Conclusions

With regards to the motor outcome, surgery for brainstem and spinal cord lesions remains challenging even in the hands of talented neurosurgeons, due to the intimate relationship between the lesion and long tracts of the motor system. The advent of MEPs has remarkably improved the reliability and efficacy of ION over the past 10 years. However, there is no scientific justification to promote a single neurophysiological technique with respect to others. Table 2 summarizes methods of ION during brainstem and spinal cord surgery.

Brainstem Surgery

Particularly for brainstem surgery, the more techniques that can be reasonably integrated, the more likely that successful monitoring can be conducted. MEPs as much as SEPs and BAERs should all be monitored. These techniques provide both specific information about a particular system (motor, sensory or auditory), and less specific information about the "well-being" of the brainstem. Yet, only MEP retain a specificity for the motor system and, in this perspective, cannot be replaced by other monitoring techniques.

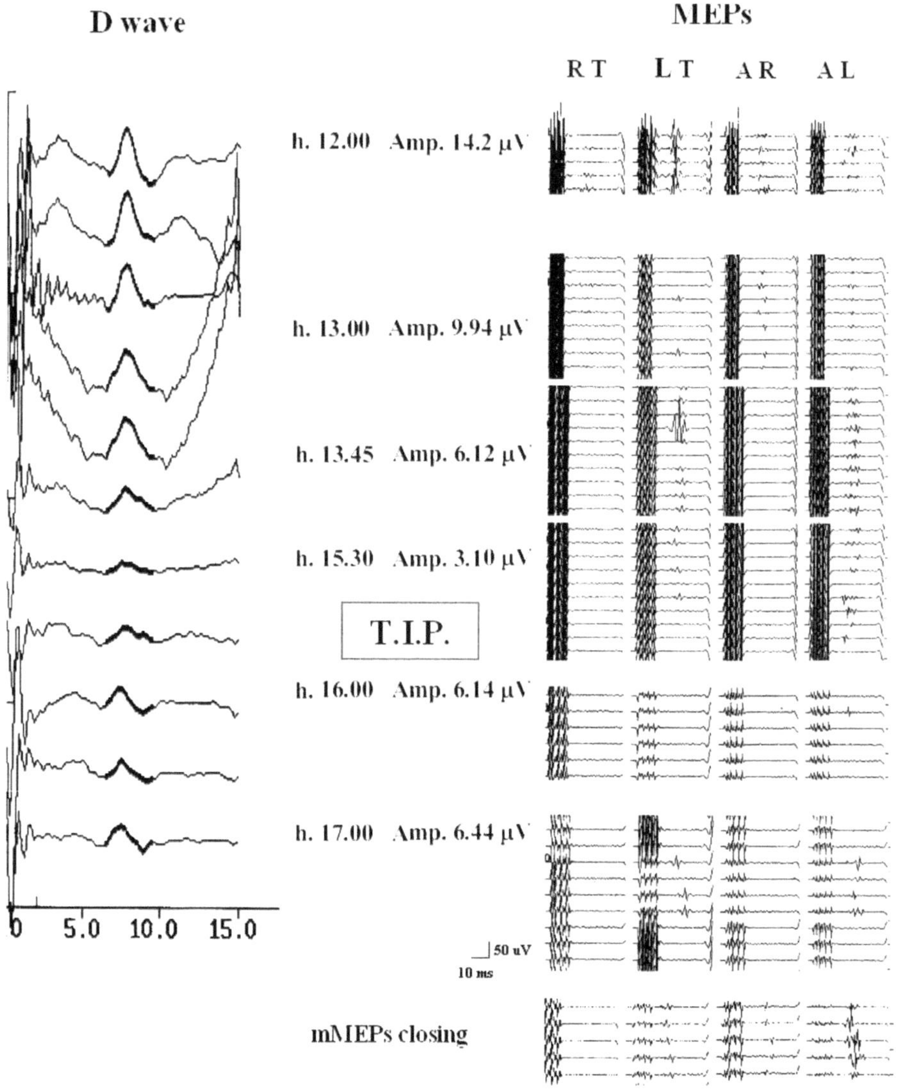

Fig. 17. Motor evoked potential monitoring during surgery for the spinal cord ependymoma illustrated in Fig. 16. (Left panel) continuous recording of the D-wave from an epidural catheter inserted below the level of the lesion; same TES stimulation parameters as for mMEPs are used. (Right panel) mMEPs recordings from the right (*RT*) and left (*LT*) tibialis anterior and from the right (*AR*) and left (*AL*) abductor hallucis muscles, after TES (see text for details)

Table 2. *When and How to Monitor in Brainstem and Spinal Cord Surgery*

Procedures	Mapping techniques	Monitoring techniques
Posterior fossa and brainstem surgery	– direct mapping of motor nuclei of the cranial nerves in the floor of the IV ventricle – direct mapping of the CT within the cerebral peduncle	– BAER – cortical SEPs – mMEPs or eMEPs – continuous EMG recording of muscle innervated by motor cranial nerves – monitoring of the corticobulbar tracts
Spinal cord surgery	– dorsal column mapping – direct mapping of the CT within the spinal cord (collision technique)	– cortical SEPs or epidural (spinal) SEPs – mMEPs and eMEPs

BAER Brainstem auditory evoked response, *EMG* electromyography, *eMEPs* epidural motor evoked potentials (D-wave), *mMEPs* muscle motor evoked potentials, *SEPs* somatosensory evoked potentials, *CT* corticospinal tract.

Additional techniques should then be tailored on the basis of the level and extension of the tumour. At the midbrain level, mapping of the cerebral peduncle can be added to MEPs, SEPs and BAERs to assist the surgeon in choosing the safest entry zone for midbrain tumors that have displaced the CT from its original position. At the level of the pons, mapping of the motor nuclei of the VII cranial nerve on the floor of the IV ventricle will decrease the risk of a postoperative facial palsy. Similarly, during surgery at the level of the medulla and cervicomedullary junction, mapping of the floor of the fourth ventricle allows for the identification of the motor nuclei of cranial nerves IX–X and XII.

Standard techniques for continuous monitoring of the integrity of brainstem motor cranial nerves are today limited to EMG recording directly from the oculomotor, facial, pharyngeal and tongue muscles. Unfortunately, the real neuroprotective role of these monitoring techniques is still controversial as is their correlation with the post-operative clinical functioning of motor cranial nerves. It is expected that the standardization of techniques currently considered "experimental", such as monitoring of corticobulbar pathways, will hopefully further improve brain stem monitoring in the near future.

Spinal Cord Surgery

During spinal cord surgery, both SEPs and mMEPs/eMEPs should be used. To favor SEPs as compared to MEPs, or vice versa, is not justified from a scientific background. Unfortunately MEPs are still not available in many intraoperative neurophysiological departments, but their use should be encouraged since MEPs represent the most appropriate and reliable tool to assess the functional integrity of descending motor tracts. The development and refinement of the so called "collision technique" is already opening new field of investigations in spinal cord neurophysiology. The possibility exists to extend mapping of the CT from the brain cortex, the subcortical motor pathways and cerebral peduncle down to the spinal cord, and this will hopefully further improve the outcome of patients operated on for ISCT. Dorsal column mapping, although not discussed in this chapter, is an "experimental" technique that will hopefully assist in the future during myelotomies for ISCT excision [28].

Finally, in Table 3, some principles and tips of ION are listed, which apply not only to brainstem and spinal cord surgery and—we hope—could be of some help to those who want to be involved in this discipline. Key to the success of ION is a keen, dedicated clinical neurophysiologist who can work in close harmony with the neurosurgeon and anaesthesiologist.

Table 3. *Tips for a Successful ION*

1) Receive a specific training for intraoperative monitoring.
2) Establish a close relationship with both the surgeon and the anesthesiologist.
3) Always use a control modality to identify specific (i.e. induced by surgical maneuvers) from non-specific (i.e. induced by technical problems or variations in anesthesia, temperature, blood pressure, etc.) changes in evoked potentials. For example, monitor SEPs and MEPs also from one upper extremity during surgery for a dorsal spinal cord tumor.
4) Adopt a multimodality INM philosophy, combining information from SEPs, MEPs and other monitoring and mapping modalities.
5) Extract neurophysiological data and immediately analyse this data in order to promptly recognize neurophysiological patterns.
6) Critically integrate neurophysiological patterns with intraoperative clinical and surgical features in order to avoid false positive or false negative results.
7) Warn the neurosurgeon in time, before irreversible injury to the nervous tissue has occured.
8) When evoked responses deteriorate, stop surgery and allow time for the nervous system to recovery.

Acknowledgments

We thank dr. Vedran Deletis for his careful review of the manuscript and David Hershberger for editorial assistance.

References

1. Berenstein A, Young W, Ransohoff J, Benjamin V, Merkin H (1984) Somatosensory evoked potentials during spinal angiography and therapeutic transvascular embolization. J Neurosurg 60: 777–785
2. Blight AR (1983) Cellular morphology of chronic spinal cord injury in the cat: analysis of myelinated axons by line-sampling. Neuroscience 10: 521–543
3. Boyd SG, Rothwell JC, Cowan JMA, Webb PJ, Morley T, Asselman P, Marsden CD (1986) A method of monitoring function in corticospinal pathways during scoliosis surgery with a note on motor conduction velocities. J Neurol Neurosurg Psychiatry 49: 251–257
4. Burke D, Hicks R, Gandevia SC, Stephen J, Woodforth I, Crawford M (1993) Direct comparison of corticospinal volleys in human subjects to transcranial magnetic and electrical stimulation. J Physiol 470: 383–393
5. Burke D, Hicks RG, Stephen JPH (1990) Corticospinal volleys evoked by anodal and cathodal stimulation of the human motor cortex. J Physiol 425: 283–299
6. Cohen AR, Young W, Ransohoff J (1981) Intraspinal localization of the somatosensory evoked potential. Neurosurgery 9: 157–162
7. Deletis V (1993) Intraoperative monitoring of the functional integrity of the motor pathways. In: Devinsky O, Beric A, Dogali M (eds) Electrical and magnetic stimulation of the brain and spinal cord. Raven Press, New York, pp 201–214
8. Deletis V (1999) Intraoperative neurophysiological monitoring. In: McLone D (ed) Pediatric neurosurgery: surgery of the developing nervous system. W.B. Saunders, Philadelphia, pp 1204–1213
9. Deletis V (2001) The 'motor' inaccuracy in neurogenic motor evoked potentials. Clin Neurophysiol 112: 1365–1366
10. Deletis V (2002) Intraoperative neurophysiology and methodologies used to monitor the functional integrity of the motor system. In: Deletis V, Shils J (eds) Neurophysiology in neurosurgery: a modern intraoperative approach. Academic Press, San Diego, pp 25–51
11. Deletis V, Bueno de Camargo A (2001) Interventional neurophysiological mapping during spinal cord procedures. Stereotact Funct Neurosurg 77: 25–28
12. Deletis V, Bueno de Camargo A (2001) Transcranial electrical motor evoked potential monitoring for brain tumor resection. Neurosurgery 49: 1488–1489 (letter)
13. Deletis V, Kothbauer K (1998) Intraoperative neurophysiology of the corticospinal tract. In: Stalberg E, Sharma HS, Olsson Y (eds) Spinal cord monitoring. Springer, Wien New York

14. Deletis V, Sala F (2001) The role of intraoperative neurophysiology in the protection or documentation of surgically induced injury to the spinal cord. Annals NYAS 939: 137–144
15. Deletis V, Sala F, Morota N (2000) Intraoperative neurophysiological monitoring and mapping during brain stem surgery: a modern approach. Operative techniques in neurosurgery 3: 109–113
16. Eisner W, Schmid UD, Reulen HJ, Oeckler R, Olteanu-Nerbe V, Gall C, Kothbauer K (1995) The mapping and continuous monitoring of the intrinsic motor nuclei during brain stem surgery. Neurosurgery 37: 255–265
17. Engler GL, Spielholz NI, Bernhard WN, Danziger F, Merkin H, Wolff T (1978) Somatosensory evoked potentials during Harrington instrumentation for scoliosis. J Bone Joint Surg 60 A: 528–532
18. Ginsburg HH, Shetter AG, Raudzens PA (1985) Postoperative paraplegia with preserved intraoperative somatosensory evoked potentials. Case report. J Neurosurg 63: 296–300
19. Grabb PA, Albright L, Sclabassi RJ, Pollack IF (1997) Continuous intraoperative electromyographic monitoring of cranial nerves during resection of fourth ventricular tumors in children. J Neurosurg 86: 1–4
20. Hicks R, Burke D, Stephen J, Woodforth I, Crawford M (1992) Corticospinal volleys evoked by electrical stimulation of human motor cortex after withdrawal of volatile anesthetics. J Physiol 456: 393–404
21. Jallo GI, Kim BS, Epstein FJ (2001) The current management of intramedullary neoplasms in children and young adults. Annals of Neurosurgery 1: http://annals-neurosurgery.org
22. Jones SJ, Harrison R, Koh KF, Mendoza N, Crockard HA (1996) Motor evoked potential monitoring during spinal surgery: responses of distal limb muscles to transcranial cortical stimulation with pulse trains. Electroencephalography Clin Neurophysiol 100: 375–383
23. Katayama Y, Tsubokawa T, Maemjima S, Hirayama T, Yamamoto T (1988) Corticospinal direct response in humans: identification of the motor cortex during intracranial surgery under general anesthesia. J Neurol Neurosurg Psychiatry 51: 50–59
24. Katsuta T, Morioka T, Fujii K, Fukui M (1993) Physiological localization of the facial colliculus during direct surgery on an intrinsic brain stem lesion. Neurosurgery 32: 861–863
25. Kearse LA, Jr, Lopez-Bresnahan M, McPeck K, Tambe V (1993) Loss of somatosensory evoked potentials during intramedullary spinal cord surgery predicts postoperative neurologic deficits in motor function [corrected]. J Clin Anesth 5: 392–398
26. Kothbauer K, Deletis V, Epstein FJ (1997) Intraoperative spinal cord monitoring for intramedullary surgery: an essential adjunct. Pediatr Neurosurg 26: 247–254
27. Kothbauer K, Deletis V, Epstein FJ (1998) Motor-evoked potential monitoring for intramedullary spinal cord tumor surgery: correlation of clinical and neurophysiological data in a series of 100 consecutive procedures. Neurosurg Focus 4: article 1

28. Krzan JM, Deletis V, Epstein FJ (1998) Intraoperative neurophysiological mapping of the dorsal columns. In: Stalberg EV, De Weerd AW, Zidar J (eds) 9th European Congress of Clinical Neurophysiology (Proceedings), Monduzzi, Bologna, pp 427–431
29. Lesser RP, Raudzens PA, Luders H, Nuwer MR, Goldie WD, Morris III HH, Dinne DS, Klem G, Hahn JF, Shetter AG, Ginsburg HH, Gurd AR (1986) Postoperative neurological deficits may occur despite unchanged intraoperative somatosensory evoked potentials. Ann Neurol 19: 22–25
30. Machida M, Weinstein SL, Yamada T, Kimura J (1985) Spinal cord monitoring. Electrophysiological measures of sensory and motor function during spinal surgery. Spine 10: 407–413
31. McCormick PC, Torres R, Post KD, Stein BM (1990) Intramedullary ependymoma of the spinal cord. J Neurosurg 72: 523–533
32. Merton PA, Morton HB (1980) Stimulation of the cerebral cortex in the intact human subject. Nature 285: 227
33. Minahan RE, Sepkuty JP, Lesser RP, Sponseller PD, Kostuik JP (2001) Anterior spinal cord injury with preserved neurogenic "motor" evoked potentials. Clin Neurophysiol 112: 1442–1450
34. Morota N, Deletis V, Constantini S, Kofler M, Cohrn H, Epstein FJ (1997) The role of motor evoked potentials during surgery for intramedullary spinal cord tumors. Neurosurgery 41: 1327–1336
35. Morota N, Deletis V, Epstein FJ, Kofler M, Abbott R, Lee M, Ruskin K (1995) Brain stem mapping: neurophysiological localization of motor nuclei on the floor of the fourth ventricle. Neurosurgery 37: 922–930
36. Nadkarni TD, Rekate HL (1999) Pediatric intramedullary spinal cord tumors: critical review of the literature. Childs Nerv Syst 15: 17–28
37. Neuloh G, Schramm J (2002) Mapping and monitoring of supratentorial procedures. In: Deletis V, Shils J (eds) Neurophysiology in Neurosurgery: a modern intraoperative approach. Academy Press (in press)
38. Nuwer MR, Dawson EG, Carlson LG, Kanim LE, Sherman JE (1995) Somatosensory evoked potential spinal cord monitoring reduces neurologic deficits after scoliosis surgery: results of a large multicenter study. Electroencephalography Clin Neurophysiol 96: 6–11
39. Owen JH (1991) Evoked potential monitoring during spinal surgery. In: Bridwell KH, DeWald RL (eds) The Textbook of Spinal Surgery. J.B. Lippincott, Philadelphia, pp 31–64
40. Patton HD, Amassian VE (1954) Single- and multiple unit analysis of cortical stage of pyramidal tract activation. J Neurophysiol 17: 345–363
41. Pechstein U, Cedzich C, Nadstawek J, Schramm J (1996) Transcranial high-frequency repetitive electrical stimulation for recording myogenic motor evoked potentials with the patient under general anesthesia. Neurosurgery 39: 335–344
42. Penfield W, Boldrey E (1937) Somatic motor and sensory representation in the cerebral cortex of man as studied by electrical stimulation. Brain 60: 389–443
43. Rhoton AL (2000) The posterior fossa veins. Neurosurgery 47: S69–S92

44. Rodi Z, Vodusek DB (2002) Normative value and variability of intraoperative muscle MEP. Clin Neurophysiol 113: S62
45. Romstock J, Strauss C, Fahlbusch R (2000) Continuous electromyography monitoring of motor cranial nerves during cerebellopontine angle surgery. J Neurosurg 93: 586–593
46. Sakuma J, Matsumoto M, Ohta M, Sasaki T, Kodama N (2002) Glossopharyngeal nerve evoked potentials after stimulation of the posterior part of the tongue in dogs. Neurosurgery 51: 1026–1032; discussion 1032–1023
47. Sala F, Basso E, Lanteri P, Faccioli F, Bricolo A (2002) The impact of intraoperative neurophysiological monitoring on early neurological outcome after surgery for intramedullary spinal cord tumors. Clin Neurophysiol 113: S44
48. Sala F, Niimi Y (2002) Neurophysiological monitoring during endovascular procedures on the spine and the spinal cord. In: Deletis V, Shils J (eds) Neurophysiology in neurosurgery: a modern approach. Academic Press, San Diego, pp 119–151
49. Sala F, Niimi Y, Krzan MJ, Berenstein A, Deletis V (1999) Embolization of a spinal arteriovenous malformation: correlation between motor evoked potentials and angiographic findings: technical case report. Neurosurgery 45: 932–937; discussion 937–938
50. Schlake HP, Goldbrunner R, Siebert M, Behr R, Roosen K (2001) Intra-Operative electromyographic monitoring of extra-ocular motor nerves (Nn. III, VI) in skull base surgery. Acta Neurochir (Wien) 143: 251–261
51. Sloan TB (2002) Intraoperative neurophysiology and anesthesia management. In: Deletis V, Shils J (eds) Neurophysiology in neurosurgery: a modern intraoperative approach. Academic Press, San Diego, pp 451–474
52. Strauss C, Romstock J, Fahlbusch R (1994) Intraoperative mapping of the floor of the fourth ventricle. In: Loftus CM, Traynelis VC (eds) Intraoperative monitoring techniques in neurosurgery. McGraw-Hill, Inc, New York, pp 213–220
53. Strauss C, Romstock J, Nimsky C, Fahlbusch R (1993) Intraoperative identification of motor areas of the rhomboid fossa using direct stimulation. J Neurosurg 79: 393–399
54. Suzuki K, Matsumoto M, Ohta M, Sasaki T, Kodama N (1997) Experimental study for identification of the facial colliculus using electromyography and antidromic evoked potentials. Neurosurgery 41: 1130–1136
55. Tamaki T, Takano H, Nakagawa T (1986) Evoked spinal cord potentials elicited by spinal cord stimulation and its use in spinal cord monitoring. In: Cracco RQ, Bodis-Wollner I (eds) Evoked Potentials. Alan Liss, New York, pp 428–433
56. Taniguchi M, Cedzich C, Schramm J (1993) Modification of cortical stimulation for motor evoked potentials under general anesthesia; technical description. Neurosurgery 32: 219–226
57. Toleikis JR, Skelly JP, Carlvin AO, Burkus JK (2000) Spinally elicited peripheral nerve responses are sensory rather than motor. Clin Neurophysiol 111: 736–742

58. Woodforth IJ, Hicks RG, Crawford MR, Stephen JP, Burke DJ (1996) Variability of motor-evoked potentials recorded during nitrous oxide anesthesia from the tibialis anterior muscle after transcranial electrical stimulation. Anesthesia Analgesia 82: 744–749
59. Young W, Koreh I (1986) Potassium and calcium changes in injured spinal cords. Brain Res 365: 42–53
60. Young W, Rosenbluth J, Wojak JC, Sakatani K, Kim H (1989) Extracellular potassium activity and axonal conduction in spinal cord of the myelin-deficient mutant rat. Exp Neurol 106: 41–51
61. Zentner J (1989) Noninvasive motor evoked potential monitoring during neurosurgical operations in the spinal cord. Neurosurgery 24: 709–712
62. Zornow MK, Grafe MR, Tybor C, Swenson MR (1990) Preservation of evoked potentials in a case of anterior spinal artery syndrome. Electroencephalogr Clin Neurophysiol 77: 137–139

Motor Evoked Potential Monitoring for the Surgery of Brain Tumours and Vascular Malformations

G. Neuloh and J. Schramm

Department of Neurosurgery, University of Bonn, Germany

With 14 Figures

Contents

Abstract

Brain surgery incurs a significant risk of a new motor deficit in lesions within or adjacent to the motor areas and pathways which, for the patient, presents one of the most disabling complications of such operations.

It is a major concern of intracranial procedures to delineate and monitor motor regions in order to preserve their structural and functional integrity, while still achieving maximal cytoreduction.

The technique of motor evoked potential recording has had to be adapted to intraoperative recording conditions under general anaesthesia, but has been available for clinical use now for almost ten years. This contribution summarizes the current technique and related methods, as well as our clinical experience in some 400 cases of MEP monitoring in supratentorial tumors, lesions in and around the brainstem, and aneurysm surgery.

Intraoperative MEP recordings have been shown to reliably reflect an impending new motor deficit. Irreversible MEP deterioration heralds new paresis, and unaltered recordings predict preserved motor function. This is also true in aneurysm surgery where conventional SEP monitoring may yield false-negative results with regard to development of a new motor deficit. Moreover, if MEP deterioration can be reversed, or halted by early surgical intervention, the presence of only a transient motor deficit, or even the lack of a new postoperative deficit, indicates the success of the MEP monitoring method in the prevention of a significant motor impairment. Certain complicated lesions can only be operated on at all because MEP monitoring is available.

In conclusion, intraoperative MEP monitoring is a useful aid in brain surgery with which to avoid a new motor deficit without compromise to the surgical result. Controlled prospective studies will be required to verify the clinical value of the method.

Keywords: Motor evoked potentials; neurosurgery; intraoperative neurophysiology; cerebral aneurysms; cerebral arteriovenous malformations.

Introduction

To minimize the risk of a new motor deficit is a major task facing many brain procedures. A variety of factors, such as resection close to motor areas and pathways, brain retraction and manipulation, and compromise of brain vessels with vascular procedures can affect the functional integrity of the motor system. The extended and partially exposed course of the motor regions contributes to the significant risk of new pareses after intracranial surgery: the primary motor cortex embraces all of the suprasylvian convexity and median surface of the hemisphere, the efferent pyramidal fibres transect the hemisphere via the corona radiata and converge within the internal capsule amidst the basal ganglia, to reach the surface of the brain in the cerebral peduncle, and enter into the ponto-medullary corticospinal tract before leaving the brain in the the ipsilateral anterior, or the contralateral lateral corticospinal tract. Seen from the patient's point of view, a persisting disabling weakness is one of the most devastating complications of brain surgery. At the same time, a maximum surgical result in terms of cytoreduction, reduced seizure frequency and elimination of bleeding risk from vascular lesions, is obviously important as well. Information about the functional state of the motor systems would clearly be an important aid to achieve both objectives without major compromise in many instances.

Direct Stimulation Mapping

Intraoperative electrical stimulation of the exposed cortex is the traditional approach, which has been applied in neurosurgery for many decades in order to delineate the motor area, and to identify the descending pyramidal tract (Horsley V, 1891, Krause F, 1911, Penfield W *et al.*, 1937, King RB *et al.*, 1987, Berger MS *et al.*, 1989, Ebeling U *et al.*, 1989, Ebeling U *et al.*, 1992, Berger MS, 1995, 1996, Berger MS *et al.*, 1997). While the same method requires awake craniotomy for speech mapping purposes, it has recently been adopted under conditions of general anesthesia if used for motor mapping alone (Yingling CD *et al.*, 1999). More recent mapping methods, in particular functional imaging tools like fMRI and PET are gaining increasing importance in surgical planning, yet are still validated by intraoperative stimulation mapping (Stapleton SR *et al.*, 1997, Vinas FC *et al.*, 1997, Fandino J *et al.*, 1999, Roux FE *et al.*, 1999, Lehericy S *et al.*, 2000, Roberts TP *et al.*, 2000, Roux FE *et al.*, 2000, Krings T *et al.*, 2001b, Romstock J *et al.*, 2002). It must be considered that delineation of eloquent brain areas allows for their structural preservation, but cannot directly reflect functional deterioration, or remote effects occurring during the actual resection procedure, such as compromise to their vascularization. Although preoperative functional imaging may help to identify the structures to be preserved, those methods do not yet provide the required and reliably proven precision, nor do they allow for continuous intraoperative functional assessment. In fact, no imaging or mapping technique is yet conceived that provides the continuous intraoperative functional information that can be obtained by recording the actual neuromuscular activity of the motor system.

Evoked Potentials

The use of evoked potential (EP) recording as an intraoperative monitoring tool was first introduced in 1972 (Croft TJ *et al.*, 1972, Tamaki T *et al.*, 1972). It was first applied to spinal surgery and to posterior fossa procedures. Early research papers about evoked potentials and experimental spinal cord injury from this group were soon followed by clinical reports, still focusing on somatosensory and auditory evoked potential (SEP and AEP) long-tract and cranial nerve monitoring for spinal and posterior fossa surgery (Schramm J *et al.*, 1977, Schramm J *et al.*, 1979, Schramm J, 1980, Schramm J *et al.*, 1983, Fischer C *et al.*, 1985, Schramm J, Jones SJ, 1985, Schramm J *et al.*, 1985a, Schramm J *et al.*, 1985b, Schramm J *et al.*, 1986, Schramm J *et al.*, 1988a, Schramm J *et al.*, 1988b, Sindou M *et al.*, 1988, Jeanmonod D *et al.*, 1989, Schramm J *et al.*, 1989). The next step was to introduce SEP monitoring into aneurysm surgery (McPherson RW *et al.*, 1983, Mooij JJ *et al.*, 1987, Buchthal A *et al.*, 1988, Schramm J, 1989, Schramm J *et al.*, 1990, Friedman WA *et al.*, 1991, Matsuda M *et al.*, 1991, Mizoi K *et al.*, 1991, Schramm J, Moeller AR, 1991, Schramm J *et al.*, 1991, Schramm J *et al.*, 1994, Sako K *et al.*, 1995, Sako K *et al.*, 1998, Lopez JR *et al.*, 1999). The latter application has been adopted in many neurosurgical centers over the last decade. It is based on the observation of a close relationship between cortical cerebral blood flow and SEP changes (Hargadine JR *et al.*, 1980, Symon L, 1980, McPherson RW *et al.*, 1983, Coyer PE *et al.*, 1987).

The initial research was devoted to working out the technical principles of evoked potential monitoring, e.g., stimulation and recording parameters, anesthesia and electrode montages. Later on with growing clinical experience, criteria of abnormality began to emerge. Those in use today are defined arbitrarily, but with regard to, and confirmed by a long-term clinical experience.

With accumulating observations of correlation of EP monitoring and clinical outcome it became clear, that SEP and AEP monitoring does not reliably reflect impending motor impairment (Deletis V, 1993, Kothbauer K *et al.*, 1997, Deletis V, Kothbauer K, 1998). Damage to the motor tract is correlated with compromise to other fiber systems in many instances, and can be reflected by SEP or AEP in such cases. However, the topographical course, and the blood supply of motor and somatosensory pathways differ increasingly in particular at the subcortical level. Therefore, SEP recording alone is not seen as a sufficient monitoring method with spinal cord surgery anymore (Deletis V, Kothbauer K, 1998), although a multicenter retrospective analysis has shown that the rate of postoperative paresis after scoliosis surgery was significantly lower after the introduction of SEP monitoring (Nuwer MR *et al.*, 1995, Nuwer MR, 1999). Likewise, there is a significant rate of new motor deficit despite unaltered intraoperative SEP recordings in most reports on SEP monitoring for cerebral aneurysm surgery (Friedman WA *et al.*, 1991, Krieger D *et al.*, 1992, Holland NR, 1998).

Motor Evoked Potentials

The advent of the motor evoked potential technique (Merton PA *et al.*, 1980) provided a method for frequent intermittent assessment of the motor systems *functional* integrity beyond mere *structural* preservation.

Soon after the introduction of motor cortex stimulation by means of transcranially applied electric current or changing magnetic fields (Merton PA *et al.*, 1980, Barker AT *et al.*, 1985), with recording of the evoked spinal or myogenic activity (motor evoked potentials), much effort was devoted to adopt this technique for intraoperative motor monitoring purposes (Levy WJ *et al.*, 1986, Zentner J, 1989, 1991, Kakimoto M *et al.*, 2000). In particular the inhibitory effect of most anesthetic regimes on the cortico-motoneuronal transmission turned out to be an obstacle for the reliable recording of the evoked muscular activity under general anesthesia. Direct recording of the descending volleys from the exposed spinal cord (epidurally or subdurally) allowed this problem to be addressed in spinal surgery (Burke D *et al.*, 1990, Deletis V, 1993, Kothbauer K *et al.*, 1997, Deletis V, Kothbauer K, 1998, Kothbauer KF *et al.*, 1998). In contrast, intracranial procedures would require recording of muscular MEPs. In 1993, our group introduced a cortical stimulation paradigm (Taniguchi M *et al.*, 1993), which turned out to be suitable for transcranial stimulation as well (Pechstein U *et al.*, 1996, Calancie B *et al.*, 1998, Deletis V, Kothbauer K, 1998). Temporal summation of excitatory postsynaptic potentials elicited by a train of descending volleys (Taylor BA *et al.*, 1993), which is directly elicited within the efferent axon by high-frequency repetitive cortical or transcranial stimulation, can overcome the inhibitory effects of anesthesia on the cortico-motoneuronal transmission.

Most of the research and clinical reports have been concerned with spinal cord motor monitoring so far. An increasing number of clinical reports converge on some general principles regarding recording technique and clinical interpretation, so that the method is now routinely applied in many institutions performing spinal, medullary, or aortic surgery (Merton PA *et al.*, 1980, Zentner J, 1989, Burke D *et al.*, 1990, Zentner J, 1991, Jones SJ *et al.*, 1996, Nagle KJ *et al.*, 1996, Kothbauer K *et al.*, 1997, Burke D *et al.*, 1998, Calancie B *et al.*, 1998, Kothbauer KF *et al.*, 1998, Meylaerts SA *et al.*, 1999, Kakimoto M *et al.*, 2000, Calancie B *et al.*, 2001).

The intraoperative monitoring of motor evoked potentials with supratentorial and brain stem procedures has only developed over the last ten years, and must still be considered to be in progress. Since the development of multi-pulse MEP stimulation as a routine procedure compatible with general anesthesia (Taniguchi M *et al.*, 1993) several preliminary reports have been published on the clinical use of MEPs with supratentorial surgery (Cedzich C *et al.*, 1996, Kombos T *et al.*, 2001, Zhou HH *et al.*, 2001).

Here we review our ten-year experience with MEP monitoring for brain surgery in over 400 cases. We describe the principles of the technique including anaesthetic techniques and discuss clinical applications along with limitations and drawbacks of the method.

MEP Technique

Electrical stimulation is performed transcranially, or directly at the cortex if the craniotomy allows the placement of an electrode grid over the central/pericentral region. High-frequency trains of 5 (4–9) short (300–10000 μs) monopolar anodal pulses are then applied with a maximum intensity of 25–30 mA for direct cortical, and 100–200 mA for transcranial stimulation. Stimulation is repeated every 10–30 s, but can be applied more frequently with critical steps of the procedure. Muscular MEP responses are recorded from contralateral facial and distal limb muscles, e.g., finger flexors, thenar, and anterior tibial muscle. MEPs occur as oligo-/polyphasic deflections at 15–30 ms after the stimulus onset, and can usually be recorded as unaveraged responses due to their high amplitude of up to 10 mV.

Total intravenous anesthesia with propofol, or anesthesia with a modern opioid (remifentanil, alfentanil, sufentanil) in combination with a halogenated agent (isoflurane, sevoflurane) is most compatible with MEP recording, and neuromuscular block should be avoided.

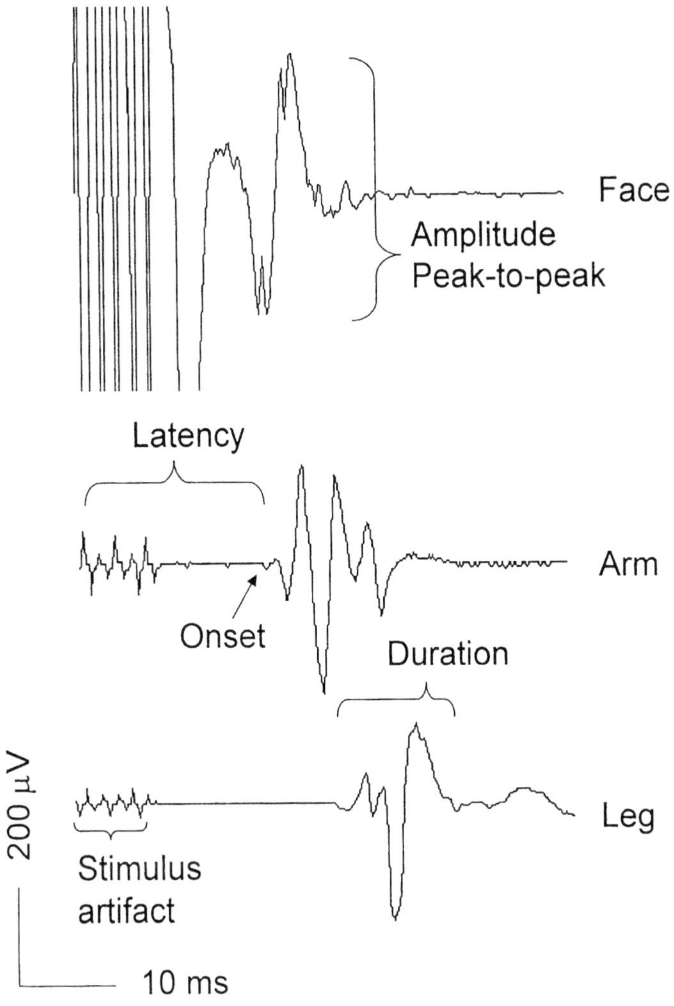

Fig. 1. Typical MEP responses. After transcranial stimulation with a train of 5 pulses at 500 Hz, MEP responses were recorded from contralateral facial (orbicularis oris and orbicularis oculi) and distal limb (thenar, anterior tibial) muscles. Important MEP parameters such as amplitude, latency, and potential duration, are indicated. Typically the responses are polyphasic, i.e. have more than two peaks

Stimulation

Transcranial stimulation is usually required for MEP monitoring with aneurysm surgery, with deeply seated lesions, or locations remote from the central area, and with infratentorial procedures. It is performed via subdermal needle electrodes, which are placed at specific stimulation sites as described below. If straight needle electrodes are used, which may easily slip from the skin, redundant placement of at least two electrodes at each site is advisable. Alternatively, spiral shaped corkscrew electrodes may be used, which cannot slip intraoperatively, but are comparably expensive.

Direct cortical stimulation is preferred if an electrode strip or grid can be placed to cross the central sulcus around the hand (or leg) area. In many instances it is necessary to push the electrode forward underneath the dura, if the craniotomy does not directly expose all of the central region. In our institution, electrode strips with one row of eight electrodes are used, or grids with two (or even four) rows of eight electrodes, depending on the size of the craniotomy, and on the cortical field to be mapped. With increasing distortion of the gyral relief from a central tumor, more extensive mapping via larger electrode grids must precede the actual monitoring procedure, as described in the section on "Neurophysiological Functional Mapping of the Pericentral Cortex" (pp. 188–196).

Stimulation site: For transcranial stimulation, the stimulating anodal electrode is typically placed at 1–2 cm anterior to the C3/C4 positions of the international 10–20-system, in order to be positioned over the hand motor cortex. The cathode is placed at the corresponding contralateral position (Jones SJ *et al.*, 1996, Pechstein U *et al.*, 1996, Deletis V, Kothbauer K, 1998). Alternatively, the anode can be placed at 1–2 cm anterior to Cz, the cathode at Fpz or Fz. A circumferential cathode from linked electrodes at Fz, F3/4 and A1/2 can be more efficient (Ubags LH *et al.*, 1996). Stimulation over the hand area (C3/4+) has proven most efficient, but produces a large twitching artifact from excitation of back musculature. The artifact is considerable lower with more medial stimulation sites C1/2 (see Fig. 2), but the stimulus efficiency is usually lower as well. If the patient is in a sitting position, subdural air may collect intracranially, preventing transcranial stimulation, given the standard electrode montages. In those cases, stimulation electrodes may be placed more laterally/basally (Kombos T *et al.*, 2000b), e.g., between C3/4+ and a position 4 cm anterior to the T3/4 sites (see Fig. 2).

If direct cortical stimulation is applied with craniotomies that enable the placement of surface electrodes over the central region, it is often impossible to identify the central cortex by mere anatomic criteria due to the limited view and the often distorted local topography with central lesions. Therefore, SEP phase reversal is always performed preceding cortical stimulation (Cedzich C *et al.*, 1996). The point of best precentral response is considered closest to the motor hand/foot area and is selected for stimulation. In most cases, the craniotomy allows for median nerve SEP recording. The electrode grid is placed over the tentative central sulcus, as close to the sensorimotor hand area as permitted by the craniotomy. It is helpful if the craniotomy can be planned to extend to the superior temporal line about 6–7 cm lateral from the midline, which lies approximately over the sensorimotor hand area. With parafalcine lesions, only recording

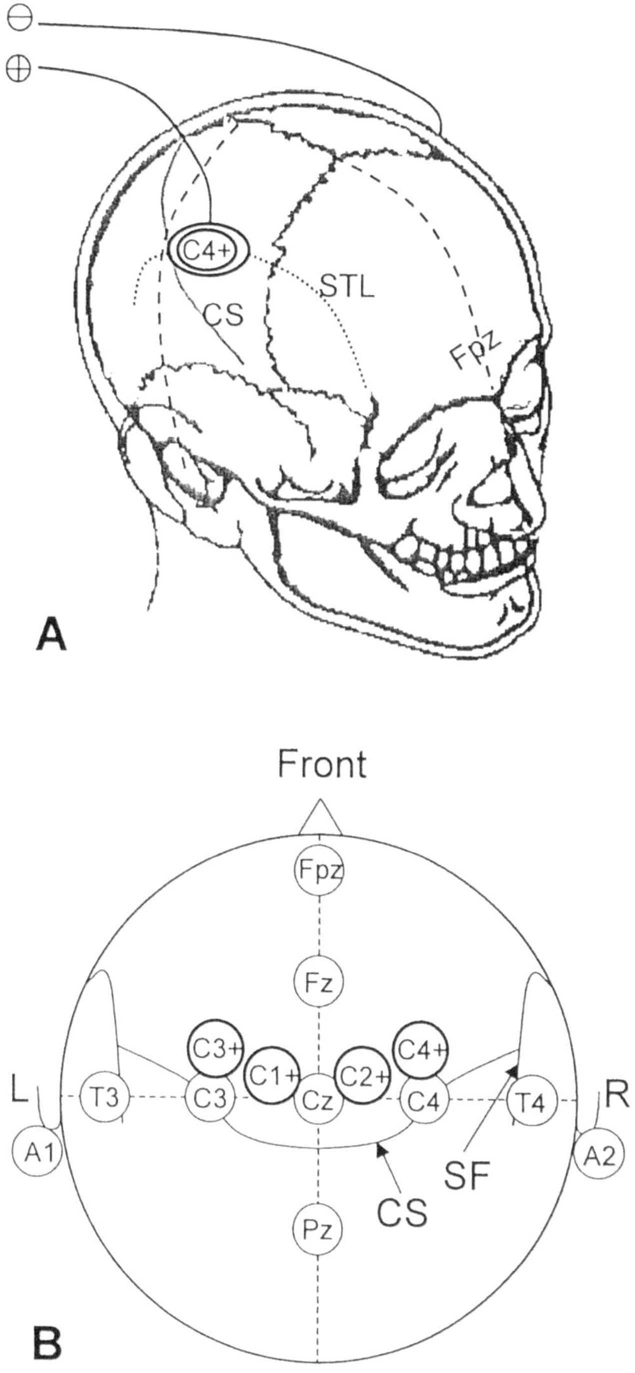

Fig. 2. Typical MEP stimulation sites. MEP stimulation sites are shown in a schematic way as related to anatomical landmarks (A), and the international 10–20 system for EEG recording (B). C3/4+ indicates a position about 1–2 cm anterior to the corresponding position C3/4 etc. of the 10–20 system. C3/4 position is easiest found at the cross-over point of the superior temporal line (about 7 cm lateral to the midline) and a line connecting the earlobes via the vertex point. C1/2+ positions are located between C3/4+ and Cz. *CS* Central sulcus, *SF* sylvian fissure

of tibial nerve SEP phase reversal from the paramedian motor cortex may be possible, and MEPs will be recorded mainly from the leg. SEP phase reversal results may be confirmed and completed by systematic MEP stimulation through all grid electrodes, as described in the section on "Neurophysiological Functional Mapping of the Pericentral Cortex" (pp. 188–196).

Extraaxial tumors like meningiomas and bridging veins may hamper subdural advancement and correct positioning of the electrode strip. In our experience, correct electrode positioning is the single most important factor for successful cortical stimulation. In a statistical analysis of failed MEP stimulation in about 8% of 180 monitoring cases, atypical electrode position after failed phase reversal mapping, or due to non-central craniotomy was significantly associated with unsuccessful stimulation, whereas anesthesia, partial neuromuscular block, or preexisting pareses did not prevent successful MEP recording.

Stimulus train parameters: A high-frequency train of *constant-current monopolar anodal* stimuli has turned out to be best suitable for MEP stimulation (Amassian VE *et al.*, 1987, Cedzich C *et al.*, 1998, Deletis V, Kothbauer K, 1998, Deletis V *et al.*, 2001). Table 1 summarizes typical stimulation and recording parameters. Fig. 3 shows the effect of varying stimulus train parameters on MEP responses.

With transcranial stimulation in particular, constant current stimulation is preferred, because the twitching artifact from inadvertent stimulation of the autochthonous neck and back muscles is less compared with

Table 1. *Typical MEP Stimulation and Recording Parameters*

MEP technique – parameters	
Stimulation	
Site	C3+2cm/C4+2cm (transcranial) or motor cortex
Polarity	monopolar anodal
Phase	monophasic
Frequency in train	250–500 Hz
Max. Amplitude	30 (cortical)–200 (transcranial)
Pulse duration	300 (50–1000) μs
Pulse number	5 (3–9)
Repetition rate	every 1–30 s
Recording	
Sites	facial and distal arm/leg muscles
Motor response	CMAP
Sweeps	50–80 ms
Averages	none/few
Filtering	10–3000 Hz

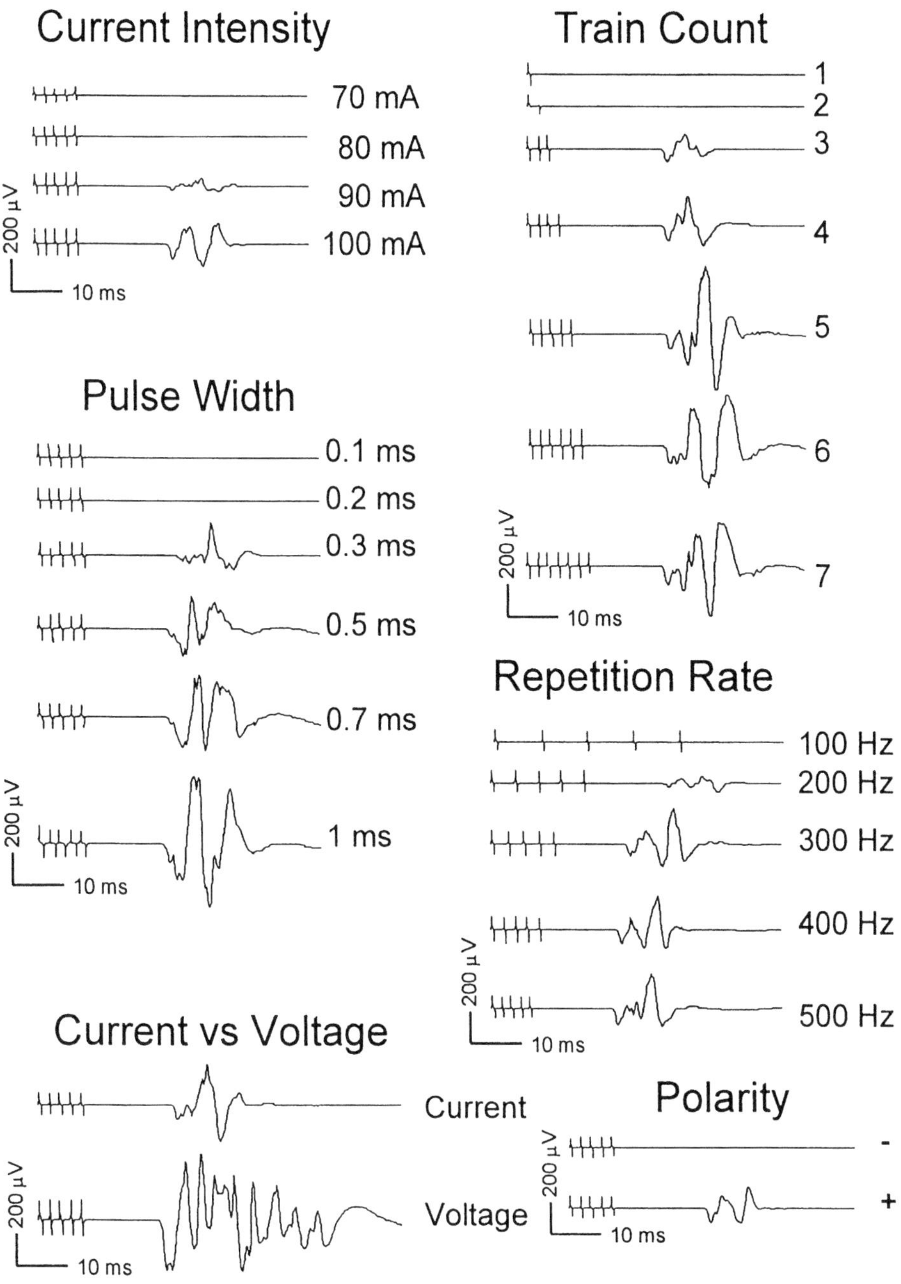

Fig. 3. Influence of varying stimulation parameters on MEPs. MEPs were recorded from the contralateral thenar muscle after transcranial stimulation at C3+2cm/C4+2cm. Based on a train of 5 anodal constant-current pulses of 300 µs duration with an interstimulus-interval of 2 ms (500 Hz) at an intensity of 100 mA, all parameters were systematically varied. Modified from (Neuloh G *et al.*, 2002)

constant voltage stimuli. However, the latter are more efficient, in particular with suboptimal placement of the stimulating electrode. With direct cortical stimulation, this twitching artifact is rarely significant, but constant current flow is preferred for safety reasons.

A higher *number of pulses in a stimulus train* leads to summation of more EPSPs at the α-motoneurons, and may yield a better motor response. With direct cortical stimulation, more than 5(4–6) pulses will rarely be necessary, with transcranial stimulation it may be helpful to increase the number of pulses to up to 9.

With transcranial stimulation, pulse duration up to 1 ms may be chosen, if a high intensity stimulation device is not available. With cortical stimulation, pulses of 300 μs (200–500 μs) duration are always sufficient, and the use of longer stimuli may mean unnecessary charge load to the brain.

The *repetition frequency within a stimulus train* is typically 250–500 Hz, corresponding to interstimulus intervals between pulse onsets (ISI) of 2–4 ms. It has been argued from experimental findings that ISIs of about 4 ms is most efficient (Deletis V, Kothbauer K, 1998). In clinical studies, however, the highest MEP amplitudes were obtained at a frequency of 2–3 ms (Taniguchi M *et al.*, 1993, Jones SJ *et al.*, 1996, Deletis V *et al.*, 2001).

A *stimulation intensity* of 50–100 mA is usually sufficient to obtain motor responses from face and upper extremities after transcranial stimulation. About 10%–20% of transcranially applied current intensity reach the cortex (Agnew WF *et al.*, 1987). Lower extremity recordings may require stimulation intensities of up to 150 mA or even higher. With direct cortical stimulation, 10–15 mA are frequently sufficient. Even with non-optimal electrode placement, preexisting paresis, partial relaxation, or other interfering factors present, 25–30 mA are not exceeded for safety reasons. For continuous monitoring, stimulation intensity somewhat above recording threshold is usually selected in order to obtain stable monitoring also with fluctuating recording conditions. Higher intensities would deliver unnecessary charge load to the brain, and might lead to excitation of the descending fibers below a superficial lesion.

Stimulus train repetition rate: In our experience it is sufficient with many procedures to repeat stimulation every 15–30 seconds, due to the slow pace at which an operation usually proceeds. Such a low repetition rate yields particularly stable MEP recordings, which may allow for a semi-quantitative interpretation of MEP amplitude changes. In critical situations, e.g. when approaching the motor tract during resection or after application of an aneurysm clip, the stimulation rate may be increased to up to more than 1 Hz, if stability of responses is not impaired. Unnecessarily frequent stimulation must be avoided in order to minimize the charge load to the cortex. Transcranial stimulation may require a fast repetition rate of 2–3 Hz in order to obtain MEP responses at a reasonable stimulation intensity. The MEP responses may only "build up" after a few stimulations. With such a fast repetition rate, the MEPs tend to vary considerably in amplitude and latency due to refractoriness phenomena, so that only a nearly complete amplitude loss can be considered significant.

Recording

Myogenic motor evoked potentials: For brain monitoring myogenic (muscular) MEP responses are recorded from facial, arm, and leg muscles. Phasic activity elicited at the neuromuscular junctions and the muscular membranes can be picked up as compound muscle action potential, CMAP, or myogenic motor evoked potential, mMEP, by surface electrodes or needles placed into the target muscle. Subdermal needle electrodes must be applied in a sterile manner. We put two electrodes in the muscle to be recorded, typically in the "belly-tendon-fashion", where one electrode is placed in the muscle itself (belly), and the other one in the region of the muscular insertion of the tendon. MEPs are most readily recorded from distal limb muscles, which are mainly supplied by mono- and oligosynaptic corticomotoneuronal connections. A typical electrode montage includes distal muscles of upper and lower extremities, e.g. forearm flexor, thenar, or hypothenar muscles and anterior tibial and possibly extensor digitorum muscles. Other muscle groups include the orbicularis oculi and orbicularis oris, the trapezoid, the biceps and triceps brachii, the quadriceps femoris and the soleus muscles. The location of the lesion, and the number of available recording channels determine the target muscles to be monitored. In our experience, recording from upper extremity muscles alone yields a significant risk for new leg and facial pareses going undetected by the monitoring procedure. Since the introduction of additional recording from facial muscles with lateral lesions, and (distal rather than proximal) lower extremity muscles with median lesions, no unexpected paresis has occurred in our institution.

MEP signal: Typically an oligophasic response of 10 mV–10 mV amplitude and 10–15 ms duration occurs with a poststimulus latency of 15–30 ms (see Fig. 1). Sweeps (recorded period of time after stimulus application) of 50 ms are sufficient to record motor responses from both upper and lower extremity muscles. With preexisting paresis, sweep length may have to be adapted to the significantly prolonged MEP latency and duration. *Latencies* are determined in a natural way as the time of first significant deflection after stimulation. *Amplitudes* are measured as the difference between the most negative and the most positive component within one response. *Neither duration* of the response, nor the *central motor conduc-*

tion time do play an important role in monitoring practice. Both latency and amplitude tend to fluctuate under general anesthesia (Woodforth IJ *et al.*, 1996, Cedzich C *et al.*, 1998) and vary greatly between individuals (Cedzich C *et al.*, 1998). With transcranial stimulation in particular, amplitudes may vary to an extent that can prevent true quantitative assessment.

Averaging and Filtering: The signal-to-noise ratio of mMEPs is usually good due to the high MEP amplitude that is often in the mV range (10 μV up to 10 mV). Therefore *averaging* is rarely required. With impaired signal-to-noise ratio due amplitude depression from neuromuscular block, inadequate anesthesia or preexisting paresis, averaging of 3–10 MEP responses may be helpful. A 10–3000 Hz *bandpass filter* is usually suitable for MEP recording. Low-pass filters below 1 kHz may be applied with a high proportion of high-frequency noise. However, myogenic MEPs may contain a high proportion of high frequency activity which is not noise. Therefore *filtering* must avoid too tight bounds in order not to filter out too much of the signal proper.

Spinal and neurogenic MEP responses: Principally, MEP responses can be recorded from any part of the cortico-muscular motor transmission pathways: from the exposed spinal cord (Burke D *et al.*, 1998, Deletis V, Kothbauer K, 1998), from peripheral nerves (Pereon Y *et al.*, 1998, Padberg AM *et al.*, 1999), and from muscles. Spinal cord and neurogenic MEPs do not depend on neuromuscular block. However, for spinal MEP monitoring with intracranial procedures, electrodes would have to be implanted invasively (Katayama Y *et al.*, 1988) as the spinal cord is not exposed for recording of descending action potentials. In addition, corticonuclear volleys cannot be recorded, preventing monitoring of facial and upper trapezoidal muscles. Moreover, spinal recordings do not reflect the somatotopy of the motor cortex as do muscle recordings. Neurogenic MEPs are difficult to record, may require averaging due to the low signal-to-noise-ratio, and have not been applied with brain surgery yet.

Adequate stimulation of the motor cortex can also elicit contralateral movement. However, reliable elicitation of movement under general anesthesia may be impossible in many cases, and does not provide quantifiable information (Yingling CD *et al.*, 1999, Kombos T *et al.*, 2000a). With direct cortical MEP stimulation, no visible movement may be elicited. Therefore, recording of peripheral electric activity after cortex stimulation has replaced the observation of movements in monitoring practice.

Safety and Anesthesia

Safety: MEP monitoring does not carry significant health risks to the patient, if it is performed according to the technical standards described in this chapter, and with current, well functioning equipment. The major concern is about charge density and charge per phase with high current electric pulses applied to the brain, and about epileptogenicity of the method. However, it has been shown that the method described here even meets the safety criteria deduced from unrealistically tight bounds established from "damage threshold" in animal models (Agnew WF *et al.*, 1987, McCreery DB *et al.*, 1990, Taniguchi M *et al.*, 1993). As a matter of fact, no neural damage has been found, neither in well over 450 cases of direct cortical or transcranial stimulation performed at our institution, nor in other large series of transcranial MEP monitoring (Jones SJ *et al.*, 1996, Kothbauer K *et al.*, 1997, Meylaerts SA *et al.*, 1999, Calancie B *et al.*, 2001). Even with the much more aggressive bipolar stimulation method, no neural injury has been reported (Woolsey CN *et al.*, 1979, King RB *et al.*, 1987, Berger MS *et al.*, 1989, Ebeling U *et al.*, 1989, Ebeling U *et al.*, 1992, Berger MS, 1995, Kombos T *et al.*, 1999, Yingling CD *et al.*, 1999). The use of electrical equipment and the application of current for stimulation purposes require optical grounding of all patient contacts and thorough grounding of the patient himself to prevent inadvertent current flow. One adverse effect is the strong muscle twitching artifact elicited by inadvertent activation of back muscles, which may interfere with the ongoing microsurgical procedure, and may require a brief halt in monitoring during dissection, or in surgery while stimulating. Overt intraoperative seizure was observed in 2 out of 400 MEP monitoring cases (0.5%) in our service: One epilepsy patient was found to have insufficient anticonvulsant level on postoperative testing, the other patient harbored a frontal AVM, with supposedly lower seizure threshold. Conversely, overt seizures were found in 11% of cases in a recently published series on the bipolar stimulation method (Yingling CD *et al.*, 1999), and subclinical autonomous EMG activity after stimulation in another 14%. The concern, that either of the methods may induce epilepsy by some kind of kindling effect (Majkowski J, 1999) has not been confirmed. In general, the epileptogenicity of the MEP stimulation method is very low due to the high-frequency repetition rate, which has a lower tendency to trigger autonomous cortical circuitry than frequencies around 50 Hz.

The placement of subdermal needle electrodes in a sterile manner for stimulation or recording has never been associated with inflammation or severe bleeding in our experience with nearly 2000 cases of intraoperative monitoring (MEP, SEP, AEP). The only adverse effect to be observed is some slight local venous bleeding after removal of the needle in some cases.

Anesthesia: Total intravenous anesthesia with propofol and modern opioids has proven suitable and is applied in many institutions performing MEP monitoring (Jones SJ *et al.*, 1996, Deletis V, Kothbauer K, 1998, Calancie B *et al.*, 2001, Kombos T *et al.*, 2001, Zhou HH *et al.*, 2001).

Volatile anesthetics, in particular halogenated agents, may have a stronger depressant effect on motor responses of up to 90% as compared with propofol, and are therefore often avoided with MEP recording (Pechstein U *et al.*, 1998). In practice, however, MEP monitoring, in particular with direct cortical stimulation, is usually still possible under inhalational anesthesia (isoflurane, sevoflurane) in combination with a modern opioid (remifentanil, alfentanil, sufentanil), if TIVA is not feasible, although the greater variability of the generally lower amplitude responses must be taken into consideration when interpreting intraoperative changes. Partial neuromuscular blockade adds to the instability of MEP responses, but recordability is preserved to at least up to 2/4 in the train of four paradigm (Lang EW *et al.*, 1996). Although neuromuscular block should be avoided with MEP monitoring in order to obtain optimum recording conditions, the muscle twitching artifact may be so disturbing that only partial relaxation allows MEP monitoring to be continued in parallel with the ongoing microsurgical procedure.

Homeostasis and adequate anesthesia are of paramount importance with myogenic MEP monitoring. Amplitudes, waveforms, and latencies may be affected by various systemic factors, including body temperature, blood pressure, blood oxygenation, and P_aCO_2 (Gugino V *et al.*, 1990, Browning JL *et al.*, 1992). Latency or amplitude alterations occurring with fluctuation of the anesthetic conditions may mask or simulate changes induced by surgical events, and may thus lead to false negative or false positive interpretations of the signal. The effects of anesthesia may be more prominent with preexisting EP alterations, e.g., due to neural damage, or with more difficult recording conditions, e.g., with lower extremity MEPs. Careful preservation of homeostasis by the anesthetist and the surgeon allows minimizing such influences.

A good cooperation of the neurophysiological/surgical and neuroanesthesiological teams is important for obtaining optimum monitoring conditions. The anesthetist may be asked to share details about the anesthesia in a given case, and to change familiar concepts and convictions in favor of an anesthetic regimen compatible with MEP monitoring.

Equipment and Staffing

The stimulation and recording devices should meet certain technical specifications to allow for efficient MEP monitoring. There are several commercially available certified machines, such as for example the Nicolet Viking IV P system (Nicolet Biomedical Instruments, Madison, Wisconsin, USA), and the Keypoint system (Medtronic, Skovlunde, Denmark), that integrate both stimulation and recording facilities, or the Inomed EWACS system (Inomed GmbH, Teningen, Germany), which works with an external stimulation device such as the Digitimer DS 7H Stimulator (Digitimer Ltd., Wellwyn Garden City, Hertfordshire, UK). Of course, there are numerous other systems available on the market for this application.

Stimulation: Remote control of the stimulating devices by a preprogrammed recording machine is desirable and available with the above mentioned machines. It helps to implement complex monitoring protocols, which may include repeated SEP phase reversal via a 2-row or 4-row grid electrode, and successive unilateral, or bilateral recording of MEPs alone, or combined SEPs and MEPs in a predefined order and frequency. For bilateral recordings, two independent stimulation devices are desirable. Maximum current output is typically 100 mA with most commercially available machines at a pulse width of more than 100 μs, or up to 1000 V in constant voltage stimulators, corresponding to about 250 mA at a typical subdermal electrode impedance below 5 kΩ. However, constant current stimulation is to be preferred, as discussed above. For stimulation intensities higher than 100 mA, which may be necessary for transcranial leg MEPs, two regular 100-mA-stimulators may be connected in series. As a general rule, it is recommendable to check the actual output profile of a particular stimulating device with an oscilloscope. Trains of 5–10 electrical pulses of 50–1000 μs duration at a repetition rate of 250–500 Hz should be available at maximum stimulation intensity. Cortical SEP phase reversal recording and MEP stimulation is performed via grid or strip electrodes with platinum, steel, or silver/silver chloride electrodes embedded in silicon, typically 8 electrodes in a row spaced at 10 mm distance. These electrode grids may be re-sterilized if the silicon embedment is changed.

Recording: For recording purposes, a minimum of 4–8 amplifiers and recording channels should be available for two reasons: First, detailed motor mapping may require recording from several adjacent muscles, and second, successive recording of bilateral SEP and MEP would require frequent intraoperative rewiring of the patient and the recording device if, e.g., only two recording channels were avialble. There should also be a "curve stack" option showing 10–30 consecutive recordings from at least one channel as a vertical waveform array at one time, allowing to easily asses changes of latency, amplitude, and potential shape in the time domain (see Figs. 8–14 for typical examples of curve stack depiction). Many important features such as response averaging, triggered versus free run recording, digital bandpass and notch (50/60 Hz) filtering, and data storage for off-line analysis are nowadays implemented in the available recording systems.

Staffing: There is permanent interaction and communication between the monitoring team, the surgeon, and the anesthesiologist during the monitoring procedure, and technical problems will frequently require immediate intervention with the stimulating or recording equipment. Therefore the presence of an experienced person performing the neurophysiological monitoring in the OR appears mandatory at this point, although remote monitoring from outside the OR via a local network is now easily available with many current monitoring programs. In our experience, a well established monitoring procedure can be performed by a technician trained in this field, if the quality of the recordings allows the strict application of clearly defined criteria with which to assess the monitoring results. However, the surgeon should be acquainted with the method, and have some physiological understanding of the monitoring results communicated by the technician. The presence of a neurophysiologist or a neurophysiologically experienced neurosurgeon will only be required if complex technical problems or ambiguous monitoring results occur.

Neurophysiological Functional Mapping of the Pericentral Cortex

MEP monitoring for supratentorial pericentral lesions is always conducted in combination with intraoperative functional mapping of the sensorimotor cortex. It is important to understand that mapping prior to resection, and continuous monitoring during resection provide complementary information: mapping identifies and delineates functional areas, whereas continuous monitoring provides information about the actual functional state of the target structure. Although this chapter is about MEP monitoring, some basic facts about functional mapping are outlined in the following section as far as they are directly related to the actual monitoring procedure.

1) Median nerve SEP phase reversal provides the fastest localization of the central sulcus and the motor strip in most cases. It helps in planning the approach, in determining the topographical relation of the lesion and the motor strip as well as the best stimulation site for continuous MEP monitoring.
2) Intraoperatively, cortex stimulation under general anesthesia allows for motor and sensory cortex mapping, whereas speech and sensory mapping requires awake craniotomy.
3) Direct cortex stimulation via implanted grid electrodes used extra-operatively can provide additional information about sensory areas and in particular speech areas. It is mainly used for epilepsy surgery planning, but also for tumor surgery.

More recent functional imaging techniques, such as functional MRI or PET, are gaining more and more relevance in surgery planning, in particular if combined with neuronavigational techniques. However, intraoperative cortex stimulation still provides the standard that serves to validate these techniques, which are therefore not further discussed in this chapter.

SEP Phase Reversal

SEP phase reversal is obtained if cortical responses after electrical stimulation of a contralateral peripheral (usually the median) nerve are recorded from an strip of electrodes positioned to cover both the postcentral and precentral gyrus and adjacent cortex areas. We use grids of 4 mm-diameter platinum or steel electrodes embedded in silicate, which are typically spaced at 1 cm within and between electrode rows, eight electrodes in a row (see Fig. 5). A subdermal reference needle electrode is placed at Fpz. Stimulation is performed according to the well-established SEP paradigm (Woolsey CN *et al.*, 1979, Mauguiere F *et al.*, 1999). We design the craniotomy in such a way that positioning of a two-row electrode grid over (or not too far from) the hand representation is possible. The electrode may have to be pushed forward under the dura (see Fig. 4). With parafalcine lesions, anterior tibial nerve SEP phase reversal can be obtained from the medial part of the hemispheric convexity. If the craniotomy is large enough, and if the local geometry is particularly distorted, a four row grid may make it easier to understand the course of the central sulcus. With small or unfavorably positioned craniotomies, single-row electrode strips can be used for phase reversal recording as well. If the trepanation does not directly expose the pericentral cortex, the recording electrode may be pushed forward under the dura, which must be done with great care not to lesion surface veins, though. Stimulation can be done via the same electrode array, by reconnecting it to a stimulating device. If the electrode grid interferes with the surgical approach, a single stimulating cup electrode may replace it. The site of best (highest amplitude) precentral response is usually chosen for MEP stimulation. The procedure of SEP phase reversal recording takes 5–15 minutes, depending on the recording conditions and the way the electrode grid is placed on the cortex.

This technique makes use of the fact that the early cortical response after peripheral nerve stimulation, which is mainly generated in the dorsal bank of the central sulcus, shows a reversal in polarity when recorded from the precentral rather than the postcentral cortex. Since some activity is also generated at the bottom and the rostral bank of the central sulcus, it is not an exact mirror image of the sensory cortex response that can be recorded precentrally, but slightly delayed and distorted (Lueders H *et al.*, 1983). Obviously, the best median nerve response can be recorded over the somatosensory hand area, but reasonable recordings are also possible somewhat laterally and medially from it. If the electrode array is positioned in strict anterior-posterior orientation, the best precentral response is obtained, marking the motor hand area. If the electrodes are positioned "off-axis", a typical deflection may only be recorded from the postcentral electrodes, also allowing identification of the central sulcus

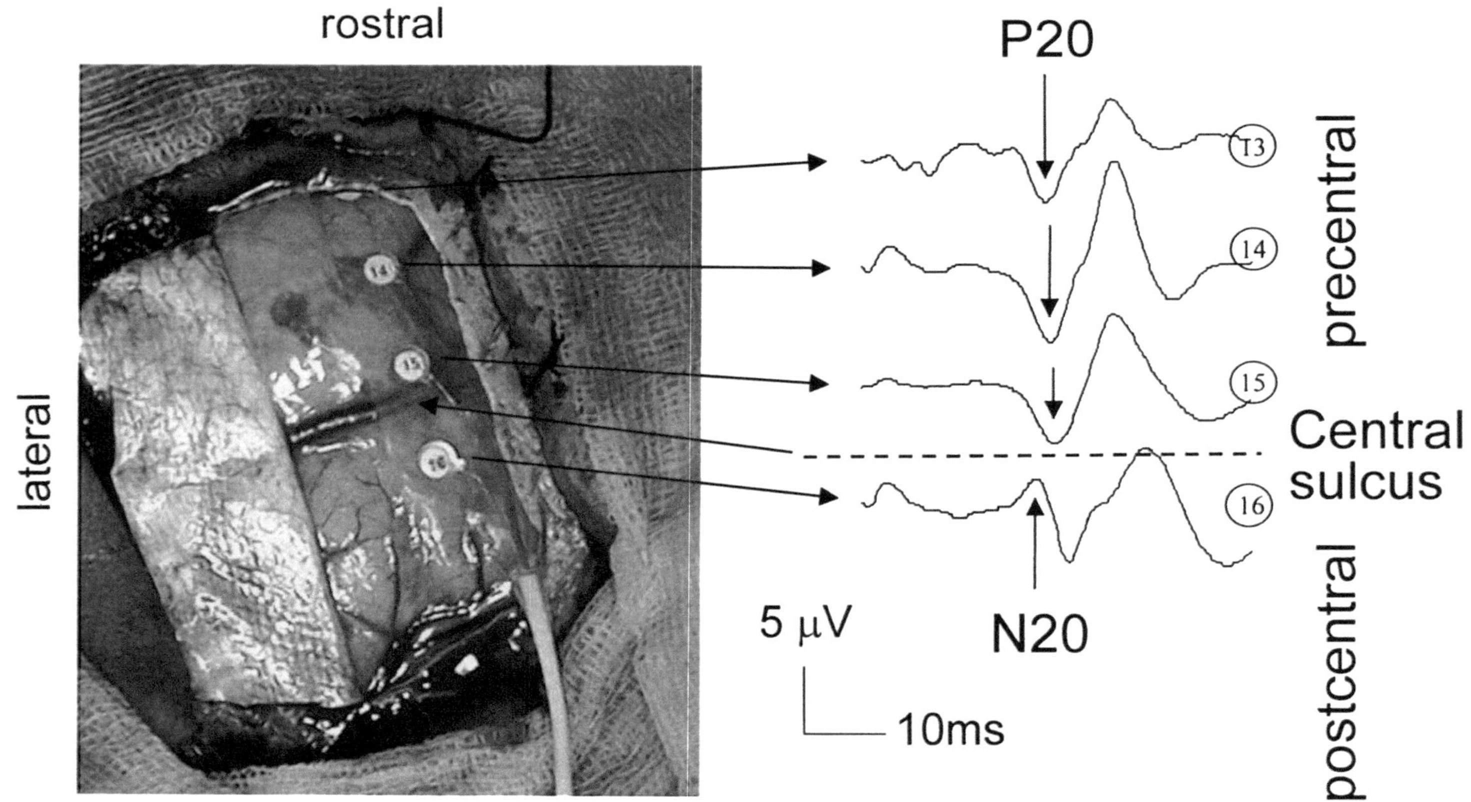

Fig. 4. Identification of central sulcus by SEP phase reversal. The case is the same as in Fig. 7. Prior to MEP recording, the position of the central sulcus and the sensorimotor cortex was confirmed by median nerve SEP phase reversal. Since the electrode would have interfered with the postcentral surgical approach, a single cup stimulation electrode was placed between electrode positions 14 and 15 for continuous MEP recording (Fig. 7)

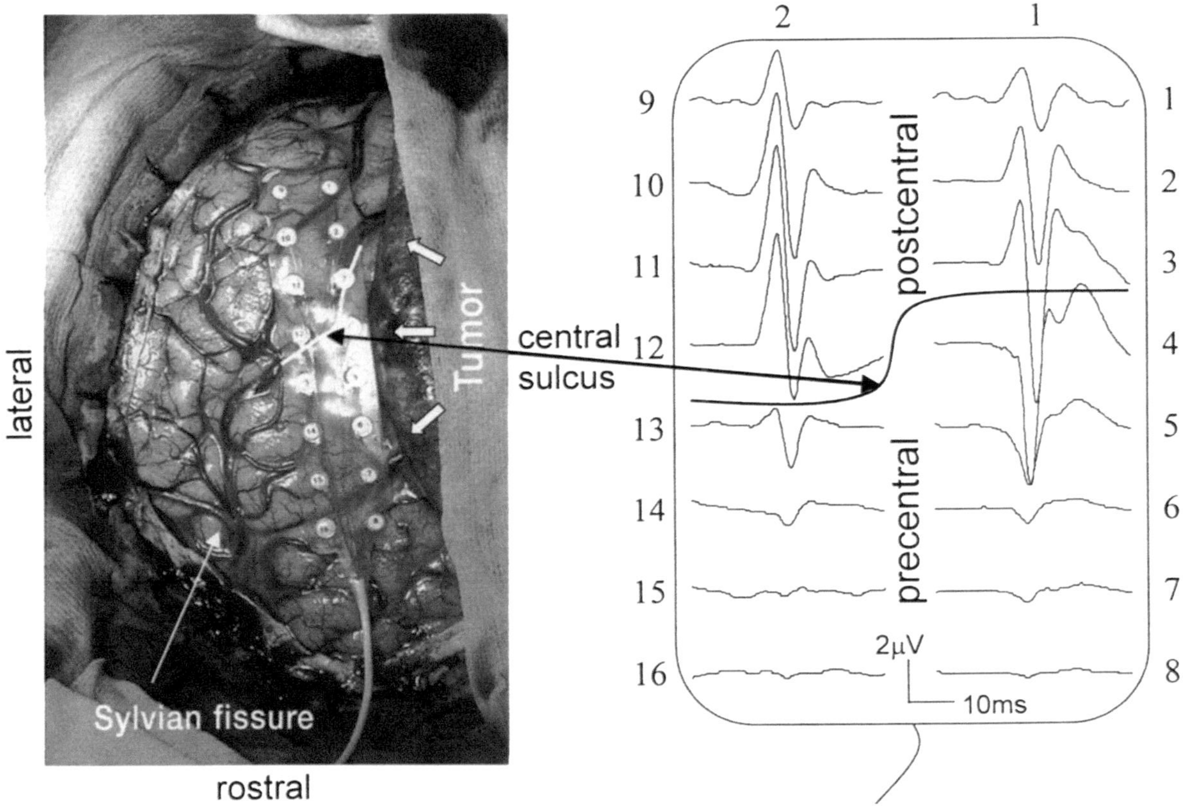

Fig. 5. SEP N20 loss. This 67 year old patient was operated for a large meningioma of the right hemispheric convexity with extensive dural infiltration beyond the tumor nodule, necessitating large dura excision. Median nerve SEPs recorded from the lateral portion of the pericentral cortex show N20 loss rather than true phase reversal. It is still possible to identify the central sulcus although the sensorimotor hand area is not accessible due to the large tumor

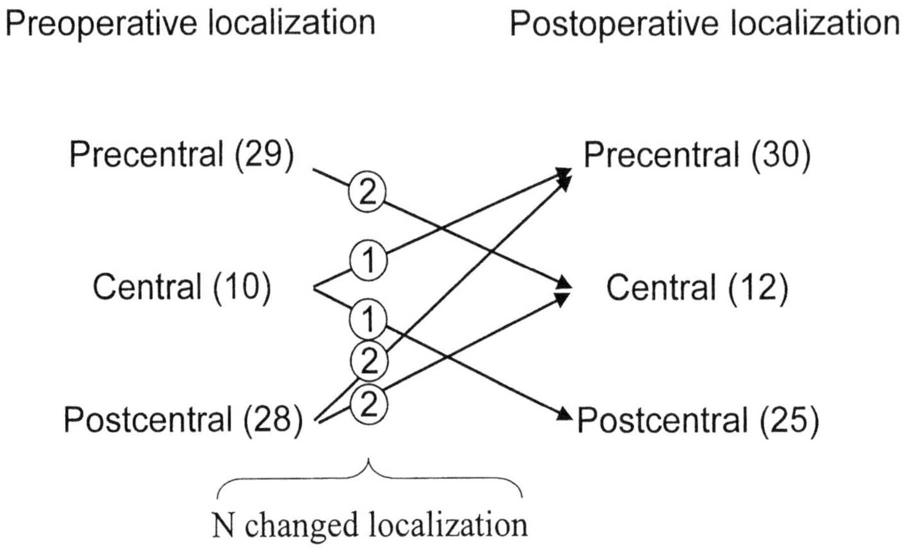

Fig. 6. Tumor localization: intraoperative SEP phase reversal vs. preoperative MRI. In a series of 67 cases, the localization of pericentral lesions was determined preoperatively by MRI and intraoperatively by SEP phase reversal. In a significant number of cases (8 patients, 12%) the localization by SEP phase reversal was found to differ from what had been concluded from preoperative MRI. This had important consequences in terms approach and extent of resection (Cedzich C *et al.*, 1996)

(Woolsey CN *et al.*, 1979, Lueders H *et al.*, 1983, Goldring S *et al.*, 1984, Cedzich C *et al.*, 1996). An excellent overview over the clinical application of the method in pericentral tumor surgery is given in (Romstock J *et al.*, 2002). The SEP phase reversal method, if combined with anatomical information and test motor stimulation over the strip electrodes, allows for reliable identification of the central sulcus, and the sensorimotor cortex nearly 100% of cases (Cedzich C *et al.*, 1996, Kombos T *et al.*, 2000a). In an earlier series from our group classification of the lesion as precentral central, or postcentral from preoperative imaging had to be corrected by intraoperative phase reversal in 12% of cases. In about 5% in this series, the tumor location was precentral rather than postcentral, or central as had been assumed preoperatively, which obviously affected the approach significantly (see Fig. 6).

Preoperative functional imaging, in particular fMRI, may of course be useful nowadays to determine the topographical relation of the lesion and the central sulcus for operation planning. However, with lesions dislocating the central sulcal cortex by bulging underneath the brain surface in a posterior or anterior direction, as well as with pathological vasculature, fMRI may yield quite misleading results. In our recent practice, fMRI has proven rather useful in positioning the electrode grid in an optimum way for phase reversal recording and MEP stimulation.

Intraoperative Stimulation Mapping

With increasing distance from the somatosensory areas, SEP phase reversal may become unreliable, or cannot even be obtained. Despite clear pragmatic advantages of the phase reversal method in general, direct motor mapping is desirable if phase reversal fails. It has been shown that this occurs in up to more than 10% of cases, while a combination of direct motor mapping and SEP phase reversal nearly always leads to identification of the central sulcus (Cedzich C *et al.*, 1996, Kombos T *et al.*, 2000a). Another important issue is detailed motor mapping beyond mere identification of the motor cortex in order to define resection borders, and the need for subcortical motor mapping in many situations. SEP phase reversal is suitable for neither application. Direct intraoperative mapping of the motor cortex via electrical stimulation is currently done in many neurosurgical centers according to the method established by Penfield, which has been used without major modifications for many decades (Woolsey CN *et al.*, 1979, King RB *et al.*, 1987, Berger MS *et al.*, 1989, Ebeling U *et al.*, 1989, Ebeling U *et al.*, 1992, Berger MS, 1995, Kombos T *et al.*, 1999, Yingling CD *et al.*, 1999). The method is summarized below. However, the more recent method of monopolar stimulation with recording of motor evoked potentials according to the technique detailed above (section 2) is preferred in our service, since it does not suffer from some of the disadvantages of the traditional method, but provides specific information otherwise not available, e.g., quantitative and, if necessary, continuous measurements.

Motor evoked potential mapping: Stimulation is done successively through all electrodes of a grid that has been positioned over the area of interest, according to the stimulation paradigm detailed in section 2. Stimulation intensity is increased stepwise until MEP response threshold. Threshold intensity is usually below 15–20 mA, but stimulation should be started at a much lower intensity, e.g. around 5 mA. If the (tentative) motor cortex has been exposed by the craniotomy, a monopolar hand-held stimulator, possibly provided with neuronavigational markers, can be used as well. This may be advantageous if the distances between grid electrodes (usually 1 cm) are too large for the required mapping resolution. Recording is done from the muscles corresponding to the cortical target area, a dense recording scheme including as many adjacent muscles as possible is desirable for detailed mapping results. Stimulation intensity at threshold is required for mapping purposes in order to preserve the resolution of the method, i.e. the specificity with regard to the target muscles. Exeedingly high stimulation intensity may lead to excitation of fibers not located directly beneath the stimulation site. Although extrapyramidal motor fibers may be activated as well, the oligosynaptic corticospinal connections originating in the primary motor cortex are obviously more sensitive to electrical cortex stimulation, if myogenic MEPs are recorded. A low recording resistance is desirable, an impedance below 1 kΩ can be achieved in most cases.

Penfield mapping: Bipolar, biphasic rectangular pulses of 1 ms duration and 2–20 mA amplitude are applied for 1–4 s at a rate of 50–60 Hz via a bipolar two-tip handheld stimulator or an electrode strip. Stimulus intensity starts at 3–5 mA, and is increased at small (1 mA) steps until a motor response is obtained. Sites of successful stimulation can be marked with sterile tokens. Instead of the traditional observation of movements of the exposed contralateral limbs, EMG recordings via electrodes from representative face-, trunk-, and limb muscle groups has been described with this method (Yingling CD *et al.*, 1999, Kombos T *et al.*, 2000a), requiring lesser stimulation intensity particularly in anesthetized patients.

In a recent series of 66 operations (Yingling CD *et al.*, 1999), motor responses to bipolar cortex stimulation could be registered by either EMG registration or observation of movement in 79% of cases. In another recent study (Kombos T *et al.*, 1999) motor responses were obtained in 95% of motor cortex stimulation, but also in almost 30% of premotor stimulation, supposedly due to spreading activation short-reaching association fibers. Neither movement observation, nor EMG recording can provide quantitative information. Another disadvantage is the high rate of clinical, and subclinical focal seizures, which was 11% plus 14%, respectively, in the series of (Yingling CD *et al.*, 1999).

Bipolar stimulation of the exposed motor cortex was introduced into neurosurgical practice by Penfield (Penfield W *et al.*, 1937), and has been advanced by several groups over the last 50 years (Woolsey CN *et al.*, 1979, King RB *et al.*, 1987, Berger MS *et al.*, 1989, Ebeling U *et al.*, 1989, Ebeling U *et al.*, 1992, Berger MS, 1995, Kombos T *et al.*, 1999, Yingling CD *et al.*, 1999). Originally tonic limb movements after cortex stimulation were observed as motor responses. The introduction of electromyographic (EMG) recording improved the sensitivity and safety of the method (Yingling CD *et al.*, 1999).

Subcortical mapping: Subcortical motor stimulation according to the MEP paradigm is possible, but may require quite high stimulation intensity as well of up to 30 mA if the distance from the motor fibers is more than a few millimeters. However, the continuous MEP monitoring paradigm with cortical stimulation has conceptual advantages over subcortical stimulation: It is performed automatically without interfering with the ongoing surgical procedure. It yields warning, or reassuring information also if motor impairment is to be expected at remote sites, or from remote factors, such as compromise to important perforating arteries, which may occur already at some distance from the motor tract.

There are also anecdotal reports on the use of the Penfield method for stimulation of subcortical motor tracts (Skirboll SS *et al.*, 1996, Berger MS *et al.*, 1997, Duffau H *et al.*, 1999, Yingling CD *et al.*, 1999), both for defining resection borders with tumors approaching deep motor (and other functional) pathways, and for determining the extent of functioning tissue within the boundaries of infiltrative gliomas. However, this was not easy to replicate at our institution. Repetitive stimulation at a frequency of 50–60 Hz leads to cortical facilitation, while subcortical stimulation must rely on spinal facilitation, which requires much higher frequencies (Taniguchi M *et al.*, 1993). It seems that the method, essentially corresponding to single-pulse stimulation at the subcortical level, leads to motor tract excitation when stimulating directly the pyramidal tract, or with only a very thin 1–2 mm layer of white substance in between. Subcortical Penfield stimulation at an intensity high enough to elicit motor responses through a thicker layer may lead to seizures by retrograde cortex activation, so that we have abandoned this method at this stage.

The need for subcortical identification of motor fibers is obvious: During tumor surgery, the resection borders with diffusely infiltrating lesions may be defined, and the proximity of motor fibres will alert the surgeon already when resecting within the pathological tissue, as traction, heat, and compromise to perforators may functionally damage those fibres already at this stage. Neuronavigation relying on preoperative imaging is not admissible with such deep-seated lesions due to the intraoperative brain shift, and cannot provide the latter functional information. Nevertheless, the recent advent of high-quality diffusion-weighted imaging of white-matter fibre tracts may be a major advance in operation planning for deep lesions (Krings T *et al.*, 2001a).

Extraoperative Mapping with Grid Electrodes

Stimulation is performed via a preimplanted (usually 8×8) electrode grid. The stimulation paradigm corresponds to the Penfield technique outlined above, although lower stimulation frequencies may be tried in some patients. Positive phenomena like elicitation of movement and sensations or auditory and visual hallucinations, from stimulation of primary cortex areas, as well as negative effects like speech arrest under stimulation of higher association areas including the speech centers, are observed and used to determine functional areas. However, functional inhibitory effects, such as impaired speech performance do not only occur from stimulation of the speech centre proper. Farther reaching inhibition, for example by short-reaching U-fibres, is possible.

This method is mostly used in surgery for chronic drug resistant epilepsy, since it allows to monitor and locate interictal spike activity over a longer period of time, and to perform functional mapping via the same electrode grid. With epilepsy cases, sensorimotor mapping is performed as often as speech mapping (Kral T *et al.*, 2002).

With non-epilepsy cases, mainly speech mapping is important with this technique, since the sophisticated motor mapping techniques outlined above make extensive extraoperative sensorimotor mapping rather dispensable. In our service, the method is applied with vascular and tumoral lesions if the risk for speech deficit is considered high in a non-epilepsy surgery.

The technique of extraoperative mapping allows the surgeon to dispose with speech cortex stimulation during awake craniotomy, which is time consuming, may be cumbersome for the patient, and prolongs surgery to many hours. This method has the major disadvantage of not allowing true verification of the mapping results after a reasonable period of time. In fact, reliable mapping typically requires two 3-hour sessions at two successive days according to our experience. Another advantage is that intraoperative seizures from the cortical stimulation are avoided. A disadvantage is of course the fact that the procedure implies two operations, first the implantation of the grid followed by the physiological evaluation, and second the resection of the lesion guided by the mapping results.

In principle, the localization of speech and sensorimotor areas can be determined noninvasively by means of functional magnetic resonance imaging with increasing validity (Binder JR *et al.*, 1996, Lehericy S *et al.*, 2000). Results from intracarotid amobarbital testing (Wada test) correspond well with electrical language mapping (Wyllie E *et al.*, 1990), but the Wada test cannot reveal the intrahemispheric topography of language areas.

Principles of Clinical Applications

Indication for MEP Monitoring

Typical lesions in which MEP monitoring has proven useful in our institution are centrally located gliomas and vascular malformations, metastases and cavernomas, and aneurysms of the anterior and the proximal posterior circulation. Insular tumors, and deep-lying vascular malformations requiring a transsylvian approach, brainstem lesions such as gliomas and cavernomas, and difficult extraaxial tumors and vascular lesions encroaching on the brainstem motor pathways are also typical indications. Although MEPs primarily serve to monitor motor pathways, indirect conclusions about other brain regions may be drawn, according to principles including local proximity and common vascular supply. However, the high sensitivity of MEPs to all kinds of interfering factors including systemic events, suggests a more specific use. Nevertheless, MEP recording can provide valuable complementary information to other monitoring data, e.g. in many cases where SEP monitoring is performed.

Interpretation of MEP Changes

In our series, significant MEP alterations occurred in about 40% of cases, compared to a rate of 11.4–56% in other studies (Deletis V, 1993, Cedzich C *et al.*, 1996, Kombos T *et al.*, 2001, Zhou HH *et al.*, 2001). MEP changes must be determined by comparison of the last recording with some recent reference measurement, e.g. a post-craniotomy recording. According to our experience, the application of latency and amplitude criteria analogous to the typical interpretation of SEP findings leads to clinically useful conclusions: Recordings are considered stable if spontaneous fluctuation is less than 50% in amplitude and 10% in latency. Only reproducible amplitude decrease or latency increase beyond these limits are considered to be a significant deterioration, whereas complete disappearance of motor responses is classified as a loss. If MEPs are recorded from several adjacent muscle groups within one limb, loss of responses if occurring not in all of these muscles are considered as deterioration rather than loss. With inconstant recordings, only impairment beyond spontaneous variation at repeated stimulation can be considered significant. Potential shape and duration are highly variable and are therefore not quantitatively evaluated. However, reproducible conspicuous changes are noted, since dispersion phenomena with impaired motor conduction may lead to a more polyphasic, broader potential, which may therefore be cautiously interpreted as a warning sign. Central motor conduction time which is routinely evaluated in perioperative and diagnostic MEP measurements, does not play a significant role in intraoperative monitoring, since it increases the complexity of the monitoring setup considerably without providing much essential additional information.

When interpreting MEP changes, normal fluctuations, technical conditions, the influence of systemic factors like anesthesia, and interindividual differences must be taken into consideration. An initial baseline potential is obtained after induction of anesthesia, before the operation actually starts. However, craniotomy and opening of the dura may already lead to significant MEP changes before the intracranial procedure proper has started, e.g., if subdural air collects over the cortex which has sagged-away from the skull. Therefore, it is advisable to update the baseline values at this point. Non-surgical factors like fluctuation of anesthesia or change in technical conditions (e.g. poorer electrode contact due to brain shift) may also lead to MEP changes during resection, necessitating the interpretation of further MEP changes with regard to the latest state considered stable rather than unimpaired.

All significant MEP changes including recovery following impairment or loss are reported to the surgeon after exclusion of technical reasons, e.g., displacement of the stimulating electrode. With persisting loss of responses from all recorded muscles in a limb, stimulus intensity may be increased up to maximum value, or until potentials reoccur. A persisting potential loss despite maximal stimulation is a particularly important warning sign. At the surgeon's request, stability of the recordings will be frequently confirmed during resection in critical areas.

In the literature, the definition of "significant MEP change" in terms of amplitude reduction was somewhat tighter (80%) in the series of (Kombos T *et al.*, 2001), whereas it was essentially identical with the 50%-criterion of the present study in (Zhou HH *et al.*, 2001). While MEP latency prolongation without amplitude reduction has hardly ever occurred in our experience, and latency changes are not considered in (Zhou HH *et al.*, 2001), isolated latency changes in 5 out of 8 patients with significant MEP alterations were observed in (Kombos T *et al.*, 2001). In our experience, exact MEP onset latency is often hard to determine with poor signal-to-noise ratio, as it occurs more frequently with low and inconstant amplitude due to inhalational narcosis.

Correlation of Intraoperative MEP Changes and Motor Outcome

Some basic principles of intraoperative MEP interpretation were observed in the initial studies (Cedzich C *et al.*, 1996), and have been confirmed in the more recent series (Kombos T *et al.*, 2001, Zhou HH *et al.*, 2001). Table 2 summarizes the current principles of clinical MEP interpretation in a simplified way, which we have found useful in 400 cases of supratentorial and infratentorial brain surgery.

Irreversible non-technical MEP loss almost invariably heralds permanent new motor deficit. Irreversible MEP alteration or reversible loss or alteration is associated with either permanent or transient motor impairment, the former being very rare with reversible MEP loss or alteration. It is important to understand that these observations do not merely allow for motor outcome *prediction*, but help to actually *prevent* new paresis. MEP loss is always preceded by at least a short period of impairment, usually amplitude reduction, and may be reversible if the cause for its deterioration has been eliminated. Only transient slight paresis is observed postoperatively in such cases. In fact, MEP recovery was usually preceded by some surgical intervention in our series. Permanent MEP deterioration or loss, frequently followed by persisting severe paresis, did mainly occur if the surgeon did not, or could not react to MEP deterioration. The rate of

Table 2. *Schematic Correlation of Intraoperative MEP Changes and Motor Outcome. These Observation Apply to the Limb/Muscle Groups Actually Monitored. Unexpected New Paresis May Occur in an Unmonitored Limb*

<table>
<tr><th rowspan="2">Intraoperative MEP findings</th><th colspan="3">Motor Outcome (monitored limb)</th></tr>
<tr><th>Permanent new deficit</th><th>Transient new deficit</th><th>No new deficit</th></tr>
<tr><td>Irreversible loss</td><td>always</td><td colspan="2">never</td></tr>
<tr><td>Irreversible deterioration</td><td>frequent</td><td>frequent</td><td>rare</td></tr>
<tr><td>Reversible loss</td><td>rare</td><td>frequent</td><td>frequent</td></tr>
<tr><td>Reversible deterioration</td><td>rare</td><td>frequent</td><td>frequent</td></tr>
<tr><td>Unaltered</td><td colspan="2">never*</td><td>always</td></tr>
</table>

* Exceptions: *SMA* syndrome, postresectional edema.

intraoperatively bleeding AVMs in this group is conspicuous. Moreover, MEP may help to determine the maximum extent of tumor resection close to motor pathways. A high rate of slight, transient deficit associated with reversible intraoperative MEP deterioration in our monitoring experience indicates that surgery was frequently completed up to the very point of imminent significant motor impairment, but not further, i.e., that the optimum extent of resection within the limits of preserved motor function could be achieved.

In our series, true new paresis occurred without preceding intraoperative MEP alteration only in specific situations. Some patients developed new postoperative paresis in a muscle group which had not been monitored intraoperatively, i.e., facial or leg pareses occurred in cases in which only arm MEPs had been recorded (see Fig. 9). Therefore, it may be concluded that recording MEPs from only one limb is not representative for the whole corticospinal tract. Consequently, our routine recording setup has been modified to include the anterior tibial muscle in all cases of medial lesions, and facial muscles are now routinely monitored with lateral lesions. No unexpected pareses have occurred since then.

With lesions involving the frontodorsal supplementary motor area (SMA), transient new postoperative weakness, in most cases also reduced speech production, frequently occurs despite intraoperatively unaltered MEPs, which resolves within a few days in most cases. This so-called SMA deficiency syndrome corresponds to an impaired initiation of movement and speech, rather than true paresis or aphasia (Zentner J *et al.*, 1996a), and thus the lack of MEP loss despite apparent new "motor deficit" does not represent a false-negative monitoring event. With SMA syndrome it has been shown that postoperative MEPs elicited by transcranial magnetic stimulation, which relies on intracortical input to the motor cortex, were abolished until functional recovery in such cases (Sala F *et al.*, 2000). Unaltered intraoperative MEPs still indicate preserved corticospinal function.

In another subgroup of patients with new paresis despite unaltered intraoperative MEPs, postoperative imaging showed significant local edema, and/or re-bleeding. Compromise of venous supply may have contributed to edema formation, particularly with meningiomas. In these cases, the new paresis may have occurred due to late postresectional effects that cannot be picked up by MEP recording during dissection. Other groups have described similar observations (Kombos T *et al.*, 2000a).

The semi-quantitative relation between MEP findings and clinical outcome schematised in table 2 corresponds with the findings of (Zhou HH *et al.*, 2001), who describe a positive correlation of MEP amplitude reduction and degree of postoperative motor deterioration. In contrast, there is convincing evidence that, with *spinal* surgery, transcranially evoked myogenic MEPs rather yield an all-or-nothing characteristic, so that only complete MEP loss indicate significant new paresis (Deletis V, 1993, Herdmann J *et al.*, 1996, Kothbauer K *et al.*, 1997, Deletis V, Kothbauer K, 1998).

Surgical Reaction to MEP Changes

Typically, reactions to MEP deterioration include temporary or definite halt of resection in the critical target area, readjustment or temporary release of retractors, application of papaverine to spastic vessels, readjustment of aneurysm clips, and release of temporary vessel clips.

In some situations, the cause for MEP loss may be detected by inspection, such as vasospasm during a transsylvian approach, and the consequence to draw, namely application of papaverine, is obvious. Most frequently, the previous course of the operation makes it possible to deduce the reason for a MEP alteration, e.g., when a retractor has been placed on the motor cortex, when a temporary clip has been applied, or when resection is approaching the motor tract. Sometimes the reason for MEP deterioration remains somewhat obscure, e.g. with frontal brain elevation during dissection of an aneurysm, and only accumulating experience of the surgical-monitoring team teaches that temporary release and readjustment of the retractor will allow for the MEPs to recover without persisting motor impairment.

The occurrence of MEP changes may also be actively excluded by preliminary occlusion of an artery with a removable clip, and testing the functional relevance of this maneuvre, for example, on a transitory artery in a case of sylvian or pericentral AVM. This highlights the fact that the non-occurrence of a MEP loss may also be valuable for the surgeon.

The repeated use of MEP monitoring is a valuable teaching experience about the danger of certain operative steps: local vasospasm or indirect retractor effects for example were not at all recognized as being dangerous until this was discovered by MEP monitoring. Similarly to BAEP monitoring in surgery for acoustic tumors, where the use of bipolar coagulation has been nearly abolished, operative methods are beginning to be influenced by the technique of MEP monitoring.

The impact of MEP monitoring on the surgical strategy is certainly a non-trivial issue. All available data are from observational rather than prospective interventional studies. Criteria for surgical intervention with MEP changes continue to emerge with growing experience of the surgical-monitoring team. MEP monitoring results provide only one piece of information in a network of intraoperative findings, including inspection, anatomical and neuronavigational information about proximity to functional areas, histopathological findings, and knowledge about the course of the operation, e.g., the approach, and previous MEP changes. The individual attitude of different surgeons towards the dangerousness of different steps of dissection introduces another variable: Dissecting close to the posterior limb of the internal capsule, one surgeon will stop resection when potential loss occurs, another surgeon will pause, and carefully resume resection when the potentials recover.

Sometimes it may even be impossible to change the course of the ongoing procedure despite significant MEP changes. In our experience for example, MEP loss occurs frequently with extensive electrocoagulation in bleeding central vascular malformations. It is obvious that the surgeon cannot react to this monitoring event in this situation other than changing to a trial of tamponade until MEPs possibly recover.

Influence of MEP Monitoring on Surgical Results and Motor Outcome

In our series of some 200 monitored supratentorial tumor surgeries, macroscopically gross total resection was achieved in about 65% of patients, and the rate of permanent new deficit was about 12%, persisting disabling deficit being even rarer (5%), despite the partially very high-risk profile of this patient group. Transient and slight new deficit was more frequent (about 20%), but had resolved by the time of discharge in most cases, with the main exception of vascular lesions, which admitted only slow recovery from new deficit. Some degree of transient new postoperative weakness is calculated in certain cases, where it is clear from the start, e.g. with infiltration of basal ganglia and internal capsule, that only subtotal resection can be achieved, and MEP deterioration is employed as a major criterion when to halt resection. On the other hand, given the closeness or the infiltration of motor pathways, surgery presumably would have remained subtotal in a larger proportion of cases without the confirmation of stable MEP recordings available during resection. Rates of postoperative motor impairment of 16% with MEP monitoring, and 12–24% with intraoperative cortical mapping (Ebeling U *et al.*, 1992, Duffau H *et al.*, 1999, Yingling CD *et al.*, 1999) have been reported in other recent studies on pericentral surgery. New neurological deficit occurred in 15.8–26% in recent series on supratentorial surgery without electrophysiological aid (Ammirati M *et al.*, 1987, Fadul C *et al.*, 1988, Vecht CJ *et al.*, 1990, Kreth FW *et al.*, 1993). The rate of permanent deficit was significantly lower in all series. Gross total resection was achieved in 29.6%–87% in recent reports on supratentorial surgery supported by cortical mapping (Duffau H *et al.*, 1999, Yingling CD *et al.*, 1999), whereas a rate of 98.6% has been reported with MEP monitoring (Kombos T *et al.*, 2001). However, any comparison of such outcome data is obviously hampered by the lack of a controlled study design with the resulting heterogeneity of patient groups and clinical criteria in different series. The true usefulness of MEP monitoring in minimizing new postoperative deficit and maximizing extent of resection will only be quantitatively assessed by means of prospectively controlled studies.

Specific Applications

Supratentorial Tumors and Vascular Malformations

For *central tumors and vascular malformations* adjacent to motor cortex areas and pathways, MEP monitoring is performed using direct cortical stimulation. In superficial lesions, the focus of recording is selected according to the motor homunculus, since only motor fibers supplying one specific circumscribed region of the body may be damaged in such lesions. With a superficial metastasis in the motor hand area, for example, recording from distal arm muscles is obviously most useful. With deeply seated lesions close to the internal capsule, a more comprehensive recording scheme, including muscles from different limbs, may be required, since the descending motor fibers converge significantly at this level. Recording from, e.g., facial, distal arm, and distal leg muscles is advisable in such cases. Concomitant SEP recording is seldom useful even with postcentral lesions, since resection will usually not be stopped because of SEP changes with preserved MEPs. Tumors in the postcentral gyrus can usually be safely resected. Sensory deficit may increase and may be disturbing due to sensory ataxia. However, this condition usually improves considerably within

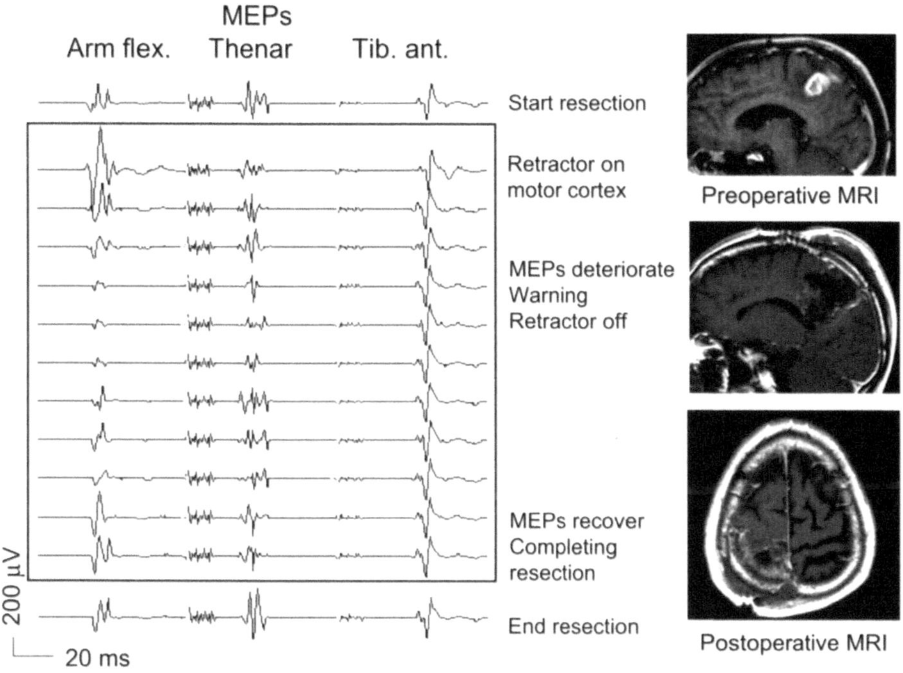

Fig. 7. MEP deterioration triggers readjustment of retractor. A 60 year old patient was operated on a postcentral glioblastoma. The postoperative MRI shows the postcentral approach. Shortly after retracting the motor cortex, upper extremity MEPs deteriorated, while lower extremity MEPs remained unaltered (box). Release and readjustment of the retractor led to quick recovery of MEPs. There was a transient mild new hemiparesis postoperatively. See also Fig. 4

weeks to months. Thus the risk in postcentral lesions, e.g. gliomas, is mainly due to the proximity to the central sulcus and the precentral cortex.

Central Tumors and AVMs

With *intrinsic central tumors*, the normal geometry may be distorted to such a degree that it is impossible to know how the motor cortex has been displaced. Intraoperative SEP phase reversal, and extraoperative sensorimotor grid mapping has been found helpful in planning the approach as well as for guidance during the resection. Even if the resection remains strictly within the tumor borders, removal of tumor tissue is indirectly exerting traction forces on the surrounding functional brain tissue, and MEP alterations may occur before resection reaches the motor fibres.

Particularly dangerous is a situation where the surgeon reaches a sulcus 2 or 2.5 cm below the surface with the risk of injuring a vessel that supplies the motor cortex. In precentral tumors the most dangerous resection area is at the posterolateral deep margin of the resection cavity, as the motor cortex lies more anterior the more it approaches the sylvian fissure. In addition, the motor fibers descending from the precentral cortex converge with increasing distance from the cortex to form the motor tract, which runs more anterior at deeper subcortical levels as well (Peraud A *et al.*, 2002). The surgeon must be particularly careful if the roof of the lateral ventricle has been opened more than 2 cm posterior to the foramen of Monro. In deep seated small lesions such as cavernomas or metastases, an oblique dorsal approach from the postcentral sulcus can be chosen in order to avoid transgressing through the converging motor fibers. Careful attention to MEP recordings may be particularly helpful in all those situations, and stable MEPs can reassure the surgeon to complete the procedure.

With *central meningiomas* bulging into the brain substance it remains an open question whether MEP monitoring is really a necessary adjunct. It may help with the preservation of anterior cerebral artery branches that have been pushed away by the meningioma, but appears less important with the dissection of the border between the meningioma and the brain, which usually is not so difficult. In fact, most new pareses after central meningioma surgery occurred despite unaltered intraoperative MEP recordings, due to postresectional edema and rebleeding, as discussed above.

With central AVMs, the distortion of the local central topography is often less pronounced than with tumors. However, the identification of the motor strip may still be difficult, and plasticity phenomena induced by the changed vasculature may lead to a changed functional organization of the pericentral cortex (Muacevic A *et al.*, 1999). Therefore, both functional mapping and continuous MEP monitoring is performed for surgery of central AVMs. During dissection of an AVM transitory branches to the motor cortex must be preserved, and test clipping may be applied under continuous monitoring. Lack of MEP changes reassures the surgeon to continue a dangerous step of dissection, or to sacrifice a vessel.

Insular Tumors and Sylvian AVMs

Insular tumors typically extend deeply towards the basal ganglia and may get very close to the internal capsule (Zentner J *et al.*, 1996b). The tumor part underneath the M2 and M3 branches that is covered by the frontal and temporal opercula and has replaced the insular cortex, the claustrum, and the external capsule and invaded the lateral putamen, is the most difficult to remove safely, especially in its most dorsal extension. Con-

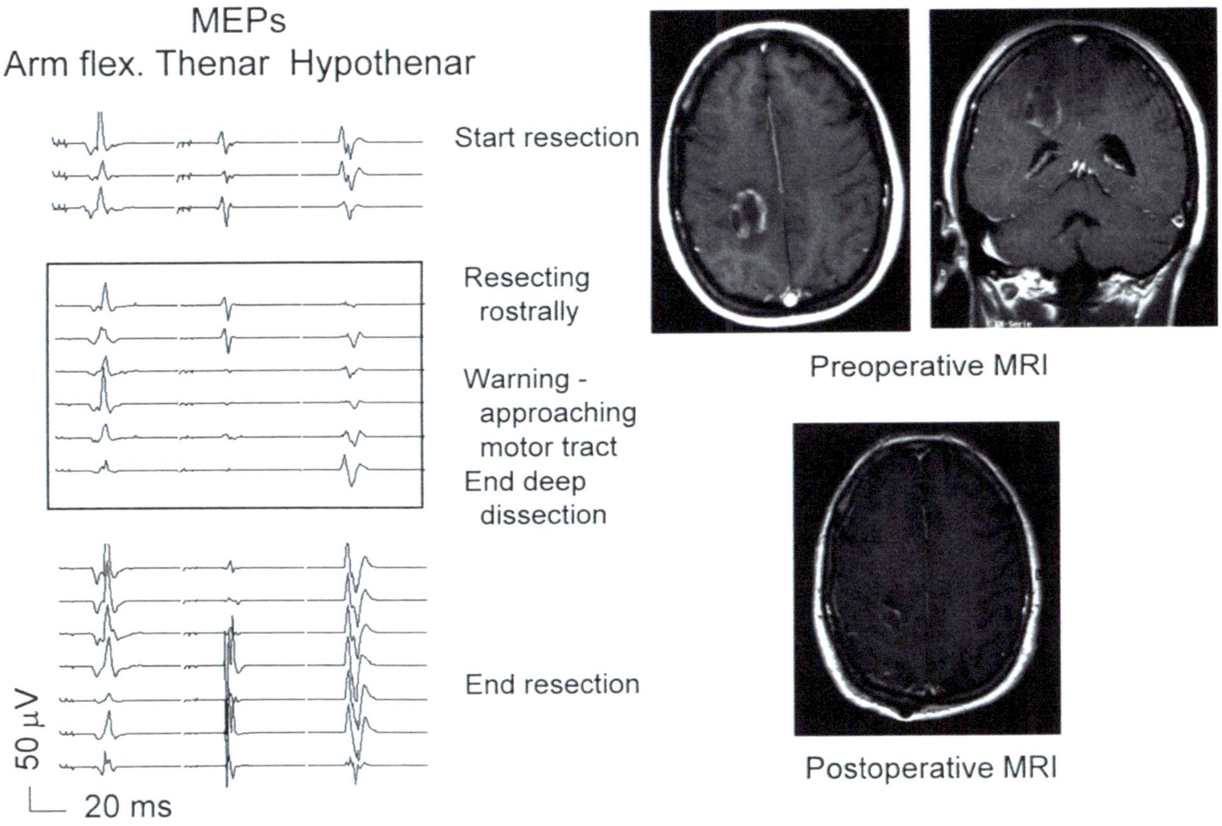

Fig. 8. MEP changes indicate closeness to motor tract. A 39 year old patient was operated on a deeply seated postcentral lesion. When resecting in the deep rostral portion of the lesion, arm MEP changes (amplitude reduction in alternating single muscles) indicated closeness of the coronal/capsular motor fibers. The resection was halted at this point, and abandoned when quick section revealed an encephalitis disseminata plaque. There was no motor deficit postoperatively

versely, resection of the temporal portion and of the frontal part on the non-dominant side poses less significant problems. With a transsylvian approach all middle cerebral artery branches have to be dissected and preserved, and the tumor resection has to proceed beyond these arteries. Ischemia from spasticity of sylvian vessels is therefore a major problem with such lesions. Another problem arises from perforating small vessels supplying the lateral putamen and parts of the tumor as well as the dorsal part of the internal capsule. Here MEP recording was found to be particularly useful in our experience. With such lesions, MEP recording provides a major criterion how far to proceed with the resection. This is of particular importance if intraoperative imaging is not available, as it is the case in most neurosurgical centers. Moreover, imaging would not reveal impending compromise to the vasculature of the internal capsule. Some 40 cases of MEP monitoring with insular tumors have shown that maximum resection without significant postoperative motor impairment is obtained if resection is stopped at first significant MEP deterioration. Postoperative imaging typically reveals resection in the putamen close to the internal capsule, and not infrequently the posterior part (2–3 cm) of the insular cortex infiltrated by the tumor has not been removed. It is our impression that persisting severe paresis can be avoided using this strategy, while slight transient paresis may be found in postoperative clinical evaluation in many instances, which resolves within days. In our institution Type B insular tumors (Yasargil MG, 1994) are only operated on when MEP monitoring is available. We are convinced that without monitoring, the removal of such gliomas would have been associated with a much higher morbidity than in our previously published paper (Zentner J *et al.*, 1996b). Since that study we have altered our strategy and no longer operate on grade IV tumors in older patients. With this new policy the incidence of motor deficits has decreased markedly, and it appears that the use of MEP monitoring has contributed to this. The size and location of the craniotomy determines whether transcranial or cortical MEPs are performed with insular and sylvian lesions. A dense recording montage including leg and arm, and facial muscles is advisable for surgery of insular lesions.

With sylvian AVMs similar considerations apply as with tumors operated via a transsylvian approach, although the scope of surgical reactions to MEP deterioration is limited, in particular if bleeding of the AVM occurs, and since incomplete removal is no option with such lesions. Identification of transitory vessels, detection of vasospasm, and retractor placement is still usefully supported by MEP monitoring (Schaller C *et al.*, 1998). With temporomesial AVMs, vascular supply of the pyramidal tract and perforating arteries that enter through the lateral-inferior part of the anterior perforated substance or branch off the anterior choroidal artery are better protected under motor monitoring conditions.

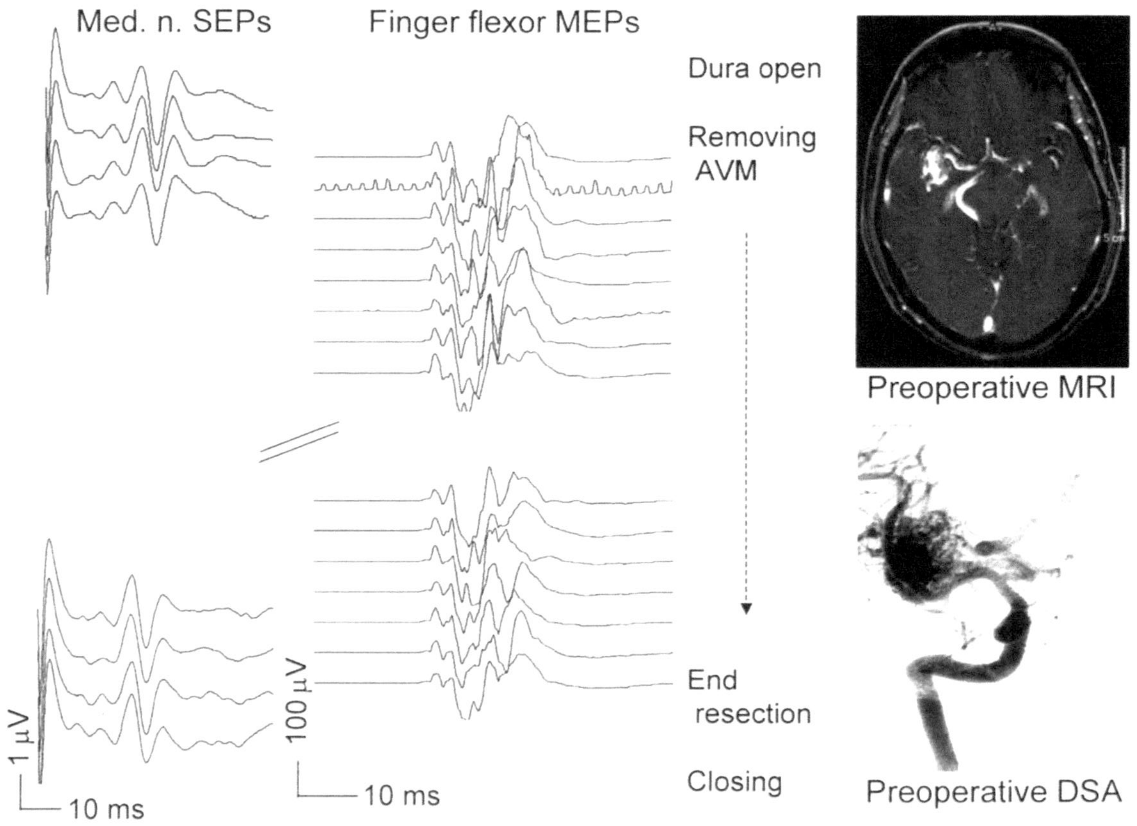

Fig. 9. Sylvian AVM – Transient new facial paresis despite unaltered arm MEPs and SEPs. This 34 year old patient was presented with symptomatic epilepsy from a right-sylvian/insular AVM. The lesion was removed via a transsylvian approach under transcranial MEP and SEP monitoring. MEPs were recorded from the contralateral arm. Despite stable intraoperative MEP and SEP recordings there was a transient new facial paresis postoperatively, i.e. in an area that had not been monitored

Infratentorial Surgery

Indications for MEP Monitoring

Long-tract monitoring by means of MEP and SEP recording is performed at our institution with lesions affecting the infratentorial sensorimotor tracts, in addition to the widely used BAEP and cranial nerve mapping and monitoring procedures (Moeller A, 1995). Obviously, surgery for brainstem astrocytomas and ependymomas as well as for metastases and cavernomas can be usefully supported by long-tract monitoring, depending on the site of the lesion. In addition, extraaxial tumors like large basal meningiomas and neuromas, may lead to a significant compression and dislocation of the brainstem and may involve its vasculature, so that MEP and SEP monitoring appears advisable in these cases. Likewise, vascular lesions like AVMs encroaching on the brainstem, and large aneurysms bulging into it, warrant long tract monitoring. As with supratentorial monitoring, only MEP recordings do reliably reflect impending motor deficit. However, SEP recording in parallel to MEP monitoring may still be advisable, since preservation of afferent pathways is almost as important at this level of the brain, taking into consideration the effect of postoperative ataxia from hemisensory loss.

Impact on Surgical Strategy

During surgery for intraaxial lesions, or extraaxial lesions which cannot be easily dissected from the surface of the brainstem, monitoring may help to decide when to stop the dissection. For example in ependymomas, resection is frequently performed without problems defining the borders. However, in some cases it may only be safely completed up to the point of MEP deterioration, with only transient postoperative paresis. Postoperative imaging has shown a thin tumor layer remaining directly on the corticospinal tract in such instances. With other lesion such as vascular malformations, cessation of surgery may not be the question, but release of a retractor compressing the brainstem or exerting traction on a perforating vessel may be triggered by monitoring results.

Clinical Experience with MEP Monitoring for Infratentorial Surgery

In a recent series of 55 operations of such lesions at our institution under long-tract monitoring, 21 patients harbored intraaxial tumors such as Gliomas and metastases, 22 patients had very large extraaxial lesions including meningiomas, neuromas, and epidermoids with significant brainstem compression, and 12 patients had vascular lesions. Continuous MEPs were monitored in 29 of these cases, SEPs in 43 cases, and both modalities in

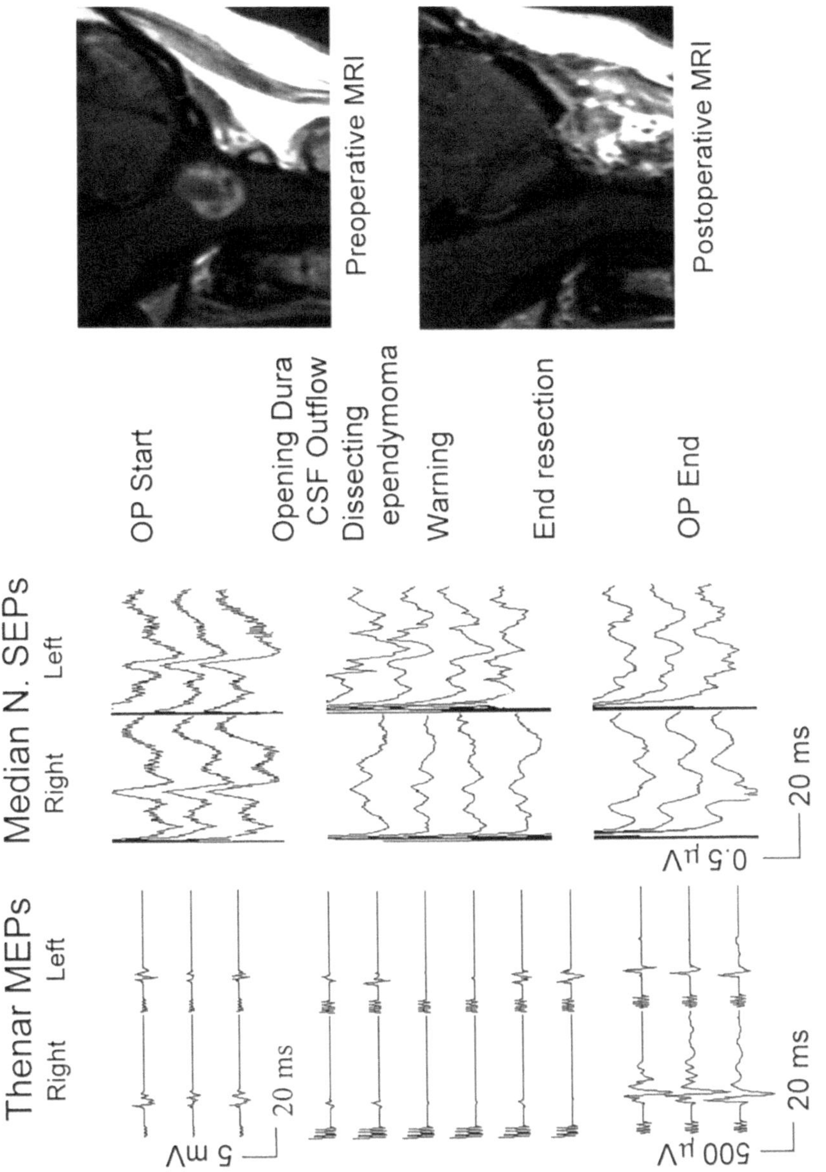

Fig. 10. EP deterioration during brainstem dissection. During surgery for a pontomedullary ependymoma, EP deterioration occurred after opening of the dura and CSF outflow. When resecting within the tumor, further MEP amplitude reduction indicated proximity of the corticospinal tract. With stabilizing MEPs resection was carefully completed, and postoperative imaging showed no residual tumor tissue. There was no new paresis, but only transient light ataxia during the postoperative course

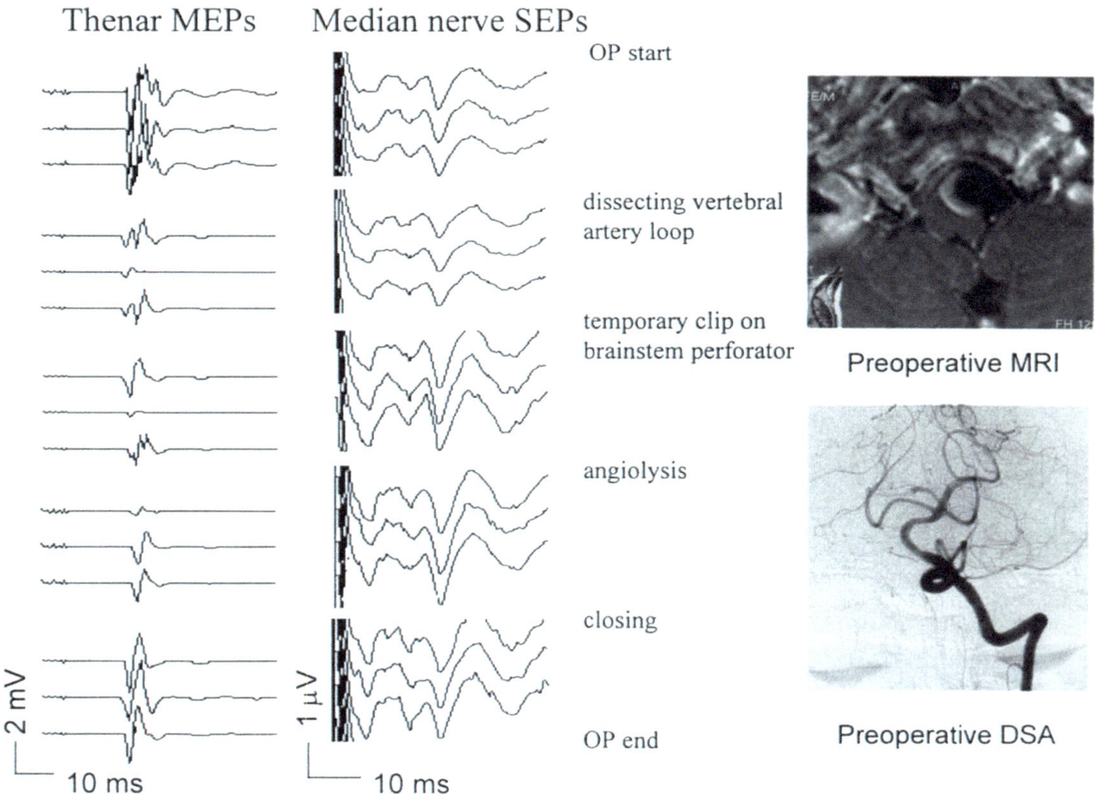

Fig. 11. Unaltered EP recordings during temporary clip on brainstem perforator. This 65 year old patient developed left hyperpathia and slight spasticity during the 13 year course after surgery for a posterior fossa arachnoid cyst. A vertebral loop compressing the medulla oblongata was found and decompression with Teflon felt implant was performed under MEP and SEP monitoring. During dissection, a ventral brainstem perforator was temporarily clipped. Unaltered MEP and SEP recordings allowed for safe completion of dissection. There was no new postoperative deficit

17 cases. With permanent MEP impairment or loss, SEPs altered reversibly, and irreversibly in one case each, and showed no alteration in one further case. With reversible MEP changes, SEPs remained always unaltered. Transient SEP impairment occurred in two cases despite unaltered MEPs. Permanent and severe new paresis occurred in two cases in this series, reversible deficit in 5 patients. New motor deficit, and unimpaired outcome was predicted by MEP monitoring in all cases, while new paresis but was missed by SEP recording in three out of four cases. MEP amplitude changes or reversible loss were associated with unimpaired motor outcome in most cases, when triggering early surgical reaction. SEP deterioration was associated with new sensory deficit or impaired coordination in several cases, also in cases with stable intraoperative MEP recordings. Conversely, when recorded in parallel, MEPs reflected, but SEPs missed both cases with transient new paresis. The case of a mid-brain cavernoma that was removed through a transventricular subforniceal approach under monitoring of MEPs and SEPs illustrates the usefulness of combined monitoring: Reversible SEP deterioration with stable MEPs was associated with transiently impaired sensation of the respective arm (Cedzich C *et al.*, 1999).

Aneurysms

SEP Monitoring for Aneurysm Surgery

The intraoperative monitoring experience with cerebral aneurysms at our institution encompasses about 120 cases of combined MEP and SEP monitoring, and another 600 previous cases of SEP recording alone. SEP monitoring for intracranial aneurysm surgery has become popular in many neurosurgical centers (Little JR *et al.*, 1987, Momma F *et al.*, 1987, Mooij JJ *et al.*, 1987, Buchthal A *et al.*, 1988, Scheepstra GL *et al.*, 1989, Schramm J *et al.*, 1990, Friedman WA *et al.*, 1991, Schramm J *et al.*, 1994, Moeller A, 1995, Sako K *et al.*, 1995, Holland NR, 1998, Sako K *et al.*, 1998, Lopez JR *et al.*, 1999). In our service, aneurysms are operated under monitoring conditions whenever possible. In certain instances, such as with very large aneurysms of the MCA or Carotid bifurcation, elective surgery may even be delayed until monitoring is available. Small peripheral MCA-aneurysms, especially if they have not bled, may be clipped without monitoring, though. In previous series with SEP aneurysm monitoring from this group (Schramm J *et al.*, 1990, Schramm J *et al.*, 1994), SEP changes occurred in 11.3%, mainly due to intentional or inadvertent vessel occlusion, and led to readjustment of clips or retractors, or removal of temporary clips in 8.1% of cases (Schramm J *et al.*, 1994), whereas the surgical strategy remained unchanged due to stable SEP recordings in 9.7%

of cases (Schramm J *et al.*, 1990). It was concluded that SEP monitoring is particularly helpful in large and multilobulated, or complicated aneurysms of the anterior circulation, with trapping procedures and during temporary vessel occlusion (Schramm J *et al.*, 1994).

However, with accumulating experience it has become clear, that SEP recording alone misses impending motor impairment in up to 10% of procedures in a series (Krieger D *et al.*, 1992, Holland NR, 1998), obviously due to a poor sensitivity for subcortical ischemia induced by compromise of perforating vessels (Holland NR, 1998). Therefore, MEP monitoring in parallel to SEP recording was introduced at our institution.

Conduction of MEP Monitoring for Aneurysm Surgery

Given the more basal location of the craniotomy required for aneurysm surgery, transcranial rather than direct cortical MEP stimulation is usually applied. Recording from upper extremities (and face) is sufficient at least with posterior fossa and middle cerebral artery aneurysms. Since the anterior cerebral artery (ACA) supplies the leg motor cortex, additional lower extremity recordings may be useful with aneurysms of the ACA or the anterior communicating artery. However, reliable leg MEP recordings may require very strong stimulation with an exceedingly strong stimulus artifact, that is frequently incompatible with microsurgical dissection. Since cortical ischemia is reliably picked up by SEP monitoring, and the perforating arteries branching from the ACA, e.g. the Heubnerian recurrent artery, supply motor pathways including corticonuclear efferents and fibers descending towards cervical motoneurons rather than lumbosacral leg motoneurons, it will usually be sufficient to monitor only the more readily recordable arm MEPs. Bilateral recordings are mandatory with aneurysms close to the midline such as anterior communicating or basilar artery aneurysms, whereas unilateral recording is sufficient with, e.g., aneurysms of the medial cerebral artery.

Clinical Experience with MEP Monitoring for Aneurysm Surgery

In a recent series of 100 aneurysm surgeries, significant intraoperative events (e.g., deliberate or inadvertent vessel occlusion) occurred in 52 cases. Twelve patients sustained a new postoperative motor deficit, which was, however, permanently disabling in only two of them. The high rate of transient slight pareses discovered in this series is presumably due to the prospective nature of the study with very precise follow-up examinations. With new paresis, early CT revealed subcortical ischemia in eight, and cortical ischemia in two patients, while only edema was found in two further cases. Combined MEP and SEP monitoring was performed in this series, with successful SEP recordings in about 95%, and recordable MEPs in 90%. However, MEP stimulation had to be terminated in several cases due to a strong stimulation twitching artifact interfering with the microsurgical procedure. MEP changes indicated vessel occlusion, inadequate retractors, vasospasm, and compromise to perforators in 21/33 cases and altered the surgical strategy in 16 cases. SEPs reflected such conditions in only 7/33 cases and altered the surgical strategy in 4 cases. MEP deterioration predicted postoperative motor impairment in 10/11 monitored cases, except for one patient displaying edema on postoperative CT. SEP indicated new paresis in only 17%. Inadvertent vessel occlusion was first reflected by micro Doppler recordings in most cases, whereas with intentional incomplete vessel occlusion, stable EP recordings predicted unimpaired postoperative outcome, and enabled the surgeon to proceed. Thus, MEP recording is clearly superior to SEPs for motor monitoring purposes, particularly in detecting impending new paresis from subcortical ischemia. The micro Doppler is the best method for detecting inadvertent vessel occlusion, but cannot assess remote flow conditions, which are better reflected by EP recordings.

Clinical Relevance of Aneurysm Monitoring

At this point it is not yet clear whether MEP monitoring can reliably help to *prevent* new deficit, which seems to occur mainly with inadvertent compromise of perforating arteries at dissection of the aneurysm, or during temporal clipping of a major artery. One might argue that the dissection will be performed with maximum care anyway, and perforators will be microsurgically inspected. However, MEP deterioration may redirect the surgeon to renewed inspection, and kinked, or spastic small arteries have been found on such occasions. Even if the vascular compromise is irreversible despite surgical intervention, e.g. with severe vasospasm, MEP and SEP monitoring can indicate to the surgeon at which step of dissection motor tract impairment occurred, which will be of educational value. Stable evoked potential recordings are of equally great value in many instances of aneurysm surgery. With temporary clipping for safer dissection, the surgeon may safely proceed with the procedure as long as stable monitoring indicates unimpaired sensorimotor function. Similarly, if inspection or micro Doppler readings suggest narrowing of the lumen of a major vessel by the definite clip, which may occur, e.g., with thrombosed aneurysms, the clip may be left in place if stable MEP and SEP recordings indicate sufficient blood flow over the target vessel or via collateral flow.

Summary and Conclusions

Why MEP Monitoring?

New postoperative paresis is one of the most important risks of brain surgery. On the one hand, it may occur relatively frequently due to the

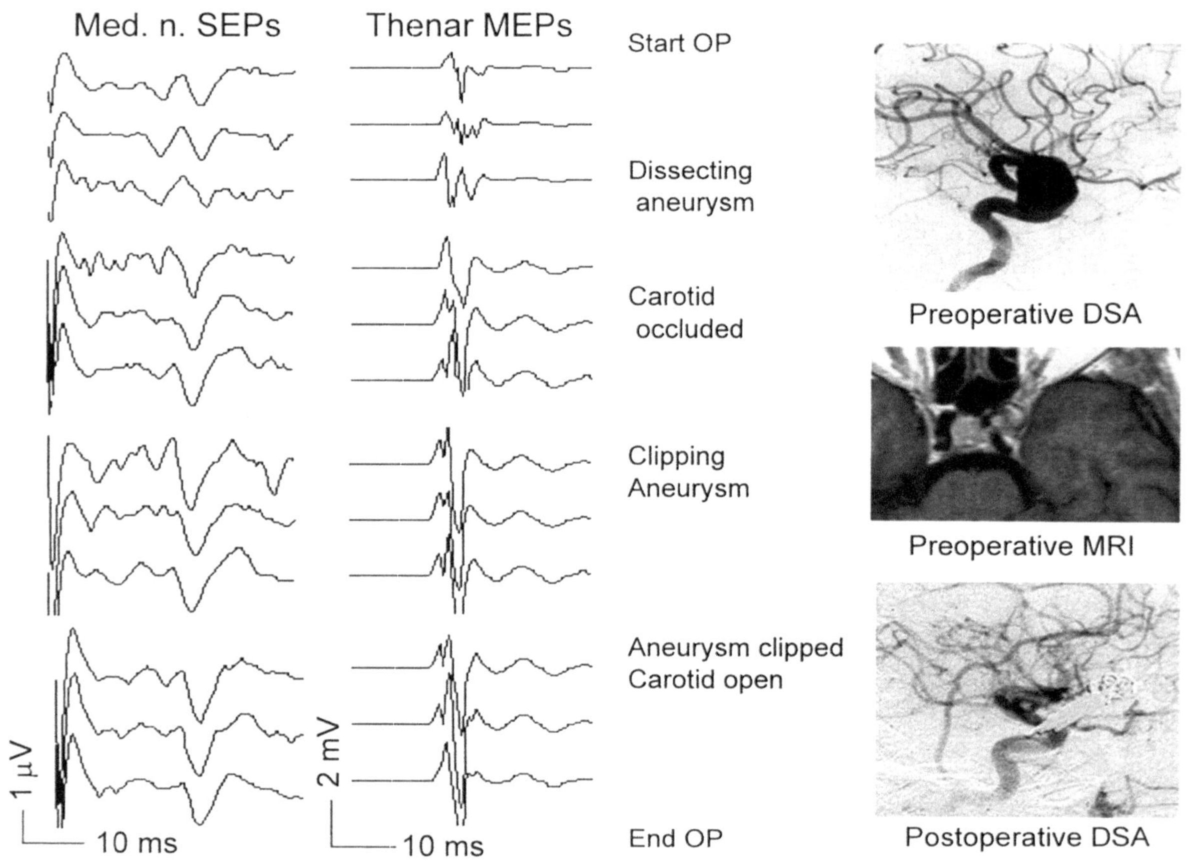

Fig. 12. Stable MEPs/SEPs during temporary carotid occlusion. This 58 year old patient was operated on a large left carotid aneurysm. Stable MEP and SEP recordings during dissection, and temporary carotid occlusion allowed safe completion of the procedure. There was no new postoperative deficit

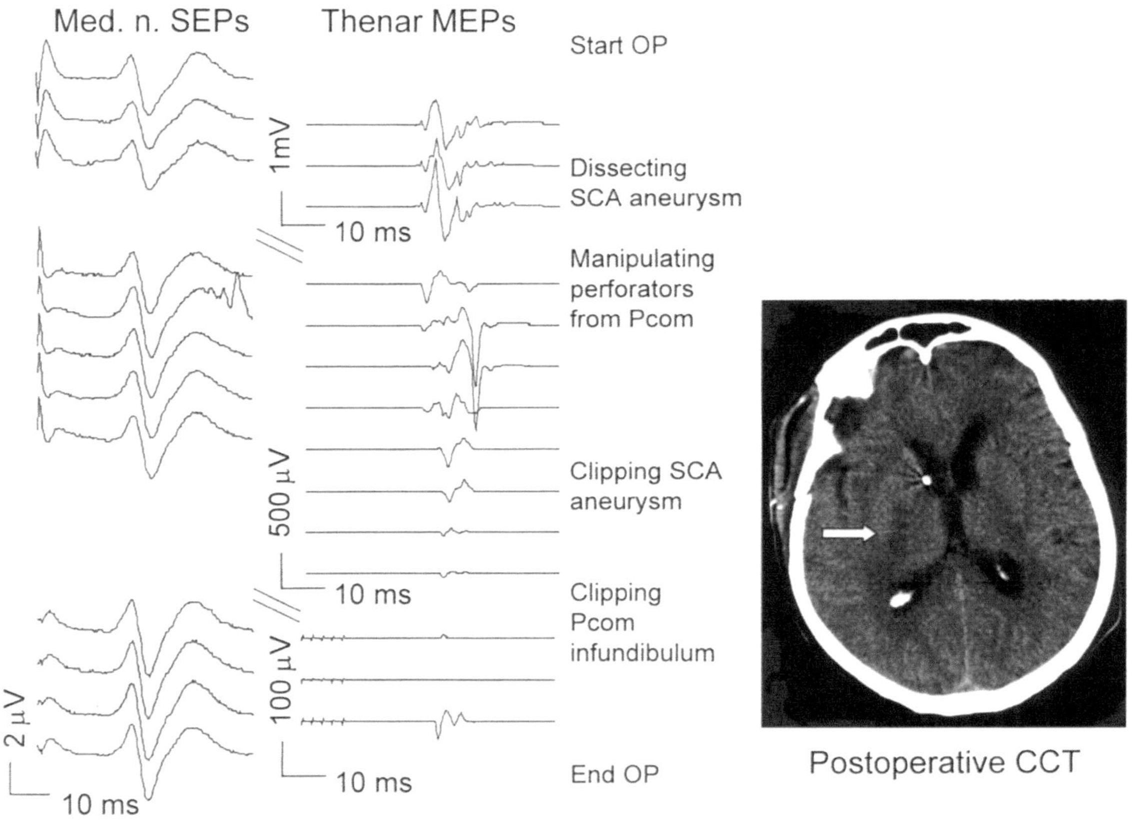

Fig. 13. MEP deterioration indicates subcortical stroke despite preserved SEPs. During dissection and clipping of a basilar tip and a posterior communicating artery aneurysm via a common pterional approach, a significant thenar MEP deterioration occurred while median nerve SEPs remained unaltered. Postoperatively, there was a new slight hemiparesis, and CT scan revealed a small basal ganglia infarction. Modified from (Neuloh G *et al.*, 2002)

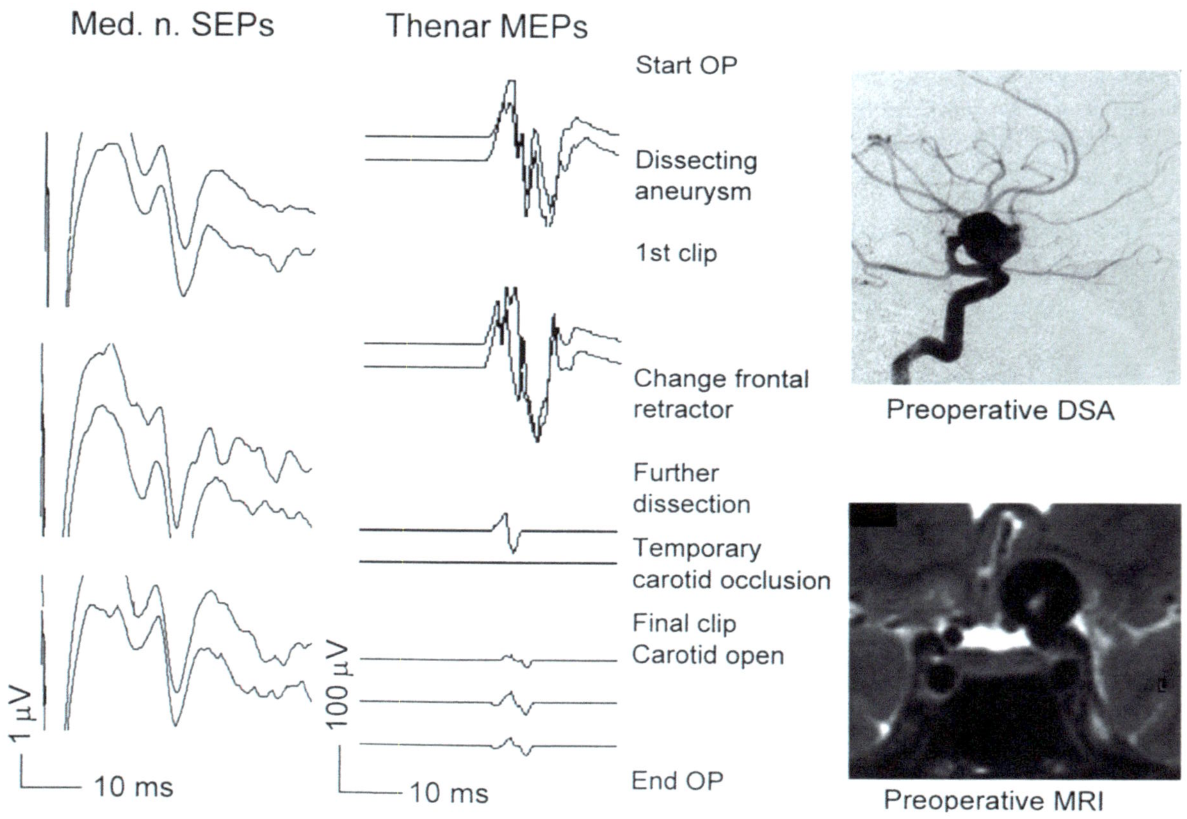

Fig. 14. MEPs reflect motor impairment despite stable SEPs. This 64 year old patient was operated on a large left carotid aneurysm. MEPs deteriorated during dissection and frontal brain elevation, and vanished during temporary carotid occlusion, but recovered partially with reopening of the carotid artery. SEPs remained unchanged. There was a transient paresis postoperatively, the CCT showed some frontobasal edema, possibly from retractor placement

extended and exposed course of the motor system within the brain. On the other hand, it is perceived as particularly incapacitating by the patient, even more so if it is severe and permanent. However, brain surgery will require resection close to motor pathways in many instances, and will benefit, e.g., in terms of extent of resection achieved, if continuous information about the motor system's functional state is available. Since voluntary movement cannot be assessed in the anesthesized patient, continuous recording of MEPs will help to assess the functional integrity of the motor areas and pathways.

When to Perform MEP Monitoring?

MEPs reflect the functional state of the motor system and are of no proven use for monitoring of other functional systems of the brain. Therefore indications for MEP monitoring will primarily include cases where the functional integrity of the motor areas and pathways proper is actually at risk. These include tumors and vascular lesions within and around the paracentral sensorimotor cortex, or adjacent to the corona radiata and the internal capsule such as insular lesions. Surgery for tumors and vascular lesions within the brainstem or encroaching on it may frequently require long-tract monitoring as well, depending on the approach and the site of the lesion as discussed in the previous section (pp. 202–213). Moreover, the motor system is at risk in surgery for cerebral aneurysms, for example from inadvertent or prolonged intentional vessel occlusion, from brain retraction, and compromise to perforating vessels. Therefore MEP recordings will usefully supplement the traditional SEP monitoring method in this field as well.

How to Achieve Good MEP Monitoring?

In many instances, functional mapping of the pericentral region must precede the actual monitoring procedure, for example with tumors invading the sensorimotor cortex. In such cases, functional mapping helps both to delimitate the eloquent areas from the lesion, and to define the optimum site for direct cortical MEP stimulation. In our experience, sensorimotor mapping is done best by recording of SEP phase reversal. However, direct motor cortex stimulation mapping, or perioperative speech mapping will be required under specific conditions, e.g., if SEP phase reversal fails, or if speech areas are also at risk due to the location of the lesion. If the pericentral cortex is not exposed by the craniotomy, transcranial MEP stimulation is performed at predefined sites. Stimulation by high-frequency anodal train stimulation yields reliable MEP responses in the anesthesized patient. Careful selection of stimulation parameters as discussed in section tree is of paramount importance in order to obtain good MEP recordings. An adequate anesthetic regimen, at this point possibly total intravenous anesthesia with propofol, without bolus injections of anesthetics and without neuromuscular block, as well as stable homeostasis conditions is important as well. Recording from at least one distal arm and leg muscle group each is recommendable in most cases in order to avoid false-negative monitoring results, and facial muscles may be included in the recording scheme with surgery at the lateral convexity of the central region. Recording from additional muscles (e.g., proximal limb muscles) may be useful for mapping purposes, but with continuous monitoring the risk of false-positive responses increases with a highly redundant recording setup.

How to Assess and Interpret MEPs?

Even with stable recordings an amplitude variation below 50% is difficult to achieve. Therefore, a pragmatic criterion of 50% amplitude decrease has proven applicable in our experience when classifying MEP changes as significant. Isolated latency changes do rarely occur, a pragmatic criterion of 10–15% latency increase at repeated recordings is applicable in these cases. Frequently, the individual level of variation has to be determined in a given case in order to define significant amplitude and latency deviations from the baseline. MEP response duration and shape is not routinely assessed for monitoring purposes. All significant changes must be reported to the surgeon, and stability of MEP responses will be of interest with critical steps of surgery as well.

The following correlation of intraoperative MEP changes and motor outcome should be considered when interpreting MEP monitoring results in terms of the motor system's functional state: Unaltered MEP responses predict unimpaired motor outcome in the monitored limbs, except in cases where dynamical processes such as postresectional edema take place after the actual surgical and monitoring procedure. Reversible MEP deterioration is associated with transient new paresis, or unimpaired outcome in most cases, whereas irreversible deterioration without complete loss of responses may be associated with permanent new paresis in a certain proportion of cases. Irreversible complete MEP loss invariably predicts permanent, or sometimes very long-lasting new motor deficit.

Does MEP Monitoring Influence Surgical Strategy and Postoperative Outcome?

A definite answer to this question will only be obtained from prospectively controlled interventional studies. However in our experience there are numerous observations in favour of a positive answer to this question. One important fact is the influence of stable recordings on the surgical procedure in critical situation where the surgeon would for example abandon resection close to the internal capsule, or remove a clip without the confirmation from MEP recordings that there is no immediate risk for motor deterioration. In insular tumors early cessation of surgery would be the only solution when approaching the motor pathways if stable MEP recordings were not available, and cytoreduction would be clearly worse as compared with resection under MEP monitoring conditions. Another important observation is the reversibility of MEP deterioration or even loss, if early surgical intervention such as cessation of surgery at a certain spot, application of papaverine to spastic vessels, or readjustment of retractors is triggered by the monitoring procedure. Since transient slight new paresis was observed in many such cases, but permanent deficit occurred with irreversible MEP loss, it can be concluded that more severe motor impairment would have occurred without the early warning from MEP deterioration and consecutive surgical intervention. Even in cases where the MEP deterioration turns out to be irreversible with permanent new paresis postoperatively, the correlation of surgical events associated with this MEP change such as severe vasospasm during a transsylvian approach will be of educational value, potentially influencing not only a single surgical procedure, but also certain aspects of a surgical approach in general, and contribute in this way to an improved outcome for the patients.

References

1. Agnew WF, McCreery DB (1987) Considerations for safety in the use of extracranial stimulation for motor evoked potentials. Neurosurgery 201: 143–147
2. Amassian VE, Stewart M, Quirk GJ, Rosenthal JL (1987) Physiological basis of motor effects of a transient stimulus to cerebral cortex. Neurosurgery 201: 74–93
3. Ammirati M, Vick N, Liao YL, Ciric I, Mikhael M (1987) Effect of the extent of surgical resection on survival and quality of life in patients with supratentorial glioblastomas and anaplastic astrocytomas. Neurosurgery 212: 201–206
4. Barker AT, Jalinous R, Freeston IL (1985) Non-invasive magnetic stimulation of human motor cortex. Lancet 18437: 1106–1107
5. Berger MS (1995) Functional mapping-guided resection of low-grade gliomas. Clin Neurosurg 42: 437–452
6. Berger MS (1996) Minimalism through intraoperative functional mapping. Clin Neurosurg 43: 324–337
7. Berger MS, Kincaid J, Ojemann GA, Lettich E (1989) Brain mapping techniques to maximize resection, safety, and seizure control in children with brain tumors. Neurosurgery 255: 786–792

8. Berger MS, Rostomily RC (1997) Low grade gliomas: functional mapping resection strategies, extent of resection, and outcome. J Neurooncol 341: 85–101
9. Binder JR, Swanson SJ, Hammeke TA, Morris GL, Mueller WM, Fischer M, Benbadis S, Frost JA, Rao SM, Haughton VM (1996) Determination of language dominance using functional MRI: a comparison with the Wada test. Neurology 464: 978–984
10. Browning JL, Heizer ML, Baskin DS (1992) Variations in corticomotor and somatosensory evoked potentials: effects of temperature, halothane anesthesia, and arterial partial pressure of CO2. Anesth Analg 745: 643–648
11. Buchthal A, Belopavlovic M, Mooij JJ (1988) Evoked potential monitoring and temporary clipping in cerebral aneurysm surgery. Acta Neurochir (Wien) 931–932: 28–36
12. Burke D, Hicks RG (1998) Surgical monitoring of motor pathways. J Clin Neurophysiol 153: 194–205
13. Burke D, Hicks RG, Stephen JP (1990) Corticospinal volleys evoked by anodal and cathodal stimulation of the human motor cortex. J Physiol 425: 283–299
14. Calancie B, Harris W, Brindle GF, Green BA, Landy HJ (2001) Threshold-level repetitive transcranial electrical stimulation for intraoperative monitoring of central motor conduction. J Neurosurg [Suppl] 952: 161–168
15. Calancie B, Harris W, Broton JG, Alexeeva N, Green BA (1998) "Threshold-level" multipulse transcranial electrical stimulation of motor cortex for intraoperative monitoring of spinal motor tracts: description of method and comparison to somatosensory evoked potential monitoring. J Neurosurg 883: 457–470
16. Cedzich C, Pechstein U, Schramm J, Schafer S (1998) Electrophysiological considerations regarding electrical stimulation of motor cortex and brain stem in humans. Neurosurgery 423: 527–532
17. Cedzich C, Pechstein U, Zentner J, Van Roost D (1999) Minimally invasive stereotactically-guided extirpation of brain stem cavernoma with the aid of electrophysiological methods. Minim Invasive Neurosurg 421: 41–43
18. Cedzich C, Taniguchi M, Schafer S, Schramm J (1996) Somatosensory evoked potential phase reversal and direct motor cortex stimulation during surgery in and around the central region. Neurosurgery 385: 962–970
19. Coyer PE, Simeone FA, Michele JJ (1987) Latency of the cortical component of the somatosensory evoked potential in relation to cerebral blood flow measured in the white matter of the cat brain during focal ischemia. Neurosurgery 214: 497–502
20. Croft TJ, Brodkey JS, Nulsen FE (1972) Reversible spinal cord trauma: a model for electrical monitoring of spinal cord function. J Neurosurg 364: 402–406
21. Deletis V (1993) Intraoperative monitoring of the functional integrity of the motor pathways. Adv Neurol 63: 201–214
22. Deletis V, Isgum V, Amassian VE (2001) Neurophysiological mechanisms underlying motor evoked potentials in anesthetized humans. Part 1. Recov-

ery time of corticospinal tract direct waves elicited by pairs of transcranial electrical stimuli. Clin Neurophysiol 1123: 438–444
23. Deletis V, Kothbauer K (1998) Intraoperative neurophysiology of the corticospinal tract. Spinal cord monitoring. In: Staalberg E, Sharma HS, Olsson Y (eds) Springer, Wien New York, pp 421–444
24. Duffau H, Capelle L, Sichez J, Faillot T, Abdennour L, Law Koune JD, Dadoun S, Bitar A, Arthuis F, Van Effenterre R, Fohanno D (1999) Intra-operative direct electrical stimulations of the central nervous system: the Salpetriere experience with 60 patients. Acta Neurochir (Wien) 14111: 1157–1167
25. Ebeling U, Huber P (1992) Localization of central lesions by correlation of CT findings and neurological deficits. Acta Neurochir (Wien) 1191–1194: 17–22
26. Ebeling U, Schmid UD, Reulen HJ (1989) Tumour-surgery within the central motor strip: surgical results with the aid of electrical motor cortex stimulation. Acta Neurochir (Wien) 1013–1014: 100–107
27. Fadul C, Wood J, Thaler H, Galicich J, Patterson RH Jr, Posner JB (1988) Morbidity and mortality of craniotomy for excision of supratentorial gliomas. Neurology 389: 1374–1379
28. Fandino J, Kollias SS, Wieser HG, Valavanis A, Yonekawa Y (1999) Intraoperative validation of functional magnetic resonance imaging and cortical reorganization patterns in patients with brain tumors involving the primary motor cortex. J Neurosurg 912: 238–250
29. Fischer C, Ibanez V, Mauguiere F (1985) [Peroperative monitoring of early auditory evoked potentials]. Presse Med 1437: 1914–1918
30. Friedman WA, Chadwick GM, Verhoeven FJ, Mahla M, Day AL (1991) Monitoring of somatosensory evoked potentials during surgery for middle cerebral artery aneurysms. Neurosurgery 291: 83–88
31. Goldring S, Gregorie EM (1984) Surgical management of epilepsy using epidural recordings to localize the seizure focus. Review of 100 cases. J Neurosurg 603: 457–466
32. Gugino V, Chabot RJ (1990) Somatosensory evoked potentials. Int Anesthesiol Clin 283: 154–164
33. Hargadine JR, Branston NM, Symon L (1980) Central conduction time in primate brain ischemia—a study in baboons. Stroke 116: 637–642
34. Herdmann J, Deletis V, Edmonds HL Jr, Morota N (1996) Spinal cord and nerve root monitoring in spine surgery and related procedures. Spine 217: 879–885
35. Holland NR (1998) Subcortical strokes from intracranial aneurysm surgery: implications for intraoperative neuromonitoring. J Clin Neurophysiol 155: 439–446
36. Horsley V (1891) The croonian lecture: on the mammalian nervous system, its functions and their localization, determined by an electrical method. Phylos trans R Soc Lond 182: 267–326
37. Jeanmonod D, Sindou M, Mauguiere F (1989) Intra-operative spinal cord evoked potentials during cervical and lumbo-sacral microsurgical DREZ-

tomy (MDT) for chronic pain and spasticity (preliminary data). Acta Neurochir (Wien) [Suppl] 46: 58–61
38. Jones SJ, Harrison R, Koh KF, Mendoza N, Crockard HA (1996) Motor evoked potential monitoring during spinal surgery: responses of distal limb muscles to transcranial cortical stimulation with pulse trains. Electroencephalogr Clin Neurophysiol 1005: 375–383
39. Kakimoto M, Inoue S, Sakamoto T, Kawaguchi M, Kitaguchi K, Furuya H (2000) [Intraoperative motor evoked potential monitoring: a review of 115 cases]. Masui 493: 240–244
40. Katayama Y, Tsubokawa T, Maejima S, Hirayama T, Yamamoto T (1988) Corticospinal direct response in humans: identification of the motor cortex during intracranial surgery under general anaesthesia. J Neurol Neurosurg Psychiatry 511: 50–59
41. King RB, Schell GR (1987) Cortical localization and monitoring during cerebral operations. J Neurosurg 672: 210–219
42. Kombos T, Suess O, Ciklatekerlio O, Brock M (2001) Monitoring of intraoperative motor evoked potentials to increase the safety of surgery in and around the motor cortex. J Neurosurg 954: 608–614
43. Kombos T, Suess O, Funk T, Kern BC, Brock M (2000a) Intra-operative mapping of the motor cortex during surgery in and around the motor cortex. Acta Neurochir (Wien) 1423: 263–268
44. Kombos T, Suess O, Kern BC, Funk T, Hoell T, Kopetsch O, Brock M (1999) Comparison between monopolar and bipolar electrical stimulation of the motor cortex. Acta Neurochir (Wien) 14112: 1295–1301
45. Kombos T, Suess O, Pietila T, Brock M (2000b) Subdural air limits the elicitation of compound muscle action potentials by high-frequency transcranial electrical stimulation. Br J Neurosurg 143: 240–243
46. Kothbauer K, Deletis V, Epstein FJ (1997) Intraoperative spinal cord monitoring for intramedullary surgery: an essential adjunct. Pediatr Neurosurg 265: 247–254
47. Kothbauer KF, Deletis V, Epstein FJ (1998) Intraoperative monitoring. Pediatr Neurosurg 291: 54–55
48. Kral T, Clusmann H, Urbach J, Schramm J, Elger CE, Kurthen M, Grunwald T (2002) Prachirurgische Abklarung bei der Epilepsiechirurgie (Bonner Algorithmus). Zentralbl Neurochir 633: 106–110
49. Krause F (1911) Chirurgie des Gehirns und Rückenmarks nach eigenen Erfahrungen. Urban & Schwarzenberg, Berlin
50. Kreth FW, Warnke PC, Scheremet R, Ostertag CB (1993) Surgical resection and radiation therapy versus biopsy and radiation therapy in the treatment of glioblastoma multiforme. J Neurosurg 785: 762–766
51. Krieger D, Adams HP, Albert F, von Haken M, Hacke W (1992) Pure motor hemiparesis with stable somatosensory evoked potential monitoring during aneurysm surgery: case report. Neurosurgery 311: 145–150
52. Krings T, Reinges MH, Thiex R, Gilsbach JM, Thron A (2001a) Functional and diffusion-weighted magnetic resonance images of space-occupying lesions

affecting the motor system: imaging the motor cortex and pyramidal tracts. J Neurosurg 955: 816–824
53. Krings T, Schreckenberger M, Rohde V, Foltys H, Spetzger U, Sabri O, Reinges MH, Kemeny S, Meyer PT, Moller-Hartmann W, Korinth M, Gilsbach JM, Buell U, Thron A (2001b) Metabolic and electrophysiological validation of functional MRI. J Neurol Neurosurg Psychiatry 716: 762–771
54. Lang EW, Beutler AS, Chesnut RM, Patel PM, Kennelly NA, Kalkman CJ, Drummond JC, Garfin SR (1996) Myogenic motor-evoked potential monitoring using partial neuromuscular blockade in surgery of the spine. Spine 2114: 1676–1686
55. Lehericy S, Duffau H, Cornu P, Capelle L, Pidoux B, Carpentier A, Auliac S, Clemenceau S, Sichez JP, Bitar A, Valery CA, Van Effenterre R, Faillot T, Srour A, Fohanno D, Philippon J, Le Bihan D, Marsault C (2000) Correspondence between functional magnetic resonance imaging somatotopy and individual brain anatomy of the central region: comparison with intraoperative stimulation in patients with brain tumors. J Neurosurg 924: 589–598
56. Levy WJ, York DH (1986) Evoked potential monitoring of the motor system. Biomed Sci Instrum 22: 51–54
57. Little JR, Lesser RP, Luders H (1987) Electrophysiological monitoring during basilar aneurysm operation. Neurosurgery 203: 421–427
58. Lopez JR, Chang SD, Steinberg GK (1999) The use of electrophysiological monitoring in the intraoperative management of intracranial aneurysms. J Neurol Neurosurg Psychiatry 662: 189–196
59. Lueders H, Lesser RP, Hahn J, Dinner DS, Klem G (1983) Cortical somatosensory evoked potentials in response to hand stimulation. J Neurosurg 586: 885–894
60. Majkowski J (1999) Kindling: clinical relevance for epileptogenicity in humans. Adv Neurol 81: 105–113
61. Matsuda M, Kidooka M, Nakazawa T, Kyoshima K, Handa J (1991) Intraoperative monitoring of somatosensory evoked potentials in patients with cerebral aneurysm—correlation between central conduction time and postoperative neurological status. Neurol Med Chir (Tokyo) 311: 13–17
62. Mauguiere F, Allison T, Babiloni C, Buchner H, Eisen AA, Goodin DS, Jones SJ, Kakigi R, Matsuoka S, Nuwer M, Rossini PM, Shibasaki H (1999) Somatosensory evoked potentials. The International Federation of Clinical Neurophysiology. Electroencephalogr Clin Neurophysiol [Suppl] 52: 79–90
63. McCreery DB, Agnew WF, Yuen TG, Bullara L (1990) Charge density and charge per phase as cofactors in neural injury induced by electrical stimulation. IEEE Trans Biomed Eng 3710: 996–1001
64. McPherson RW, Niedermeyer EF, Otenasek RJ, Hanley DF (1983) Correlation of transient neurological deficit and somatosensory evoked potentials after intracranial aneurysm surgery. Case report. J Neurosurg 591: 146–149
65. Merton PA, Morton HB (1980) Stimulation of the cerebral cortex in the intact human subject. Nature 2855762: 227

66. Meylaerts SA, Jacobs MJ, van Iterson V, De Haan P, Kalkman CJ (1999) Comparison of transcranial motor evoked potentials and somatosensory evoked potentials during thoracoabdominal aortic aneurysm repair. Ann Surg 2306: 742–749
67. Mizoi K, Yoshimoto T (1991) Intraoperative monitoring of the somatosensory evoked potentials and cerebral blood flow during aneurysm surgery—safety evaluation for temporary vascular occlusion. Neurol Med Chir (Tokyo) 316: 318–325
68. Moeller A (1995) Intraoperative neurophysiologic monitoring. Harwood Academic Publishers, Luxembourg
69. Momma F, Wang AD, Symon L (1987) Effects of temporary arterial occlusion on somatosensory evoked responses in aneurysm surgery. Surg Neurol 274: 343–352
70. Mooij JJ, Buchthal A, Belopavlovic M (1987) Somatosensory evoked potential monitoring of temporary middle cerebral artery occlusion during aneurysm operation. Neurosurgery 214: 492–496
71. Muacevic A, Steiger HJ (1999) Computer-assisted resection of cerebral arteriovenous malformations. Neurosurgery 455: 1164–1170; discussion 1170–1171
72. Nagle KJ, Emerson RG, Adams DC, Heyer EJ, Roye DP, Schwab FJ, Weidenbaum M, McCormick P, Pile-Spellman J, Stein BM, Farcy JP, Gallo EJ, Dowling KC, Turner CA (1996) Intraoperative monitoring of motor evoked potentials: a review of 116 cases. Neurology 474: 999–1004
73. Neuloh G, Schramm J (2002) Intraoperative neurophysiological mapping and monitoring for supratentorial procedures. Neurophysiology in neurosurgery. In: Deletis SJ (ed) Academic Press, Amsterdam Boston London, pp 339–401
74. Nuwer MR (1999) Spinal cord monitoring. Muscle Nerve 2212: 1620–1630
75. Nuwer MR, Dawson EG, Carlson LG, Kanim LE, Sherman JE (1995) Somatosensory evoked potential spinal cord monitoring reduces neurologic deficits after scoliosis surgery: results of a large multicenter survey. Electroencephalogr Clin Neurophysiol 961: 6–11
76. Padberg AM, Bridwell KH (1999) Spinal cord monitoring: current state of the art. Orthop Clin North Am 303: 407–433, viii
77. Pechstein U, Cedzich C, Nadstawek J, Schramm J (1996) Transcranial high-frequency repetitive electrical stimulation for recording myogenic motor evoked potentials with the patient under general anesthesia. Neurosurgery 392: 335–343; discussion 343–344
78. Pechstein U, Nadstawek J, Zentner J, Schramm J (1998) Isoflurane plus nitrous oxide versus propofol for recording of motor evoked potentials after high frequency repetitive electrical stimulation. Electroencephalogr Clin Neurophysiol 1082: 175–181
79. Penfield W, Boldrey E (1937) Somatic motor and sensory representation in the cerebral cortex of amn as studied by electrical stimulation. Brain 60: 389–443
80. Peraud A, Meschede M, Eisner W, Ilmberger J, Reulen HJ (2002) Surgical resection of grade II astrocytomas in the superior frontal gyrus. Neurosurgery 505: 966–975; discussion 975–977

81. Pereon Y, Bernard JM, Fayet G, Delecrin J, Passuti N, Guiheneuc P (1998) Usefulness of neurogenic motor evoked potentials for spinal cord monitoring: findings in 112 consecutive patients undergoing surgery for spinal deformity. Electroencephalogr Clin Neurophysiol 1081: 17–23
82. Roberts TP, Ferrari P, Perry D, Rowley HA, Berger MS (2000) Presurgical mapping with magnetic source imaging: comparisons with intraoperative findings. Brain Tumor Pathol 172: 57–64
83. Romstock J, Fahlbusch R, Ganslandt O, Nimsky C, Strauss C (2002) Localisation of the sensorimotor cortex during surgery for brain tumours: feasibility and waveform patterns of somatosensory evoked potentials. J Neurol Neurosurg Psychiatry 722: 221–229
84. Roux FE, Boulanouar K, Ibarrola D, Tremoulet M, Chollet F, Berry I (2000) Functional MRI and intraoperative brain mapping to evaluate brain plasticity in patients with brain tumours and hemiparesis. J Neurol Neurosurg Psychiatry 694: 453–463
85. Roux FE, Boulanouar K, Ranjeva JP, Tremoulet M, Henry P, Manelfe C, Sabatier J, Berry I (1999) Usefulness of motor functional MRI correlated to cortical mapping in Rolandic low-grade astrocytomas. Acta Neurochir (Wien) 1411: 71–79
86. Sako K, Nakai H, Kawata Y, Takizawa K, Satho M, Yonemasu Y (1998) Temporary arterial occlusion during anterior communicating or anterior cerebral artery aneurysm operation under tibial nerve somatosensory evoked potential monitoring. Surg Neurol 493: 316–322; discussion 322–333
87. Sako K, Nakai H, Takizawa K, Tokumitsu N, Satho M, Katho M (1995) [Aneurysm surgery using temporary occlusion under SEP monitoring]. No Shinkei Geka 231: 35–41
88. Sala F, Krzan MJ, Jallo G, Epstein FJ, Deletis V (2000) Prognostic value of motor evoked potentials elicited by multipulse magnetic stimulation in a surgically induced transitory lesion of the supplementary motor area: a case report. J Neurol Neurosurg Psychiatry 696: 828–831
89. Schaller C, Zentner J (1998) Vasospastic reactions in response to the transsylvian approach. Surg Neurol 492: 170–175
90. Scheepstra GL, de Lange JJ, Booij LH, Ros HH (1989) Median nerve evoked potentials during propofol anaesthesia. Br J Anaesth 621: 92–94
91. Schramm J (1980) Clinical experience with the objective localization of the lesion in cervical myelopathy. Adv Neurosurg 8: 26–32
92. Schramm J (1989) Intraoperative monitoring with evoked potentials in cerebral vascular surgery and posterior fossa surgery. Neuromonitoring in surgery. Elsevier Science Publishers B.V. (Biomedical Division), pp 243–262
93. Schramm J, Hashizume K (1977) Somatosensory evoked potentials (SEP) in patients with peripheral, spinal and supraspinal lesions of the semsory system. Adv Neurosurg 4: 250–256
94. Schramm J, Hashizume K, Fukushima T, Takahashi H (1979) Experimental spinal cord injury produced by slow, graded compression. Alterations of cortical and spinal evoked potentials. J Neurosurg 501: 48–57

95. Schramm J, Jones SJ (1985) (eds) Spinal cord monitoring. Springer, Wien New York
96. Schramm J, Koht A, Schmidt G, Pechstein U, Taniguchi M, Fahlbusch R (1990) Surgical and electrophysiological observations during clipping of 134 aneurysms with evoked potential monitoring. Neurosurgery 261: 61–70
97. Schramm J, Krause R, Shigeno T, Brock M (1983) Experimental investigation on the spinal cord evoked injury potential. J Neurosurg 593: 485–492
98. Schramm J, Moeller AR (1991) (eds) Intraoperative neurophysiologic monitoring in neurosurgery. Springer, Berlin Heidelberg New York Tokyo
99. Schramm J, Mokrusch T, Fahlbusch R, Hochstetter A (1985a) [Intra- and perioperative acoustic evoked brain stem potentials in cerebellopontile angle operations]. Hno 3311: 495–498
100. Schramm J, Mokrusch T, Fahlbusch R, Hochstetter A (1988a) Detailed analysis of intraoperative changes monitoring brain stem acoustic evoked potentials. Neurosurgery 224: 694–702
101. Schramm J, Romstöck J, Thurner F, Fahlbusch R (1985b) Variance of latency and amplitude in SEPs monitored during spinal operations with and without cord manipulation. Spinal cord monitoring. In: Schramm J, Jones SJ (eds) Springer, Berlin Heidelberg New York Tokyo, pp 186–196
102. Schramm J, Romstock J, Watanabe E (1986) [Intraoperative monitoring of the spinal cord. Personal results and an inventory of the current status]. Z Orthop Ihre Grenzgeb 1246: 726–731
103. Schramm J, Taniguchi M (1991) Value of stable and changing somatosensory evoked potentials (SSEP) During aneurysm surgery. Intraoperative neurophysiologic monitoring in neurosurgery. In: Schramm J, Moller AR (eds) Springer, Berlin Heidelberg New York Tokyo, pp 151–161
104. Schramm J, Watanabe E (1988b) [Possibilities and limits of intraoperative functional assessment of the spinal cord]. Klin Wochenschr 66 [Suppl] 14: 19–26
105. Schramm J, Watanabe E, Strauss C, Fahlbusch R (1989) Neurophysiologic monitoring in posterior fossa surgery. I. Technical principles, applicability and limitations. Acta Neurochir (Wien) 981–982: 9–18
106. Schramm J, Zentner J, Pechstein U (1994) Intraoperative SEP monitoring in aneurysm surgery. Neurol Res 161: 20–22
107. Sindou M, Keravel Y (1988) Microsurgical procedures in the peripheral nerves and the dorsal root entry zone for the treatment of spasticity. Scand J Rehabil Med [Suppl] 17: 139–143
108. Skirboll SS, Ojemann GA, Berger MS, Lettich E, Winn HR (1996) Functional cortex and subcortical white matter located within gliomas. Neurosurgery 384: 678–684; discussion 684–685
109. Stapleton SR, Kiriakopoulos E, Mikulis D, Drake JM, Hoffman HJ, Humphreys R, Hwang P, Otsubo H, Holowka S, Logan W, Rutka JT (1997) Combined utility of functional MRI, cortical mapping, and frameless stereotaxy in the resection of lesions in eloquent areas of brain in children. Pediatr Neurosurg 262: 68–82

110. Symon L (1980) The relationship between CBF, evoked potentials and the clinical features in cerebral ischaemia. Acta Neurol Scand [Suppl] 78: 175–190
111. Tamaki T, Yamashita T, Kobayashi H (1972) Spinal cord monitoring. Jpn J Electroencephalogr Electromyograph 1: 196
112. Taniguchi M, Cedzich C, Schramm J (1993) Modification of cortical stimulation for motor evoked potentials under general anesthesia: technical description. Neurosurgery 322: 219–226
113. Taylor BA, Fennelly ME, Taylor A, Farrell J (1993) Temporal summation—the key to motor evoked potential spinal cord monitoring in humans. J Neurol Neurosurg Psychiatry 561: 104–106
114. Ubags LH, Kalkman CJ, Been HD, Drummond JC (1996) The use of a circumferential cathode improves amplitude of intraoperative electrical transcranial myogenic motor evoked responses. Anesth Analg 825: 1011–1014
115. Vecht CJ, Avezaat CJ, van Putten WL, Eijkenboom WM, Stefanko SZ (1990) The influence of the extent of surgery on the neurological function and survival in malignant glioma. A retrospective analysis in 243 patients. J Neurol Neurosurg Psychiatry 536: 466–471
116. Vinas FC, Zamorano L, Mueller RA, Jiang Z, Chugani H, Fuerst D, Muzik O, Mangner TJ, Diaz FG (1997) [^{15}O]-water PET and intraoperative brain mapping: a comparison in the localization of eloquent cortex. Neurol Res 196: 601–608
117. Woodforth IJ, Hicks RG, Crawford MR, Stephen JP, Burke DJ (1996) Variability of motor-evoked potentials recorded during nitrous oxide anesthesia from the tibialis anterior muscle after transcranial electrical stimulation. Anesth Analg 824: 744–749
118. Woolsey CN, Erickson TC, Gilson WE (1979) Localization in somatic sensory and motor areas of human cerebral cortex as determined by direct recording of evoked potentials and electrical stimulation. J Neurosurg 514: 476–506
119. Wyllie E, Luders H, Murphy D, Morris H 3rd, Dinner D, Lesser R, Godoy J, Kotagal P, Kanner A (1990) Intracarotid amobarbital (Wada) test for language dominance: correlation with results of cortical stimulation. Epilepsia 312: 156–161
120. Yasargil MG (1994) Microneurosurgery. Georg Thieme Verlag, Stuttgart New York
121. Yingling CD, Ojemann S, Dodson B, Harrington MJ, Berger MS (1999) Identification of motor pathways during tumor surgery facilitated by multichannel electromyographic recording. J Neurosurg 916: 922–927
122. Zentner J (1989) Noninvasive motor evoked potential monitoring during neurosurgical operations on the spinal cord. Neurosurgery 245: 709–712
123. Zentner J (1991) Motor evoked potential monitoring during neurosurgical operations on the spinal cord. Neurosurg Rev 141: 29–36
124. Zentner J, Hufnagel A, Pechstein U, Wolf HK, Schramm J (1996a) Functional results after resective procedures involving the supplementary motor area. J Neurosurg 854: 542–549

125. Zentner J, Meyer B, Stangl A, Schramm J (1996b) Intrinsic tumors of the insula: a prospective surgical study of 30 patients. J Neurosurg 852: 263–271
126. Zhou HH, Kelly PJ (2001) Transcranial electrical motor evoked potential monitoring for brain tumor resection. Neurosurgery 485: 1075–1080; discussion 1080–1081

Functional Neuronavigation and Intraoperative MRI

C. Nimsky, O. Ganslandt, and R. Fahlbusch

Department of Neurosurgery, University Erlangen-Nürnberg, Erlangen, Germany

With 9 Figures

Contents

Abstract

Our concept of computer assisted surgery is based on the combination of intraoperative magnetic resonance (MR) imaging with microscope-based neuronavigation, providing anatomical and functional guidance simultaneously. Intraoperative imaging evaluates the extent of a resection, while the additional use of functional neuronavigation, which displays the position of eloquent brain areas in the operative field, prevents increasing neurological deficits, which would otherwise result from extended resections.

Up to mid 2001 we performed intraoperative MR imaging using a low-field 0.2 Tesla scanner in 330 patients. The main indications were the evaluation of the extent of resection in gliomas, pituitary tumours, and in epilepsy surgery. Intraoperative MR imaging proved to serve as intraoperative quality control with the possibility of an immediate modification of the surgical strategy, i.e. extension of the resection. Integrated use of functional neuronavigation prevented increased neurological deficits. Compared to routine pre- or postoperative imaging being performed with high-Tesla machines, intraoperative image quality and sequence spectrum could not compete. This led to the development of the concept to adapt a high-field MR scanner to the operating environment, preserving the benefits of using standard microsurgical equipment and microscope-based neuronavigational guidance with integrated functional data, which was successfully implemented by April 2002. Up to the end of 2002, 95 patients were investigated with the new setup. Improved image quality, intraoperative workflow, as well as enhanced sophisticated intraoperative imaging possibilities are the major benefits of the high-field setup.

Keywords: Epilepsy surgery; functional imaging; functional MR imaging; functional neuronavigation; glioma surgery; intraoperative imaging; high-field MR imaging; low-field MR imaging; magnetoencephalography; pituitary tumour surgery.

Introduction

Intraoperative imaging has attracted increasing interest in recent years. The ability to objectively determine the extent of tumour removal during surgery is highly advantageous. If the resection is incomplete, one can attempt to remove the tumour residues that were missed during the same operation. In contrast to subjective estimation by the surgeon, intraoperative imaging allows for objective evaluation, thus acting as quality control during surgery.

In addition to intraoperative imaging, an integral part of our concept of computer aided surgery is the possibility of using neuronavigation simultaneously. In standard neuronavigation, also known as frameless stereotaxy, the real space of the surgical field is registered to the 3-dimensional image space, which is based on anatomical data only. We prefer the application of microscope-based neuronavigation, where the extent and localization of a tumour is superimposed on the microscope field of view through contours. Aside from standard neuronavigation based on anatomical information only, which has become a routine tool in many neurosurgical departments, we integrate information from preoperative magnetoencephalographic (MEG) measurements and functional magnetic resonance imaging (fMRI) to define the localization of eloquent brain areas, such as the motor and speech areas, in individual patients [28, 29, 53, 54, 68, 80]. This method of functional neuronavigation allows more thorough resection of tumours at risk zones with lower morbidity [28, 80].

Functional Neuronavigation

Besides anatomical landmarks and patterns, different functional imaging modalities have been used to define the central region as part of the preoperative assessment, especially MEG [27, 48, 69, 87], fMRI [2, 5, 19, 20, 26, 42, 45, 57, 63, 70, 90, 92, 121], and PET [5, 26, 89] and were integrated into MR-images to allow the preoperative localization of the somatosensory cortex. The increasing application of neuronavigation has allowed the integration of this functional information into the operating room, leading to so-called functional neuronavigation. This was demonstrated initially with a pointer-based navigation system for MEG [29, 91, 94] and later in the same manner for fMRI [64, 80, 101].

The method of MEG-based functional neuronavigation has been applied routinely in surgery around the motor strip in our department for the last six years in more than 140 patients (Fig. 1). But this approach is limited since the availability of MEG is restricted and not amenable to common use in the neurosurgical community. Thus we additionally made efforts to integrate fMRI data into the functional neuronavigation setup as well. fMRI can detect subtle regional changes in blood flow and levels of deoxygenated hemoglobin associated with increased neuronal activity. In echo-planar imaging technique, image contrast is mainly generated by changes in deoxyhemoglobin levels known as blood oxygenation level-dependent (BOLD) contrast [56, 82, 107]. fMRI can be performed with standard MRI equipment such as a 1.5 T MR scanner, which is widely available, making fMRI relatively easily available in comparison to MEG. In accordance with one of our previous papers [109] we reported in a pilot study of eight subjects and six patients that the activation sites for fMRI and MEG sensorimotor activity may differ [80]. This did not lead to a mislocalization of the central sulcus in any case. However, it is conceivable that in patients with large space occupying lesions and complex vessel geometry, fMRI and MEG localizations might be ambiguous. Therefore, we investigated a larger number of patients to check the clinical applicability of fMRI and MEG for the delineation of the central sulcus in presurgical functional imaging. Special interest was laid upon the different spatial activation sites of the primary sensorimotor cortex in MEG and fMRI (see below) [54].

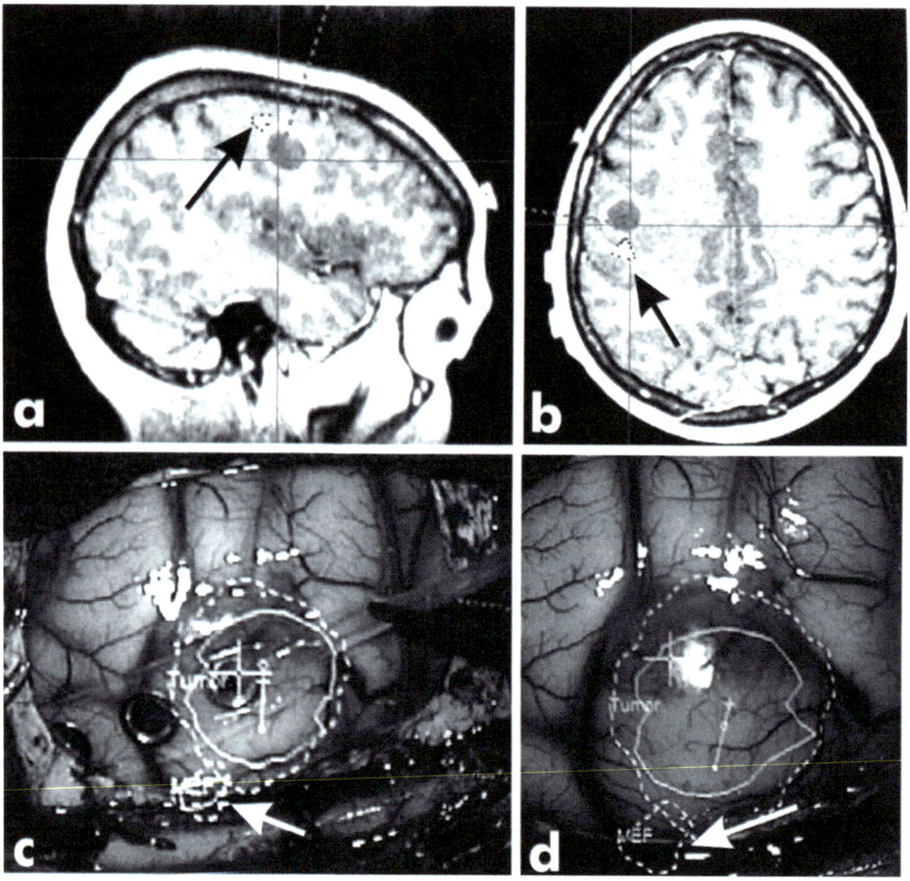

Fig. 1. MEG-based functional neuronavigation in a 31-years old female patient with a left precentral glioblastoma. A white pyramid represents the motor evoked fields (*MEF*) (black arrow in the navigation images a/b and white arrow in the microscope views c/d, where the MEF is displayed in the microscope view besides the tumour contour using heads-up display technique (a/b: sagittal and coronal view of the navigation screen, c: the identification of the motor area is verified by phase reversal, d: enlarged view without strip electrode)

Magnetoencephalography

MEG was recorded using a 2 × 37 channel biomagnetometer (MAGNES II, Biomagnetic Technologies Inc., San Diego, USA) placed inside a magnetically shielded room. Each patient's headshape was scanned relative to the MEG sensor using an electromagnetic 3-D-digitizer (Isotrak, Polhemus, Vermont, USA). For identification of the central region, different activations were evoked. Motor activation was performed by brisk flexion movements of the index finger contralateral to the lesion every 3 to 5 seconds. Sensory activity was evoked with a pneumatically driven tactile stimulator applying tactile stimuli with an interstimulus interval of 800 ms to the index finger contralateral to the lesion using a fingerclip. The MEG sources were localized applying a single current dipole model with a locally fitted sphere as head model. The anatomical source location was performed with a contour fit algorithm, fitting the digitally scanned head-shape including the MEG dipole to the individual MR dataset [52]. Recently, MEG-data representing language related areas were included in the functional neuronavigation setup, calculated using the method of current density distribution by spatial filtering [53, 68].

Functional MRI

fMRI was performed with a 1.5 T MR scanner (Magnetom Symphony, Siemens AG Medical Solutions, Erlangen, Germany) using Echo Planar Imaging (EPI) technique. Functional images were measured using a T2* weighted EPI sequence (TE: 62 ms, TR: 114 ms, flip angle: 90°, slice thickness: 3 mm, FOV: 200 mm, matrix: 64×64, interpolated 128*128) resulting in an acquisition time of 1.82 s for each fMRI volume. Anatomical images were collected in identical positions using a T1 weighted spin echo sequence (TE: 15 ms, TR: 450 ms, flip angle: 90°, slice thickness: 3 mm, FOV: 200 mm, matrix: 256×256). The fMRI experiments consisted of 3 activation and 3 baseline (rest) periods. During each activation and rest period, 10 EPI volumes were collected, yielding a total of 60 functional volume data sets. Motor activation was performed by iterative clenching of the hand contralateral to the lesion once per second. Sensory activity was evoked using the identical stimulation paradigm as in the MEG experiments. Functional activation maps were calculated with a cross correlation analysis [4, 120] between measured and expected activation time course for each voxel using the AFNI (Analysis of Functional NeuroImages) software package [21] omitting the first two EPI volumes of each period.

On the day before surgery, 8 to 10 adhesive skin fiducials were attached on the patient's head in a scattered pattern for referencing. For anatomical neuronavigation, a T1-weighted 3-D gradient echo sequence (FLASH: fast low angle shot, TE: 7.0 ms, TR: 16.1 ms, flip angle: 30°, slab 168 mm, 112 slices, FOV: 250 mm, matrix: 256×256) was measured with a 0.2 T Magnetom Open (Siemens AG Medical Solutions, Erlangen, Germany). All the functional data were registered with the anatomical 3-D dataset using a contour fit algorithm [52]. The functional data were displayed in the anatomical dataset by markers placed into the MR images. A pyramid or cube with white or black intensity accordingly represented a functional modality. The patient was registered with the navigation microscope (MKM or NC4-microscope with SMN-system, navigational software STP4.0; Zeiss, Oberkochen, Germany). The contours of the lesion, the predefined surgical approach and optionally also the markers for functional data, were displayed using the heads-up display of the microscopes.

Clinical Impact of Functional Imaging

Modern brain mapping techniques such as fMRI, MEG, and PET, which were recently mostly applied in a research setting, have gradually emerged in the clinical armamentarium to supplement clinical and radiological findings for preoperative planning of brain surgery. In a retrospective study we focused on how the decision to resect a glioma was influenced by MEG. In a time period of 5 consecutive years we have investigated every patient proposed for surgery, who harboured a lesion adjacent to an eloquent brain area. Altogether 191 patients were examined, 119 of them harboured supratentorial gliomas. About every forth patient (26.8%) revealed a significant possible risk of postoperative neurological morbidity according to MEG and thus was not considered to be a candidate for surgery. This corresponds to published data where 12 out of 40 investigated patients (30%) with tumours and vascular malformations underwent non-surgical therapy according to the MEG results [40]. When functional data was used in combination with frameless stereotactic devices the postoperative morbidity was as low as 2.3%. Overall morbidity however was 6.8%. These data reflect the beneficial effects of functional neuronavigation in comparison to data of other studies with morbidity rates varying from 6 to 31.7% [1, 8, 16, 17, 23]. These figures can also be interpreted as a result of a more careful patient selection through the help of preoperative brain mapping. Preoperative identification of eloquent brain areas has an impact on the risk evaluation in glioma surgery, where functional neuronavigation reduces the risk for postoperative neurological deficits. Besides identification of the motor strip, the localisation of language areas is of great clinical impact. Initial results based on smaller patient numbers indicate positive effects as well for MEG-based identification of Broca and Wernicke areas [53, 68].

Comparing MEG/fMRI

Interestingly, in all comparative studies, both MEG as well as fMRI measurements were able to locate the motor cortex. MEG is measuring neuronal activity and fMRI is measuring changes in blood oxygenation levels which are not directly located at the same place. Since the fMRI BOLD signal change not only shows activation in the parenchyma of the activated cortical structures but also in the draining veins [105] it is difficult to differentiate between activated cortex and venous drainage especially in patients with morphological malformations. This is important when these data are not just used in a preoperative assessment, but are transferred into the operating room to localize the central region or other eloquent cortex areas (e.g., in language mapping) in 3-D space. In studies which compare MEG directly with fMRI, both could correctly identify the anatomical cortex, but distances of more than 10 mm between the major MEG and fMRI activity were measured [109]. Others have even reported a distance of up to 16 mm [7], which reflects the different underlying substrates being measured. A previous study investigating voluntary movements with high resolution EEG and fMRI in healthy volunteers yielded exactly the same difference of 10 mm as in our study between movement related activation sites [3]. Although we used slightly different activation patterns for the motor task and an identical paradigm for the sensory stimulation the difference between maximum MEG and fMRI activation was larger for the somatosensory stimulation (15 mm) than for the motor task (10 mm). In the motor task MEG and fMRI activations projected onto the same neuroanatomical structure the hand knob of the precentral gyrus. Although the motor fMRI activation was significantly more posterior (5 mm) and more lateral (3 mm) adjacent to the hand knob in the central sulcus following the topology of the draining veins in normal anatomy (Fig. 2).

Depending on the distance and direction in 3-D space a difference of up to 15 mm between the major MEG and fMRI-activity may result in fatal errors when trying to use these data in the operating theatre. Therefore some groups emphasize the dual use and integration of both modalities [6, 41, 95]. Due to our experience the spatial difference between fMRI activation and MEG localization did not result in an erroneous estimation of the central sulcus in any case. Thus we indicate no preference concerning the intraoperative usability of either modality, MEG or fMRI. For validation we measured the phase reversal of somatosensory evoked potentials and could confirm the identification of the central sulcus in all cases which were operated with neuronavigation [96].

There are sources for inaccuracy in the transformation of the anatomically distorted T2*-weighted images, which have to be matched on the anatomical T1-weighted images to be used in neuronavigation [110]. Especially at the brain air barrier and at the skull base, both not so relevant in fMRI around the central sulcus, distortions are considerable and have to be corrected by special algorithms describing a field map. Furthermore, fMRI data may be compromised by motion artefacts and not only by measuring the BOLD effect in the capillary vessels of the activated cortex, but also measuring activity by the contributions of large vessels [6, 42, 105]. The accuracy of the functional method itself concerning spatial resolution is in the range of about 3–5 mm for MEG [58] and at least 3 mm

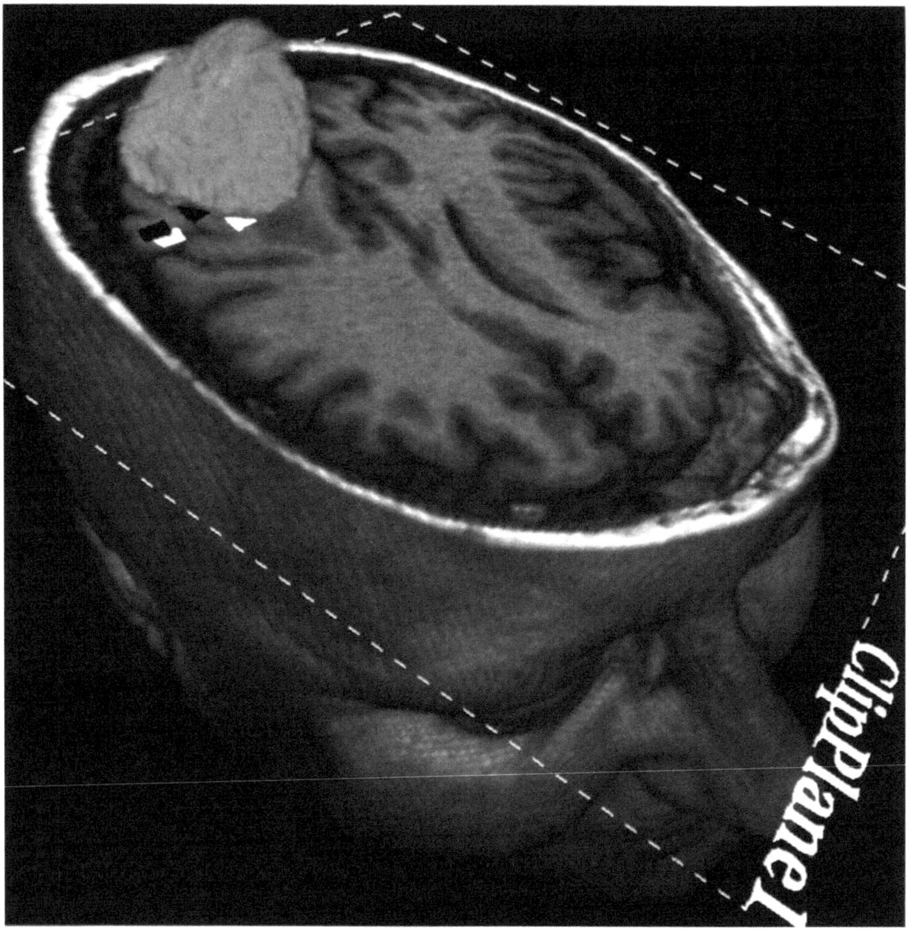

Fig. 2. Volume rendering of a right parietal meningioma. Markers representing functional data (white: MEG, black: fMRI) delineate the central sulcus. The precentral gyrus is identified by the white triangle (motor evoked fields MEG); the motor-fMRI activity (black triangle) is localized in the central sulcus. The postcentral gyrus is identified by both square markers, which are visualized more laterally

for fMRI, resulting from the voxel size of 3 mm of the T2* images. The overall accuracy is influenced by further factors: the accuracy of the matching algorithm, being below 2 mm for the contour fit we used [52] and image distortion of the anatomical MR-images [110]. Considering that fMRI equals MEG in correct identification of the central region, as is our impression by the current data, and as it is stated by others [101], it will help to initiate a common application of functional neuronavigation due to wide availability of fMRI in contrast to the restricted availability of MEG. Integration of language mapping with MEG and fMRI as well as

MEG- and fMRI-based localization of foci in epilepsy surgery [104] with integration of these data into functional neuronavigation will open new therapeutic options.

Intraoperative Imaging

The first attempts to use ultrasound and computed tomography during a neurosurgical operation were made in the 1980s. At that time, though, image quality was quite limited, so that neither method gained wide use in neurosurgery. Since then, MRI has become the method of choice for pre-operative diagnosis of brain tumours and epilepsy. However, the closed-bore design and the strong fringe fields of the first MR scanners prevented their use in the operating room. Then, in the mid-1990s, the concept of intraoperative imaging experienced a renaissance with the development of open MR systems. For the first time, these systems made it possible to use MR for intraoperative imaging.

The 0.2 T Magnetom Open (Siemens AG Medical Solutions, Erlangen, Germany), initially designed solely for diagnostic purposes, was adapted to surgical use in 1995 in close cooperation between the Neurosurgical departments of the Erlangen and Heidelberg Universities and Siemens [108, 113]. In addition to intraoperative imaging, an integral part of our concept was the possibility of simultaneously using microscope-based neu-ronavigation with integrated functional data from MEG or fMRI. However, the only navigation microscope available at that time, the robotic MKM (Zeiss, Oberkochen, Germany), had to be spatially separated from intraoperative imaging. As a result, the original concept required transportation of the patient during surgery. For intraoperative imaging, the patient had to be moved on an air-cushion table, with the open skull, over a distance of 5 meters from an adjacent operating room reserved for the actual surgery to the MR scanner.

The most important indications for intraoperative imaging included: resection control during surgery of gliomas [9, 11, 51, 99, 117], hormonally inactive pituitary tumours [24, 65, 88], and epilepsy surgery [13, 14, 46, 103]. Intraoperative MRI also allowed compensation for brain shift [71, 74, 78, 116]. We found that the support of intraoperative MR allowed for more thorough resections with less morbidity, while simultaneous use of functional neuronavigation allowed for conservation of eloquent brain areas despite extended resections.

Low-Field MR Imaging

As demonstrated by other groups [9, 11, 60, 113, 122] the use of MR scanners in the operating environment is safe and reliable, as well as applicable to neurosurgical procedures, even if these procedures have to a certain extent to be adapted to the MR environment. Even intraoperative patient transport, which was necessary in our primary conceptual design combining intraoperative imaging with microscope-based neuronavigation, did not cause any problems. The development of new navigation microscopes, which can be used in the fringe field of the MR scanner [79], allowed us to abandon the cumbersome and time-consuming intraoperative patient transport. Then it was possible to integrate full microscope-based neuronavigational support to the concept of surgery in the fringe-field, which was initially applied in transsphenoidal surgery [24, 108] and afterwards proposed for open cranial surgery [98].

Comparing the low-field MR systems available for intraoperative use, up to now there are two main concepts for intraoperative MRI: placing the patient directly inside the scanner, as implemented in the systems which are specially designed for intraoperative use, like the Signa SP 'double-doughnut' scanner (General Electric Medical Systems, Milwaukee, Wisconsin, USA) [10] offering nearly real-time imaging but with the drawback of restricted patient access. The 0.12 T PoleStar N-10 (Odin Technologies, Yokneam, Israel) has a similar design, where the disk-shaped magnets are placed below the operating table and are moved upwards for scanning [31, 100]. On the other hand, standard diagnostic scanners adapted to the operating environment necessitate some kind of patient transport. Either the patient is operated in the fringe field of the scanner, with the patient placed on the movable MR tray [79], or on a rotating operating table [60]. Alternatively, surgery is performed in a separate operating room, like in the 'twin-operating room' setup, necessitating a longer intraoperative patient transport [108, 113], which has been implemented with the 0.3 T AIRIS II scanner (Hitachi Medical corporation, Tokyo, Japan) [11].

Clinical Experience

Between March 1996 and July 2001 we performed intraoperative MRI in 330 patients. Histopathological examinations revealed pituitary adenoma in 61, craniopharyngioma in 26, glioma in 106, cavernoma in 18, and other diagnoses in 54 patients. The remaining 65 patients were either non-lesional cases in whom resective or disconnective epilepsy surgery was carried out ($n = 40$), or procedures in which intraoperative MR imaging was used as an online control during catheter or electrode placements ($n = 25$). Among the 330 procedures there were 240 craniotomies, 59 transsphenoidal approaches, and 31 burr hole procedures [75, 81].

We did not observe any adverse effects of intraoperative imaging on the postoperative course of the patients. Intraoperative patient transport (applied in 166 patients) did not cause any harm. On average a delay of some 10 minutes occurred until imaging could start. The same time was necessary until surgery could be continued after imaging, so that intra-operative patient transport was time consuming and cumbersome. In the patients being operated directly on the MR-tray at the 5 G line ($n = 164$), scanning could be started on average after one minute. The overall complication rate in respect to wound infection ($n = 3$) and re-bleeding ($n = 1$) was a bit lower than the average number in our department.

Pituitary Tumour Surgery

Among 59 patients who underwent transsphenoidal pituitary surgery (52 pituitary adenomas, 7 craniopharyngiomas), image evaluation was limited in 16 cases either due to drilling artefacts with metal debris disturbing image interpretation in the sella region, or blood in the resection cavity mimicking tumour remnants. With increasing experience in image interpretation, as well as repeated intraoperative scanning and the application of T2-weighted sequences further evaluation was possible in most of the cases. In 30 patients a repeat inspection of the surgical field was performed. This led to further tumour removal in 17 out of the total 59 patients. Among a subgroup ($n = 44$) of resectable pituitary macroadenomas with a distinct suprasellar extension the further resection of tumour increased the rate of complete removal from 45% (21/44) to 75% (33/44). In the remaining patients an extensive supra- and parasellar tumour extension did not allow complete transsphenoidal removal, necessitating surveillance, secondary transcranial surgery, or radiation therapy. Only in one patient was there a transient postoperative visual impairment, caused by a secondary empty sella after complete tumour removal. In none of the 59 patients was endocrine function disturbed due to the surgical intervention [24].

In transsphenoidal pituitary surgery intraoperative MRI provides in cases of residual tumour the possibility for further tumour removal, thus increasing the rate of complete tumour removals significantly [65, 88]. The removal of the suprasellar tumour portion could be evaluated reliably. However, low-field MRI seemed not to be suitable for the evaluation of the removal of intrasellar microadenomas. Also the interpretation of the extent of removal of parasellar, i.e. intracavernous tumour parts, was limited. Image artefacts caused by metal debris from drilling or by a blood accumulation in the resection cavity were further challenges. In case of incomplete tumour removal, intraoperative MRI enabled additional therapy planning at a much earlier state, since normally further diagnostics studies have to be delayed for two to three months after surgery, to obtain artefact-free postoperative images [24].

Glioma Surgery

In glioma surgery intraoperative low-field MRI showed remaining tumour in a high percentage (63%). In 42% of these patients we continued tumour resection and tried to remove as much tumour as possible: in more than half of them neuronavigation was updated by the intraoperative image data to localize the tumour remnants. Further tumour removal increased the gross total removal rate especially in the low-grade gliomas. However in some patients despite further tumour removal no complete resection was performed since some tumour remnants invaded eloquent brain areas, which were identified by functional neuronavigation [73].

In 34 patients, i.e. in 51% (34 out of 67) of the patients with incomplete resection, the resection cavity was inspected again. In 28 (42% of the patients with tumour remnants, i.e. 26% of all glioma patients) of them the extent of the resection could be expanded based on the results of intraoperative MR scanning. This increased the rate of gross total removal in the WHO grade I tumours ($n = 15$) from 87% to 100%. In the grade II tumours ($n = 37$) further tumour removal was performed in 19 patients. In 10 of them, a finally complete tumour removal could be achieved. Thus, the gross total removal rate in the grade II increased from 27% to 54%. In the remaining 9 patients, despite further tumour removal, complete resection was not possible due to small tumour remnants infiltrating eloquent brain areas. In the high grade astrocytomas the extent of resection was enlarged in 7 patients only. This resulted in an increased gross total removal of 55% versus 41% in the grade III and 22% versus 19% in the grade IV patients. In 8 of the low grade gliomas (WHO grade II) the resection could not be extended primarily, because eloquent brain areas were infiltrated. In the majority (82%: 32 out of 39) of the high grade gliomas (grade III & IV), where intraoperative imaging had depicted incomplete removal, it was the policy above all to avoid new neurological deficits.

45 glioma patients presented preoperative neurological deficits, which resolved at least partially in 32, worsened only in two patients. In the group of patients without preoperative neurological deficits ($n = 61$) three developed a postoperative neurological deterioration (twice paresis, once aphasia). In total, in 5 glioma patients (4.7%) a neurological postoperative deficit occurred, while only in two of them functional neuronavigation had been applied.

We used functional neuronavigation if the tumour was located near eloquent areas to preserve neurological function [28, 53, 54, 80], as a limitation for total resection. In 66 patients in addition to anatomical data, functional data from MEG ($n = 64$) or fMRI ($n = 15$), identifying either motor, sensory, or language related cortex, were integrated into the neuronavigation which allowed identification and avoided damaging of these eloquent brain areas during surgery. The identification of these cortical areas just at the beginning of tumour removal was not affected by brain shift, since the functional markers from MEG or fMRI were used to identify the eloquent cortex just after dural opening. The application of functional neuronavigation was associated with low postoperative morbidity. Only in two of the 66 patients (3%) in whom functional neuronavigation was applied a persistent neurological deterioration occurred, which in both patients fortunately improved in the further postoperative course. Anatomical and functional neuronavigation were used as guidance to identify relevant structures. Intraoperative MR imaging allowed a delineation of the extent of the resection, so that the combination of both allowed a maximum possible resection with least neurological deficits, while taking incomplete tumour removal into account, when eloquent brain areas were infiltrated.

Besides the possibility of evaluating the completeness of a tumour resection, intraoperative MRI not only allows the delineation and visualization of the extent of brain shift, which results in great inaccuracies of neuronavigation systems in an ongoing operation, but it also offers the possibility to compensate for the effects of brain shift. This could be achieved reliably by an intraoperative update of the neuronavigation system with the intraoperative 3-D image data [74, 78, 116]. Out of the 28 patients, in whom resection was continued after intraoperative imaging, an intraoperative update of neuronavigation was performed in 18 (64%, 17% of the total 106 gliomas). The update procedure was performed only when it was difficult to localize the remaining tumour in the surgical field. In all of these cases the intraoperative update of neuronavigation was technically successful. Image transfer, new segmentation, and re-referencing added another 15 to 20 minutes to the whole procedure. The registration error, calculated as the root mean square error was low (mean: 0.85 mm, standard deviation: 0.44 mm, maximum error: 2.3 mm), resulting in a reliable navigation with high accuracy that quickly led to the suspicious areas. In 3 patients the histopathological examination revealed no tumour in the tissue that was removed additionally due to intraoperative imaging. Complete tumour removal was confirmed by early scanning in all three of them.

One of the general limitations of glioma surgery is that by histological definition a 100% tumour removal is not possible. Furthermore, the extension of the tumour into eloquent brain areas will also result in only partial removal. According to our experience with intraoperative MR evaluation of glioma removal, we are not sure about the long-term benefit of intraoperative MRI in surgery of high grade gliomas, though first reports published on this topic claim a benefit even in high grade tumours [9, 51, 117]. In the high-grade tumours the differentiation between tumour remnant, surgically induced imaging changes, oedema, and normal brain may be very difficult. Controversial reports on life expectancy in high grade gliomas emphasize that life expectancy depends more on low postoperative deficits than 'macroscopic' total tumour removal [55]. On the other hand in the low grade gliomas [49], at a microscopic level complete resection also will not be possible, but since survival of patients seems to be correlated with the extent of tumour resection [72], we believe that especially surgery of these tumours will benefit more from intraoperative imaging [99]. Up to now, it is still too early to determine the effects on life expectancy in these patients, but it can be stated that more radical resections are possible without increasing morbidity, especially when intraoperative imaging is supported by the use of functional neuronavigation. Long-term survival studies are still needed.

Epilepsy Surgery

Epilepsy surgery is another indication for intraoperative MRI, where it enabled reliable evaluation of the extent of resection or disconnection procedure. It not only allowed evaluation of the completeness of the resection in the lesional cases (e.g. glioma, cavernoma), but also in the non-lesional cases the exact extent of a callosotomy ($n = 6$) or the size and shape of a tailored temporal lobe resection ($n = 34$) in its neocortical and hippocampal extent could be documented [13, 14]. Increased knowledge of structure-function relationships as partially defined by intraoperative imaging may reduce adverse neuropsychological sequels of epilepsy surgery in the future [13, 14, 46, 103]. Frameless-stereotactic techniques have now been applied in epilepsy surgery for more than 10 years [85, 86, 93]. We used neuronavigation for the intraoperative localization and delineation of the size of a lesion, thus minimizing the surgical approach. Especially in cases of small, deep-seated cavernomas, neuronavigation proved to be very helpful in achieving a minimally invasive approach, thus minimizing damage to normal brain, especially near functionally eloquent brain areas. In callosotomies the extent of sectioning is well documented. Additionally, the placement of depth electrodes and hippocampal electrodes is facilitated, as well as the positioning of electrodes for electrocorticopgraphy can be documented for spatial correlation.

Limitations and Imaging Alternatives

Some of the limitations of intraoperative MR imaging are due to the MR technique itself. In contrast to computed tomography, MR imaging per se poses the risk of geometrical distortion depending on the sequences applied, gradient field nonlinearities, and field inhomogeneities. Inaccuracies result when MR images are used in a stereotactic coordinate system without minimizing the spatial distortion [110]. Furthermore, bony structures are better visualized using computed tomography. Of course there are limitations in image resolution, especially when low-field systems are applied, so e.g. the thalamic nuclei can not be differentiated with the low-field magnets.

Besides the limitations of our special setup implementing low-field intraoperative MR imaging necessitating either intraoperative cumbersome patient transport or the lack of a possibility for head adjustments during surgery, when operating at the 5 G line with the Magnetom Open scanner [79], the use of intraoperative low-field MR imaging has to address the problem of image quality obtained by adapting the low-field MR scanners to the operative environment. Intraoperative image quality was sufficient to evaluate tumour resection in the majority of cases. Image quality of the intraoperative low-field systems can of course not compete with the quality of high-field scanners. But image quality of the low-field systems in intraoperative use is in the majority of cases sufficient to evaluate the extent of a tumour resection. In cases with residual tumour even images, which were severely disturbed by artefacts, could advice the neurosurgeon to continue the resection.

Adequate evaluation of intraoperative MR images may be disturbed by artefacts. The external factors causing artefacts range from anaesthesia devices to the overall electrical environment. Also the specific design of separable coils may be a source for additional pitfalls, since it is mandatory that the electrical contacts in the adapters, which are near the operative field, are shielded from contamination with fluids. Imaging artefacts due to technical failures and different external influences, such as the anaesthesia equipment, prevented sufficient image evaluation in 12 patients. In 5 of them a short circuit in the adapters of the unsterile to the sterile part of the intraoperative coil occurred or an instability of the magnetic field due to a cooling failure prevented tuning and image generation. In the other 7 patients shielding defects in electrical devices resulted in severe image artefacts, disturbing image interpretation. Even in case of poor image quality, which occurred in a total of 6.5% of all patients remaining tumour, when present, could be identified.

Another source for artefacts was the operative field itself, i.e. artefacts caused by metal drill debris or the air-brain surface. Artefacts, especially at the resection border, when contrast media was applied, could be diagnosed, by comparing the intraoperative scans with the preoperative scans in a side-by-side display. Exact knowledge of the neurosurgical procedure and situation was necessary to evaluate the intraoperative images. It was important to know e.g. whether special hemostatic materials were applied or 'contrast media loaded'-blood could be detected at the resection border or in the resection cavity [50].

Comparing intraoperative MR imaging regardless of which system is used with other imaging modalities, MR presently without doubt is the most advanced intraoperative imaging modality available. Ultrasound [36, 59, 119] on the other hand provides real-time data at lower costs, but despite recent advances in image quality, soft tissue contrast and signal to noise ratio in MR imaging are clearly superior to ultrasound. The difficulties of 3-D orientation of ultrasound imaging may be resolved soon by the integration of ultrasound in neuronavigation systems, and by the application of 3-D ultrasound transducer abilities, which possibly can also compensate for the effects of brain shift [38, 43]. High-end ultrasound systems are not only able to visualize tissue, but also vessels and flow in real-time and therefore will be an important imaging modality addition in the operating room setup in the future [97].

Intraoperative CT, either with fixed systems [61, 62, 83, 84], as originally introduced, or on tracks [44], or the recently developed system of mobile CT [15, 30, 66] does not require special instruments, except for radiation-compatible head fixation. However, a low soft tissue resolution, as well as missing free slice orientation, and the need for x-ray protection requirements, are major drawbacks compared to intraoperative MR imaging. Intraoperative CT has its advantages in the evaluation of bony structures, making it highly suitable for imaging in spine surgery [39].

High-Field MR Imaging

The image quality generated by various intraoperative low-field systems, however, does not compare well with the diagnostic image quality obtained with high-field scanners, which are used in the routine neuroradiological diagnostics. In addition, these scanners did not allow for other examination modalities, such as functional imaging.

Only two centres have previously attempted to use high-field MR scanners in the operating room. The group of W. Hall and C. Truwit in Minneapolis adapted a standard Philips 1.5 T scanner to the operating room environment [33, 35]. At Calgary, G. Sutherland's group, together with IMRIS Co., took an alternate path and designed a 1.5 T scanner specifically for the requirements of an operating room. In their design, the magnet is mounted on the ceiling and moved into the appropriate imaging position [47, 111]. The active magnetic shielding of modern high-field magnets results in the 5 G zone being relatively close to the scanner, so that it appeared possible to integrate the scanner into the operating room. In this way, our concept of intraoperative MRI with integration of microscope-based neuronavigation in the low magnetic field can now be applied with high-field magnets [76]. The operating room was appropriately reconstructed between August 2001 and March 2002. We have been able to operate on the first patients using intraoperative high-field MRI and integrated microscope-based neuronavigation since the end of April 2002.

Integration of a High-Field Scanner into the Operating Room

The 1.5 T Magnetom Sonata Maestro Class scanner is installed in an operating room with radio-frequency shielding. A rotatable surgical table is adapted to the scanner to allow for a special surgical-MR tabletop. The surgical table can be locked into various positions. The principal surgical position is at 160°, with the patient's head at the 5 G line (Fig. 3). As soon as the rotating mechanism has been locked, the height of the table, the angle of tilt as well as the lateral tilt can be modified. The table movements

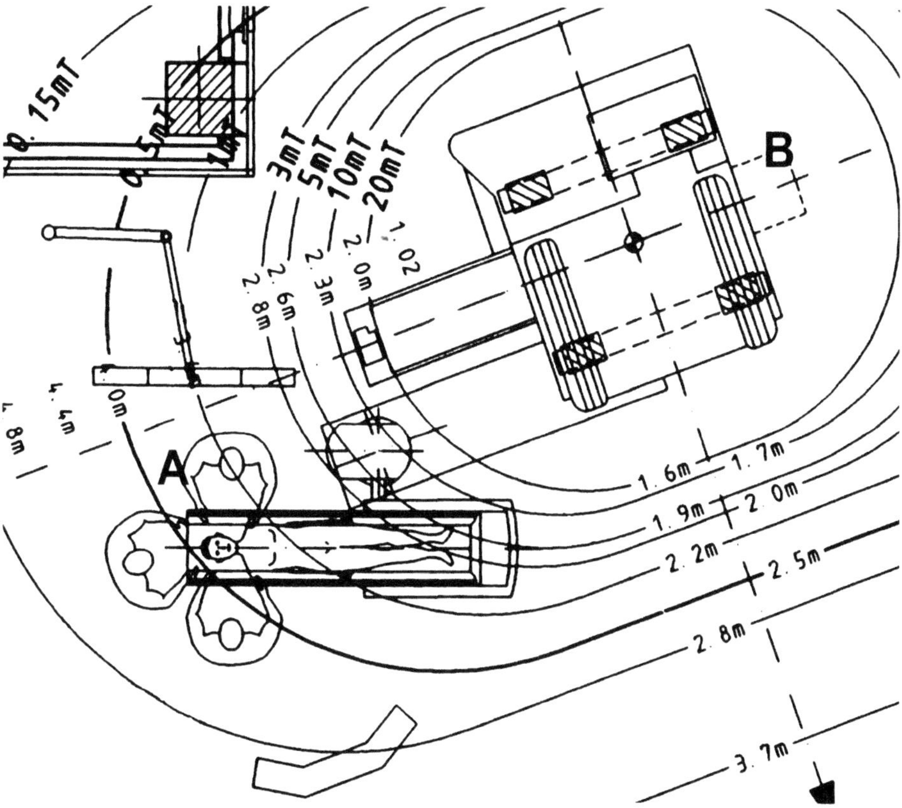

Fig. 3. Operating room setup with the high-field MR scanner. There are two operating scenarios possible: at the 5 G line with standard microsurgical equipment supported by functional microscope-based neuronavigation (position A). Alternatively surgery can be performed in the high-magnetic field with fully MR-compatible instruments (e.g. for biopsies, position B)

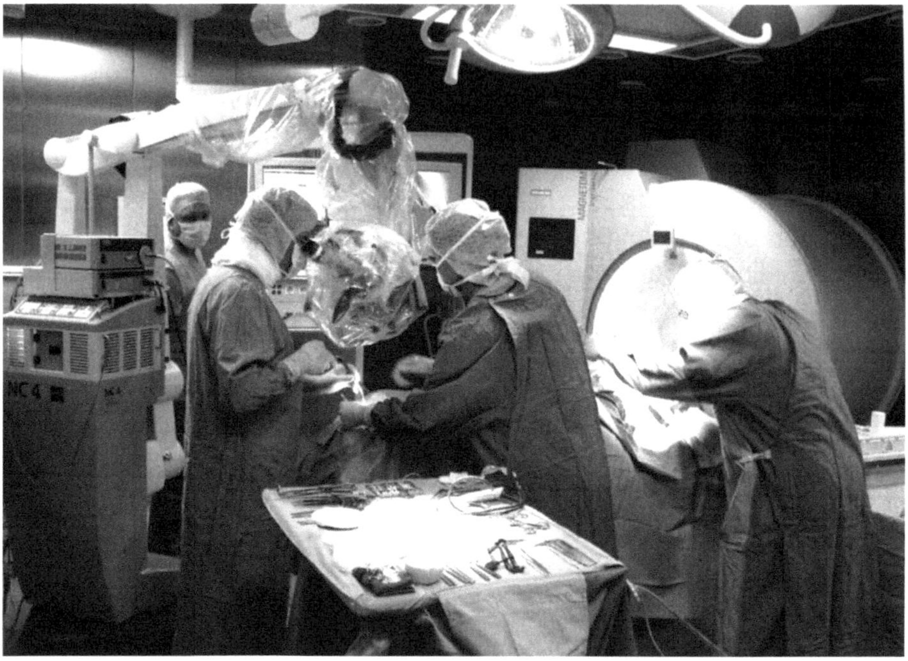

Fig. 4. OR scene during transsphenoidal surgery. The patient's head is placed at the 5 G line, the OR table is locked at the 160° position, the neurosurgeon is standing behind the head of the patient, standard microinstruments are used, only the nasal speculum is fully MR-compatible

are controlled remotely. Only the rotation about the table axis to turn the table into the axis of the scanner is performed manually for safety reasons. A MR-compatible 4-point head holder is integrated into the head coil for head fixation during craniotomy [77]. The upper part of the head coil may be sterilized using plasma sterilization. Sterile adapters screwed to the lower part of the head coil assure the possibility of sterile draping. In transsphenoidal operations that do not require head stabilization, imaging is performed using a U-shaped flexible coil adapted to the head (Fig. 4).

The ceiling outlet for laminar airflow is located above the main OR position (Luwa, Frankfurt, Germany). The laminar airflow output is surrounded by a band of fluorescent lamps for optimal illumination. For scanning, the illumination can be turned off from the control room. The entire operating theatre has MR-compatible spot-room lighting. Two ceiling-mounted surgical lamps (Heraeus Med, Hanau, Germany) are installed at the main surgical position (Fig. 3 A). Movable MR-compatible surgical lamps (Heraeus Med) can be used at the second surgical position in the high magnetic field (Fig. 3 B). An MR-compatible Servo 900C (Siemens AG Medical Solutions, Erlangen, Germany) and MR-compatible monitoring (Invivo research, Orlando, USA) are available for control of anaesthesia and for wireless 2.4 GHz data transfer from the RF cabin. The perfusors are shielded for MR compatibility. All standard gas lines are available on service outlets at different places in the RF cabin. Compressed air is integrated to operate drills. The service outlets include sockets connected to different electrical circuits so that selected sockets can be switched off from a switchboard in the MR control room to prevent artefacts generated by individual devices.

The NC4 microscope (Zeiss, Oberkochen, Germany) is installed at the left side of the head outside the 5 G line. It (along with other potentially interfering devices) is automatically switched off for the MR measurements. Navigation support is provided by the VectorVision Sky Navigation System (BrainLab, Heimstetten, Germany). A fibre optic connection ensures MR-compatible integration into the RF room. The camera used to monitor the positions of the microscope and other instruments is ceiling-mounted, as is the touch screen which is used to operate the navigation system. The microscope video is documented using Medimage software (Vepro, Pfungstadt, Germany) and in parallel as a recording on commercial S-VHS tapes. Both systems are installed in the MR control room. Two flat monitor screens (17.4″, AS4431ID, Iyamo, Nagano-Shi, Japan) mounted on a ceiling arm (Ondal, Hünfeld, Germany) display the microscope image, the image from the MR console, or various PC applications. There is also a wall mounted PC console using Autoview200 (Avocent, Munich), with which the various PC systems, such as the video documentation and the navigation system, can be operated from inside the RF cabin. There is an identical console in the MR control room. A mobile in-room MR console is available in the RF room for operating the scanner. Both the 5 G and 200 G lines are marked on the floor. The 200 G line is also marked by a raised stainless steel strip as a mechanical threshold. All equipment not completely MR compatible, such as the navigation microscope and the height-adjustable surgeon's chair are mechanically secured to the wall of the RF room. The instrument table and the various rotating chairs are fully MR-compatible (Trumpf, Saalfeld, Germany).

First Clinical Results

The first 95 patients have been examined with high-field MR by the end of 2002. These are essentially patients with gliomas, pituitary tumours, or pharmaco-resistant epilepsy. Intraoperative imaging was technically possible in all cases. Only one case encountered technical difficulties. In one of the first transsphenoidal procedures a coil cable broke during use of the flexible coil, so that the coil had to be replaced. No artefacts due to the operating room environment were observed. The ability to turn off the fluorescent lamps and specific power outlets from the MR control room proved very helpful.

As a general rule it took less than 2 minutes from the time the neurosurgeon decided to use intraoperative imaging until the imaging was actually started. With all the anaesthesia lines to and from the patient passing through the surgical table's centre of rotation, there were no delays due to anaesthesia.

The flexibility of the rotatable and adjustable surgical MR table is similar to that of the surgical table normally used. The head holder integrated into the head coil allows great variability of access; however, access for craniotomy is limited because of the greater distance of the head from the edge of the table and the coil adapter. Some modifications are certainly still needed. The integrated microscope-based neuronavigation was employed without problems and with very good clinical accuracy. Navigation could be updated during surgery without difficulty.

The intraoperative image quality obtained is clearly superior to that of low-field systems. Hardly any difference can be observed between the quality of preoperative and intraoperative imaging. Aside from improved image quality, the high-field scanner allows significantly shorter examination times, so that a more detailed sequence can be measured within the same time frame. When T2-weighted HASTE sequences (half Fourier single shot turbo spin echo) are used to check on resection of pituitary adenomas, the extent of the resection can often be determined after only a few seconds (Fig. 5). Craniopharyngioma cyst puncture can be easily monitored by intraoperative MR imaging (Fig. 6). Fig. 7 shows an example of intraoperative high-field MR imaging during removal of a ventricular meningioma. Fig. 8 depicts an example of intraoperative imaging in glioma surgery.

Summary & Future Developments

In the mid-90s a limited number of hospitals worldwide were attempting to generate MR images of the brain during surgery [10, 108, 113]. The first designs were based on low-field magnets with field strengths up to 0.5 T. In the meantime, initial reports on their use for large numbers of patients have been published [9, 99, 102, 106, 117]. Since March 1996, we have examined 330 patients with low-field intraoperative MRI. In the great majority of cases, intraoperative MRI provided reliable data regarding resections, and is gaining increased importance as intraoperative quality control [12].

However, the diagnostic quality of the intraoperative low-field system cannot compete with the image quality of routine neuroradiologic diagnosis generally performed with high-field magnets. Advances in scanner

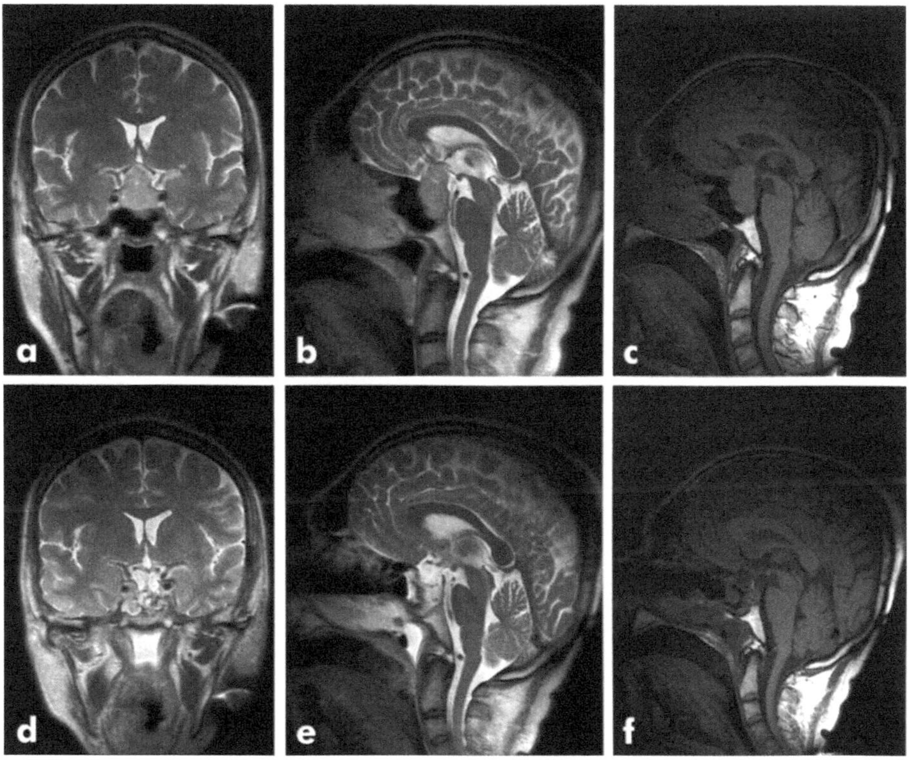

Fig. 5. 62-years old male patient with a large hormonally inactive intra- and suprasellar pituitary adenoma (a–c: preoperative, d–f: intraoperative images), intraoperative imaging depicts complete removal, which could be stated after some seconds imaging time applying the T2-weighted HASTE-sequence (a/b & d/e) (c/f: T1-weighted images)

design, including those due to active magnetic shielding, have made it possible to adapt a modern high-field scanner to the surgical environment without loss of the simultaneous capability of microscope-aided neuronavigation with integrated functional data [76]. So far, two different high-field concepts had been realized [33, 111].

Since April 2002 we have examined the first 95 patients with a 1.5 T Magnetom Sonata. Even this pilot series clearly confirms distinct improvements in intraoperative image quality. Comparison with the preoperative images indicates little significant limitation. We hope that the clearly improved image quality will result in greater reliable information regarding the extent of a resection, i.e., about the presence of residual tumour tissue and its exact location (Fig. 9).

With respect to intraoperative practicability, we found a distinct improvement over our previous designs [79, 108]. The costs of intraoperative

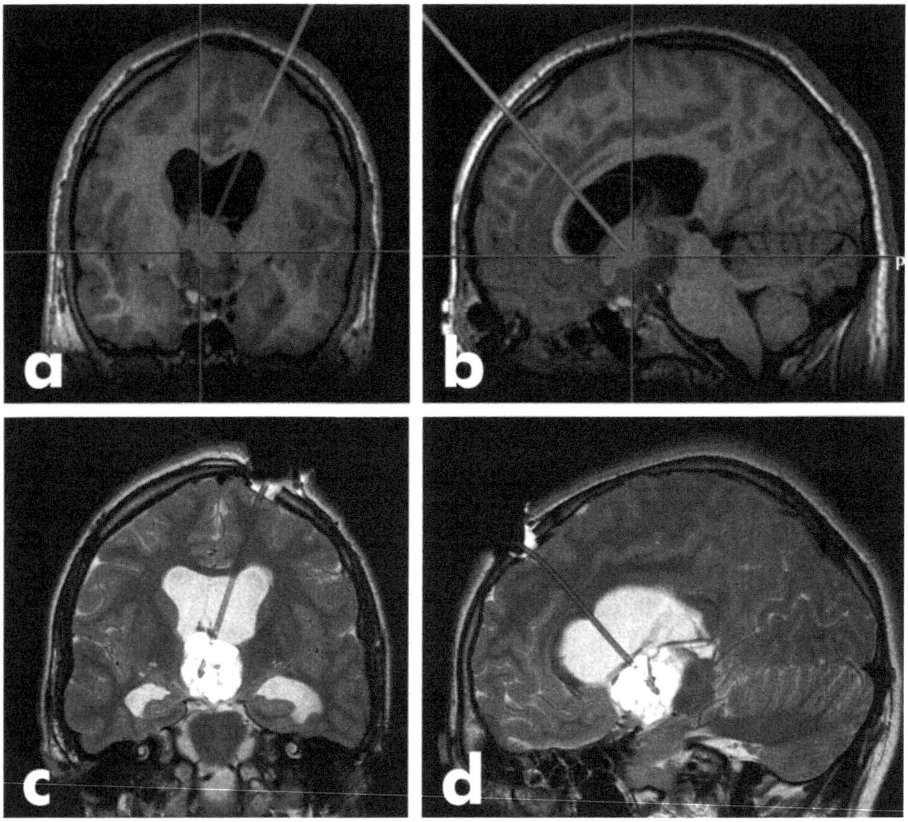

Fig. 6. 26-years old male patient with a recurrent suprasellar craniopharyngioma with acute hydrocephalus, a cysto-ventricular catheter was placed with navigation guidance (a/b: coronal and sagittal navigation screens), intraoperative high-field MR imaging with a high-resolution T2-sequence confirms the catheter placement (note that even the holes in the first 2 cm of the catheter are visualized)

imaging have been reduced for both preparation as well as intraoperative procedures. Additionally, the use of microscope-supported neuronavigation has resulted in substantially shorter preparation times. We expect that the possibility of automatic registration of images will result not only in further timesaving, but also in improved accuracy during clinical applications. Other improvements relate to head fixation during craniotomy and to the use of different coils to improve ergonomics for the surgeon.

The use of MR spectroscopy for improved delineation of a brain tumour from healthy tissue, resulting in improved definition of the tumour boundaries, opens up new dimensions in operative tumour therapy [22, 34, 114, 115]. Integration of the spectroscopic data into neuronavigation is a critical step, as is the possibility of intraoperative spectroscopic measure-

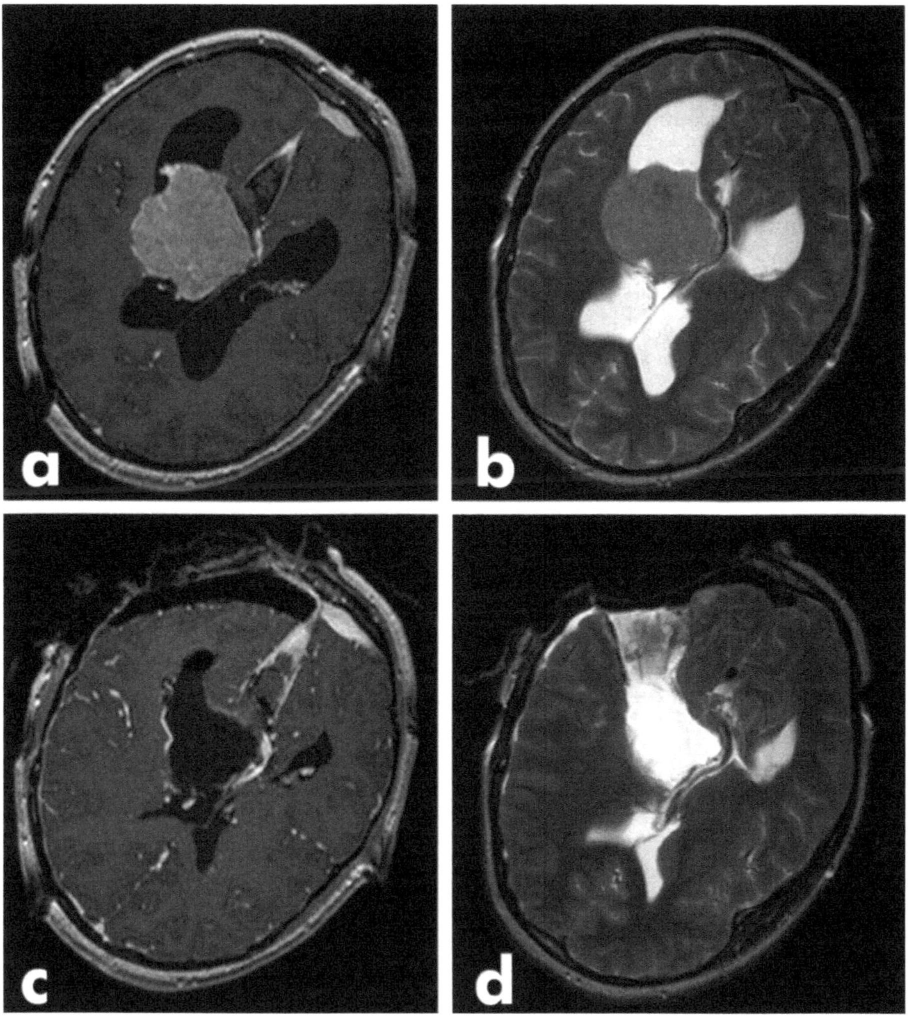

Fig. 7. 31-years old female patient with a large intraventricular meningioma, which was removed completely, confirmed by intraoperative MR imaging (a/b: preoperative, c/d: intraoperative, a/c T1-weighted, b/d: T2-weighted, note the visualization of the veins in the intraoperative T2-weighted images)

ments. The visualization of the course of brain pathways by diffusion tensor imaging and their integration in the navigational setup, as demonstrated for the pyramidal tract [18], is a further requirement. We must also attempt to compensate for the effect of brain shift not only with the anatomical data from intraoperative MR measurements [37, 78, 116] but also by localizing during surgery the important functional areas [118] which as of yet have been definable only with preoperative MEG or fMRI mea-

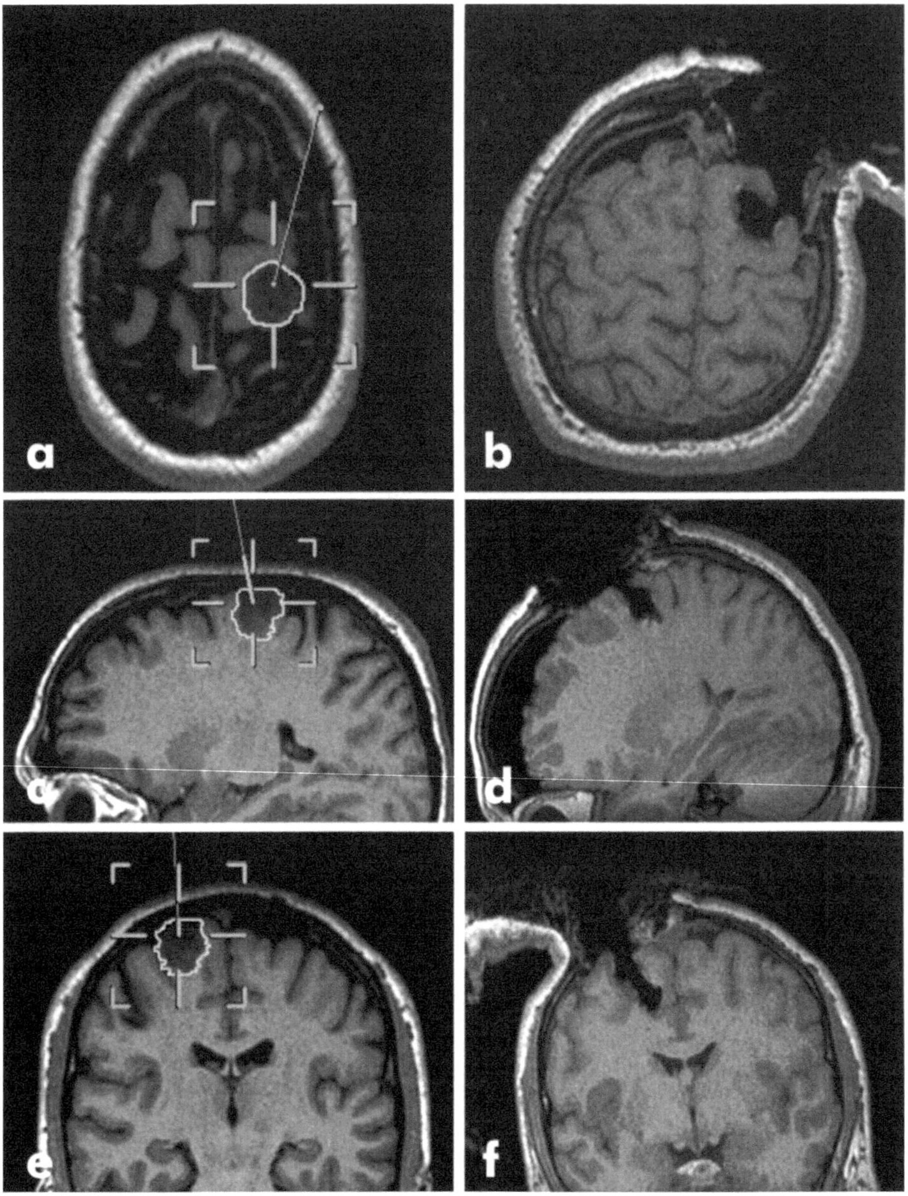

Fig. 8. 34-years old male patient with a right precentral WHO grade II astrocytoma (a/c/e: preoperative, b/d/f: intraoperative images – all images are reformatted from 3-D T1-weighted MPRAGE dataset with an isotropic voxel size of 1.0 mm, in the preoperative data the tumour border is segmented for use in microscope-based neuronavigation), intraoperative imaging depicts complete removal

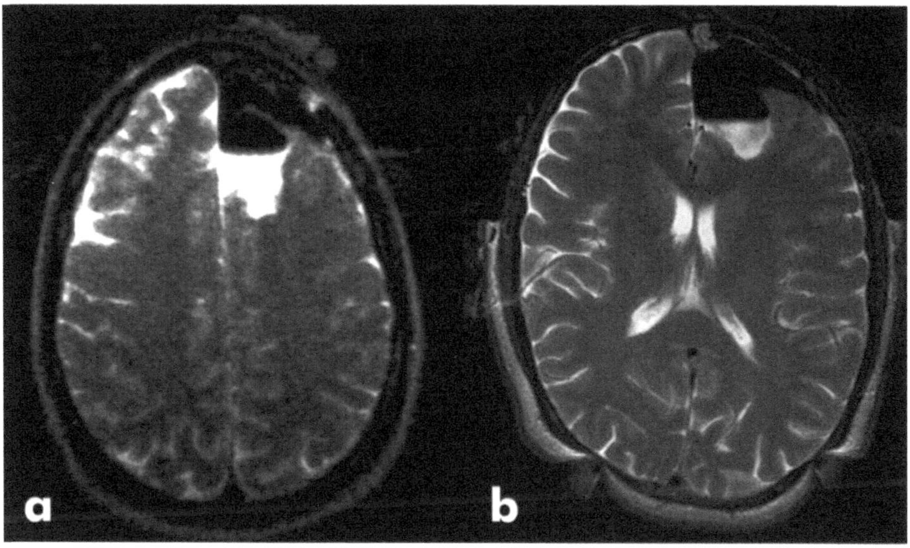

Fig. 9. Comparison of intraoperative image quality of the low-field with the high-field system clearly depicts superior quality of the high-field setup: resembling cases of intraoperative imaging of right frontal tumours, which were investigated by T2-weighted imaging with the 0.2 Tesla (a) or with the 1.5 Tesla system (b)

surements. Mathematical models that describe the brain shift phenomenon using finite elements will be helpful in this respect [25, 67].

Aside from the indications currently defined by the experience with use of low-field MR systems, intraoperative high-field MR imaging is also important in, e.g., vascular surgery. MR angiography cannot only show the complete clipping of an aneurysm; use of diffusion-weighted imaging should also allow for intraoperative diagnosis of blood supply to the respective brain areas, and thus perhaps prevent reduced perfusion [112]. The concept may also prove useful in spinal surgery, such as resection of complex intramedullary tumours or drainage of syringomyelias. Also integration of robot-assisted surgery will be a future application.

At present, intraoperative high-field MR imaging with integrated microscope-based neuronavigation is certainly the most highly developed technical method providing reliable intraoperative checking of the immediate effects of an operation so that conclusions can be drawn during the same surgery, as with respect to extended resection. However, in addition to these developments, we must still emphasize e.g. for the treatment of gliomas that critical advances are expected particularly in molecular biology. Aside from carefully weighing the indications for use of this complex as well as expensive technology, accurate cost-benefit analyses [32] will certainly be essential in the future.

Acknowledgments

We are especially grateful to Dr. Helmut Kober for his outstanding work in functional imaging. We want to thank the whole team of physicists and computer specialists of our Neurocenter helping in image processing, namely Dr. P. Grummich, Dr. P. Hastreiter, and Dr. A. Stadlbauer. Furthermore, we acknowledge continuing support by Dr. A. Oppelt and Dr. T. Vetter (Siemens Medical Solutions, Erlangen, Germany). This work is supported by the Deutsche Forschungsgemeinschaft and the Wilhelm-Sander-Stiftung.

References

1. Ammirati M, Galicich JH, Arbit E, Liao Y (1987) Reoperation in the treatment of recurrent intracranial malignant gliomas. Neurosurgery 21: 607–614
2. Atlas SW, Howard RS, Maldjian J, Alsop D, Detre JA, Listerud J, D'Esposito M, Judy KD, Zager E, Stecker M (1996) Functional magnetic resonance imaging of regional brain activity in patients with intracerebral gliomas: findings and implications for clinical management. Neurosurgery 38: 329–338
3. Ball T, Schreiber A, Feige B, Wagner M, Lucking CH, Kristeva-Feige R (1999) The role of higher-order motor areas in voluntary movement as revealed by high-resolution EEG and fMRI. Neuroimage 10: 682–694
4. Bandettini PA, Jesmanowicz A, Wong EC, Hyde JS (1993) Processing strategies for time-course data sets in functional MRI of the human brain. Magn Reson Med 30: 161–173
5. Baumann SB, Noll DC, Kondziolka DS, Schneider W, Nichols TE, Mintun MA, Lewine JD, Yonas H, Orrison WW Jr, Sclabassi RJ (1995) Comparison of functional magnetic resonance imaging with positron emission tomography and magnetoencephalography to identify the motor cortex in a patient with an arteriovenous malformation. J Image Guid Surg 1: 191–197
6. Beisteiner R, Erdler M, Teichtmeister C, Diemling M, Moser E, Edward V, Deecke L (1997) Magnetoencephalography may help to improve functional MRI brain mapping. Eur J Neurosci 9: 1072–1077
7. Beisteiner R, Gomiscek G, Erdler M, Teichtmeister C, Moser E, Deecke L (1995) Comparing localization of conventional functional magnetic resonance imaging and magnetoencephalography. Eur J Neurosci 7: 1121–1124
8. Black PM (1999) Surgery for cerebral gliomas: past, present and future. In: Howard III MA, Elliott JP, Haglund MM, McKhann II GM (Eds) Clinical neurosurgery. Lippincott Williams & Wilkins, Boston, pp 21–45
9. Black PM, Alexander III E, Martin C, Moriarty T, Nabavi A, Wong TZ, Schwartz RB, Jolesz F (1999) Craniotomy for tumor treatment in an intraoperative magnetic resonance imaging unit. Neurosurgery 45: 423–433
10. Black PM, Moriarty T, Alexander III E, Stieg P, Woodard EJ, Gleason PL, Martin CH, Kikinis R, Schwartz RB, Jolesz FA (1997) Development and implementation of intraoperative magnetic resonance imaging and its neurosurgical applications. Neurosurgery 41: 831–845

11. Bohinski RJ, Kokkino AK, Warnick RE, Gaskill-Shipley MF, Kormos DW, Lukin RR, Tew JM Jr (2001) Glioma resection in a shared-resource magnetic resonance operating room after optimal image-guided frameless stereotactic resection. Neurosurgery 48: 731–744
12. Bradley WG (2002) Achieving gross total resection of brain tumors: intraoperative MR imaging can make a big difference. AJNR Am J Neuroradiol 23: 348–349
13. Buchfelder M, Fahlbusch R, Ganslandt O, Stefan H, Nimsky C (2002) Use of intraoperative magnetic resonance imaging in tailored temporal lobe surgeries for epilepsy. Epilepsia 43: 864–873
14. Buchfelder M, Ganslandt O, Fahlbusch R, Nimsky C (2000) Intraoperative magnetic resonance imaging in epilepsy surgery. J Magn Reson Imaging 12: 547–555
15. Butler WE, Piaggio CM, Constantinou C, Niklason L, Gonzalez RG, Cosgrove GR, Zervas NT (1998) A mobile computed tomographic scanner with intraoperative and intensive care unit applications. Neurosurgery 42: 1304–1311
16. Cabantog AM, Bernstein M (1994) Complications of first craniotomy for intra-axial brain tumour. Can J Neurol Sci 21: 213–218
17. Ciric I, Ammirati M, Vick N, Mikhael M (1987) Supratentorial gliomas: surgical considerations and immediate postoperative results. Gross total resection versus partial resection. Neurosurgery 21: 21–26
18. Coenen VA, Krings T, Mayfrank L, Polin RS, Reinges MH, Thron A, Gilsbach JM (2001) Three-dimensional visualization of the pyramidal tract in a neuronavigation system during brain tumor surgery: first experiences and technical note. Neurosurgery 49: 86–93
19. Connelly A, Jackson GD, Frackowiak RS, Belliveau JW, Vargha-Khadem F, Gadian DG (1993) Functional mapping of activated human primary cortex with a clinical MR imaging system. Radiology 188: 125–130
20. Cosgrove GR, Buchbinder BR, Jiang H (1996) Functional magnetic resonance imaging for intracranial navigation. Neurosurg Clin N Am 7: 313–322
21. Cox RW, Hyde JS (1997) Software tools for analysis and visualization of fMRI data. NMR Biomed 10: 171–178
22. Dowling C, Bollen AW, Noworolski SM, McDermott MW, Barbaro NM, Day MR, Henry RG, Chang SM, Dillon WP, Nelson SJ, Vigneron DB (2001) Preoperative proton MR spectroscopic imaging of brain tumors: correlation with histopathologic analysis of resection specimens. AJNR Am J Neuroradiol 22: 604–612
23. Fadul C, Wood J, Thaler H, Galicich J, Patterson RH Jr, Posner JB (1988) Morbidity and mortality of craniotomy for excision of supratentorial gliomas. Neurology 38: 1374–1379
24. Fahlbusch R, Ganslandt O, Buchfelder M, Schott W, Nimsky C (2001) Intraoperative magnetic resonance imaging during transsphenoidal surgery. J Neurosurg 95: 381–390
25. Ferrant M, Warfield SK, Nabavi A, Jolesz F, Kikinis R (2000) Registration of 3D intraoperative MR images of the brain using a finite element bio-

mechanical model. In: Delp SL, DiGioia AM, Jaramaz B (Eds) Medical image computing and computer-assisted intervention – MICCAI 2000. Springer, Berlin Heidelberg New York Tokyo, pp 19–28
26. Fried I, Nenov VI, Ojemann SG, Woods RP (1995) Functional MR and PET imaging of rolandic and visual cortices for neurosurgical planning. J Neurosurg 83: 854–861
27. Gallen CC, Schwartz BJ, Bucholz RD, Malik G, Barkley GL, Smith J, Tung H, Copeland B, Bruno L, Assam S *et al* (1995) Presurgical localization of functional cortex using magnetic source imaging. J Neurosurg 82: 988–994
28. Ganslandt O, Fahlbusch R, Nimsky C, Kober H, Möller M, Steinmeier R, Romstöck J, Vieth J (1999) Functional neuronavigation with magnetoencephalography: outcome in 50 patients with lesions around the motor cortex. J Neurosurg 91: 73–79
29. Ganslandt O, Steinmeier R, Kober H, Vieth J, Kassubek J, Romstock J, Strauss C, Fahlbusch R (1997) Magnetic source imaging combined with image-guided frameless stereotaxy: a new method in surgery around the motor strip. Neurosurgery 41: 621–628
30. Grunert P, Muller-Forell W, Darabi K, Reisch R, Busert C, Hopf N, Perneczky A (1998) Basic principles and clinical applications of neuronavigation and intraoperative computed tomography. Comput Aided Surg 3: 166–173
31. Hadani M, Spiegelman R, Feldman Z, Berkenstadt H, Ram Z (2001) Novel, compact, intraoperative magnetic resonance imaging-guided system for conventional neurosurgical operating rooms. Neurosurgery 48: 799–809
32. Hall WA, Kowalik K, Liu H, Truwit CL, Kucharezyk J (2003) Costs and benefits of intraoperative MR-guided brain tumor resection. Acta Neurochir (Wien) [Suppl] 85: 137–142
33. Hall WA, Liu H, Martin AJ, Pozza CH, Maxwell RE, Truwit CL (2000) Safety, efficacy, and functionality of high-field strength interventional magnetic resonance imaging for neurosurgery. Neurosurgery 46: 632–642
34. Hall WA, Martin A, Liu H, Truwit CL (2001) Improving diagnostic yield in brain biopsy: coupling spectroscopic targeting with real-time needle placement. J Magn Reson Imaging 13: 12–15
35. Hall WA, Martin AJ, Liu H, Nussbaum ES, Maxwell RE, Truwit CL (1999) Brain biopsy using high-field strength interventional magnetic resonance imaging. Neurosurgery 44: 807–814
36. Hammoud MA, Ligon BL, elSouki R, Shi WM, Schomer DF, Sawaya R (1996) Use of intraoperative ultrasound for localizing tumors and determining the extent of resection: a comparative study with magnetic resonance imaging. J Neurosurg 84: 737–741
37. Hastreiter P, Engel K, Soza G, Bauer M, Wolf M, Ganslandt O, Fahlbusch R, Greiner G, Ertl T, Nimsky C (2001) Remote Analysis for Brain Shift Compensation. In: Niessen W, Viergever M (Eds) Medical image computing and computer assisted intervention 4th international conference. Springer, Berlin Heidelberg New York Tokyo, pp 1248–1249
38. Hata N, Dohi T, Iseki H, Takakura K (1997) Development of a frameless

and armless stereotactic neuronavigation system with ultrasonographic registration. Neurosurgery 41: 608–614
39. Hum B, Feigenbaum F, Cleary K, Henderson F (2000) Intraoperative computed tomography for complex craniocervical operations and spinal tumor resections. Neurosurgery 47: 374–381
40. Hund M, Rezai AR, Kronberg E, Cappell J, Zonenshayn M, Ribary U, Kelly PJ, Llinas R (1997) Magnetoencephalographic mapping: basic of a new functional risk profile in the selection of patients with cortical brain lesions. Neurosurgery 40: 936–943
41. Inoue T, Shimizu H, Nakasato N, Kumabe T, Yoshimoto T (1999) Accuracy and limitation of functional magnetic resonance imaging for identification of the central sulcus: comparison with magnetoencephalography in patients with brain tumors. Neuroimage 10: 738–748
42. Jack CR Jr, Thompson RM, Butts RK, Sharbrough FW, Kelly PJ, Hanson DP, Riederer SJ, Ehman RL, Hangiandreou NJ, Cascino GD (1994) Sensory motor cortex: correlation of presurgical mapping with functional MR imaging and invasive cortical mapping. Radiology 190: 85–92
43. Jödicke A, Deinsberger W, Erbe H, Kriete A, Böker DK (1998) Intraoperative three-dimensional ultrasonography: an approach to register brain shift using multidimensional image processing. Minim Invas Neurosurg 41: 13–19
44. Kabuto M, Kubota T, Kobayashi H, Handa Y, Sato K, Ishii H, Takeuchi H, Uno H, Arishima H, Ido K, Ueda Y, Adachi M, Ishida M, Hasegawa Y, Yanagimoto M, Goto Y (1998) Intraoperative CT imaging system using a mobile CT scanner gantry mounted on floor-embedded rails for neurosurgery. No To Shinkei 50: 1003–1008
45. Kahn T, Schwabe B, Bettag M, Harth T, Ulrich F, Rassek M, Schwarzmaier HJ, Modder U (1996) Mapping of the cortical motor hand area with functional MR imaging and MR imaging-guided laser-induced interstitial thermotherapy of brain tumors. Work in progress. Radiology 200: 149–157
46. Kaibara T, Myles ST, Lee MA, Sutherland GR (2002) Optimizing epilepsy surgery with intraoperative MR imaging. Epilepsia 43: 425–429
47. Kaibara T, Saunders JK, Sutherland GR (2000) Advances in mobile intraoperative magnetic resonance imaging. Neurosurgery 47: 131–138
48. Kamada K, Takeuchi F, Kuriki S, Oshiro O, Houkin K, Abe H (1993) Functional neurosurgical simulation with brain surface magnetic resonance images and magnetoencephalography. Neurosurgery 33: 269–273
49. Keles GE, Lamborn KR, Berger MS (2001) Low-grade hemispheric gliomas in adults: a critical review of extent of resection as a factor influencing outcome. J Neurosurg 95: 735–745
50. Knauth M, Aras N, Wirtz CR, Dorfler A, Engelhorn T, Sartor K (1999) Surgically induced intracranial contrast enhancement: potential source of diagnostic error in intraoperative MR imaging. AJNR Am J Neuroradiol 20: 1547–1553
51. Knauth M, Wirtz CR, Tronnier VM, Aras N, Kunze S, Sartor K (1999) Intraoperative MR imaging increases the extent of tumor resection in patients with high-grade gliomas. AJNR Am J Neuroradiol 20: 1642–1646

52. Kober H, Grummich P, Vieth J (1995) Fit of the digitized head surface with the surface reconstructed from MRI tomography. In: Baumgartner C (Ed) Biomagnetism: fundamental research and clinical applications. Elsevier Science, IOS Press, Amsterdam, pp 309–312
53. Kober H, Möller M, Nimsky C, Vieth J, Fahlbusch R, Ganslandt O (2001) New approach to localize speech relevant brain areas and hemispheric dominance using spatially filtered magnetoencephalography. Hum Brain Mapp 14: 236–250
54. Kober H, Nimsky C, Möller M, Hastreiter P, Fahlbusch R, Ganslandt O (2001) Correlation of sensorimotor activation with functional magnetic resonance imaging and magnetoencephalography in presurgical functional imaging: a spatial analysis. Neuroimage 14: 1214–1228
55. Kowalczuk A, Macdonald RL, Amidei C, Dohrmann III G, Erickson RK, Hekmatpanah J, Krauss S, Krishnasamy S, Masters G, Mullan SF, mundt AJ, Sweeney P, Vokes EE, Weir BKA, Wollmann RL (1997) Quantitative imaging study of extent of surgical resection and prognosis of malignant astrocytomas. Neurosurgery 41: 1028–1038
56. Kwong KK, Belliveau JW, Chesler DA, Goldberg IE, Weisskoff RM, Poncelet BP, Kennedy DN, Hoppel BE, Cohen MS, Turner R, Cheng HM, Brady TJ, Rosen BR (1992) Dynamic magnetic resonance imaging of human brain activity during primary sensory stimulation. Proc Natl Acad Sci USA 89: 5675–5679
57. Latchaw RE, Hu X, Ugurbil K, Hall WA, Madison MT, Heros RC (1995) Functional magnetic resonance imaging as a management tool for cerebral arteriovenous malformations. Neurosurgery 37: 619–625
58. Leahy RM, Mosher JC, ME S, MX H, Lewine JD (1998) A study of dipole localization accuracy for MEG and EEG using a human skull phantom. Electroenceph Clin Neurophysiol 107: 159–173
59. LeRoux PD, Winter TC, Berger MS, Mack LA, Wang K, Elliott JP (1994) A comparison between preoperative magnetic resonance and intraoperative ultrasound tumor volumes and margins. J Clin Ultrasound 22: 29–36
60. Lewin JS (1999) Interventional MR imaging: concepts, systems, and applications in neuroradiology. AJNR Am J Neuroradiol 20: 735–748
61. Lunsford LD, Parrish R, Albright L (1984) Intraoperative imaging with a therapeutic computed tomographic scanner. Neurosurgery 15: 559–561
62. Lunsford LD, Rosenbaum AE, Perry J (1982) Stereotactic surgery using the "therapeutic" CT scanner. Surg Neurol 18: 116–122
63. Maldjian J, Atlas SW, Howard RS, Greenstein E, Alsop D, Detre JA, Listerud J, D'Esposito M, Flamm ES (1996) Functional magnetic resonance imaging of regional brain activity in patients with intracerebral arteriovenous malformations before surgical or endovascular therapy. J Neurosurg 84: 477–483
64. Maldjian JA, Schulder M, Liu WC, Mun IK, Hirschorn D, Murthy R, Carmel P, Kalnin A (1997) Intraoperative functional MRI using a real-time neurosurgical navigation system. J Comput Assist Tomogr 21: 910–912
65. Martin CH, Schwartz R, Jolesz F, Black PM (1999) Transsphenoidal resec-

tion of pituitary adenomas in an intraoperative MRI unit. Pituitary 2: 155–162
66. Matula C, Rossler K, Reddy M, Schindler E, Koos WT (1998) Intraoperative computed tomography guided neuronavigation: concepts, efficiency, and work flow. Comput Aided Surg 3: 174–182
67. Miga MI, Paulsen KD, Hoopes PJ, Kennedy FE, Hartov A, Roberts DW (2000) In vivo modeling of interstitial pressure in the brain under surgical load using finite elements. J Biomech Eng 122: 354–363
68. Möller M, Kober H, Ganslandt O, Nimsky C, Vieth J, Fahlbusch R (1999) Functional mapping of speech evoked brain activity by magnetoencephalography and its clinical application. Biomed Tech (Berl) 44: 159–161
69. Morioka T, Yamamoto T, Katsuta T, Fujii K, Fukui M (1994) Presurgical three-dimensional magnetic source imaging of the somatosensory cortex in a patient with a peri-Rolandic lesion: technical note. Neurosurgery 34: 930–934
70. Mueller WM, Yetkin FZ, Hammeke TA, Morris GL, Swanson SJ, Reichert K, Cox R, Haughton VM (1996) Functional magnetic resonance imaging mapping of the motor cortex in patients with cerebral tumors. Neurosurgery 39: 515–520
71. Nabavi A, Black PM, Gering DT, Westin CF, Mehta V, Pergolizzi RS Jr, Ferrant M, Warfield SK, Hata N, Schwartz RB, Wells WM, 3rd, Kikinis R, Jolesz FA (2001) Serial intraoperative magnetic resonance imaging of brain shift. Neurosurgery 48: 787–798
72. Nicolato A, Gerosa MA, Fina P, Iuzzolino P, Giorgiutti F, Bricolo A (1995) Prognostic factors in low-grade supratentorial astrocytomas: a uni-multivariate statistical analysis in 76 surgically treated adult patients. Surg Neurol 44: 208–223
73. Nimsky C, Ganslandt O, Buchfelder M, Fahlbusch R (2002) Glioma surgery evaluated by intraoperative low-field magnetic resonance imaging. Acta Neurochir (Wien) [Suppl] 85: 55–63
74. Nimsky C, Ganslandt O, Cerny S, Hastreiter P, Greiner G, Fahlbusch R (2000) Quantification of, visualization of, and compensation for brain shift using intraoperative magnetic resonance imaging. Neurosurgery 47: 1070–1080
75. Nimsky C, Ganslandt O, Fahlbusch R (2002) From intraoperative patient transport to surgery in the fringe field – intraoperative application of magnetic resonance imaging using a 0.2-Tesla scanner: the Erlangen experience. Tech Neurosurg 7: 265–273
76. Nimsky C, Ganslandt O, Fahlbusch R (2002) How to implement high-field intraoperative magnetic resonance imaging. In: Lemke HU, Vannier MW, Inamura K, Farman AG, Doi K, Reiber JHC (Eds) CARS2002. Springer, Berlin Heidelberg New York Tokyo, pp 139–143
77. Nimsky C, Ganslandt O, Fischer H, Oppelt A, Vetter T, Distler P, Kuth R (2000) Kombination aus Kopffixation und Kopfspule für neurochirurgische Operationen. Siemens Technik Report 3: 64–65
78. Nimsky C, Ganslandt O, Hastreiter P, Fahlbusch R (2001) Intraoperative compensation for brain shift. Surg Neurol 56: 357–364

79. Nimsky C, Ganslandt O, Kober H, Buchfelder M, Fahlbusch R (2001) Intraoperative magnetic resonance imaging combined with neuronavigation: a new concept. Neurosurgery 48: 1082–1091
80. Nimsky C, Ganslandt O, Kober H, Möller M, Ulmer S, Tomandl B, Fahlbusch R (1999) Integration of functional magnetic resonance imaging supported by magnetoencephalography in functional neuronavigation. Neurosurgery 44: 1249–1256
81. Nimsky C, Ganslandt O, Tomandl B, Buchfelder M, Fahlbusch R (2002) Low-field magnetic resonance imaging for intraoperative use in neurosurgery: a 5-year experience. Eur Radiol 12: 2690–2703
82. Ogawa S, Lee CC, Kay AR, Tank DW (1990) Brain magnetic resonance imaging with contrast dependent on blood oxygenation. Proc Natl Acad Sci USA 87: 9868–9872
83. Okudera H, Kyoshima K, Kobayashi S, Sugita K (1994) Intraoperative CT scan findings during resection of glial tumours. Neurol Res 16: 265–267
84. Okudera H, Takemae T, Kobayashi S (1993) Intraoperative computed tomographic scanning during transsphenoidal surgery: technical note. Neurosurgery 32: 1041–1043
85. Olivier A, Alonso-Vanegas M, Comeau R, Peters TM (1996) Image-guided surgery of epilepsy. Neurosurg Clin N Am 7: 229–243
86. Olivier A, Germano IM, Cukiert A, Peters T (1994) Frameless stereotaxy for surgery of the epilepsies: preliminary experience. Technical note. J Neurosurg 81: 629–633
87. Orrison WW Jr, Rose DF, Hart BL, Maclin EL, Sanders JA, Willis BK, Marchand EP, Wood CC, Davis LE (1992) Noninvasive preoperative cortical localization by magnetic source imaging. AJNR Am J Neuroradiol 13: 1124–1128
88. Pergolizzi RS Jr, Nabavi A, Schwartz RB, Hsu L, Wong TZ, Martin C, Black PM, Jolesz FA (2001) Intra-operative MR guidance during transsphenoidal pituitary resection: preliminary results. J Magn Reson Imaging 13: 136–141
89. Poline JB, Vandenberghe R, Holmes AP, Friston KJ, Frackowiak RS (1996) Reproducibility of PET activation studies: lessons from a multicenter European experiment. EU concerted action on functional imaging 4: 34–54
90. Pujol J, Conesa G, Deus J, Vendrell P, Isamat F, Zannoli G, Marti VJ, Capdevila A (1996) Presurgical identification of the primary sensorimotor cortex by functional magnetic resonance imaging. J Neurosurg 84: 7–13
91. Rezai AR, Hund M, Kronberg E, Zonenshayn M, Cappell J, Ribary U, Kall B, Llinas R, Kelly PJ (1996) The interactive use of magnetoencephalography in stereotactic image-guided neurosurgery. Neurosurgery 39: 92–102
92. Righini A, de DO, Prinster A, Spagnoli D, Appollonio I, Bello L, Scifo P, Tomei G, Villani R, Fazio F, Leonardi M (1996) Functional MRI: primary motor cortex localization in patients with brain tumors. J Comput Assist Tomogr 20: 702–708
93. Roberts DW, Jobst BC, Siegel AM, Lewis PJ, Darcey TM, Thadani VM,

Williamson PD (2001) Investigation of extra-temporal epilepsy. Stereotact Funct Neurosurg 77: 216–218

94. Roberts TP, Zusman E, McDermott M, Barbaro N, Rowley HA (1995) Correlation of functional magnetic source imaging with intraoperative cortical stimulation in neurosurgical patients. J Image Guid Surg 1: 339–347
95. Roberts TPL, Rowley HA (1997) Mapping of the sensorimotor cortex: functional MR and magnetic source imaging. AJNR Am J Neuroradiol 18: 871–880
96. Romstöck J, Fahlbusch R, Ganslandt O, Nimsky C, Strauss C (2002) Localisation of the sensorimotor cortex during surgery for brain tumours: feasibility and waveform patterns of somatosensory evoked potentials. J Neurol Neurosurg Psychiatry 72: 221–229
97. Rubin JM, Quint DJ (2000) Intraoperative US versus intraoperative MR imaging for guidance during intracranial neurosurgery. Radiology 215: 917–918 (letter)
98. Rubino GJ, Farahani K, McGill D, Van De Wiele B, Villablanca JP, Wang-Mathieson A (2000) Magnetic resonance imaging-guided neurosurgery in the magnetic fringe fields: the next step in neuronavigation. Neurosurgery 46: 643–654
99. Schneider JP, Schulz T, Schmidt F, Dietrich J, Lieberenz S, Trantakis C, Seifert V, Kellermann S, Schober R, Schaffranietz L, Laufer M, Kahn T (2001) Gross-total surgery of supratentorial low-grade gliomas under intraoperative MR guidance. AJNR Am J Neuroradiol 22: 89–98
100. Schulder M, Liang D, Carmel PW (2001) Cranial surgery navigation aided by a compact intraoperative magnetic resonance imager. J Neurosurg 94: 936–945
101. Schulder M, Maldjian JA, Liu WC, Holodny AI, Kalnin AT, Mun IK, Carmel PW (1998) Functional image-guided surgery of intracranial tumors located in or near the sensorimotor cortex. J Neurosurg 89: 412–418
102. Schwartz RB, Hsu L, Wong TZ, Kacher DF, Zamani AA, Black PM, Alexander III E, Stieg PE, Moriarty TM, Martin CA, Kikinis R, Jolesz FA (1999) Intraoperative MR imaging guidance for intracranial neurosurgery: experience with the first 200 cases. Radiology 211: 477–488
103. Schwartz TH, Marks D, Pak J, Hill J, Mandelbaum DE, Holodny AI, Schulder M (2002) Standardization of amygdalohippocampectomy with intraoperative magnetic resonance imaging: preliminary experience. Epilepsia 43: 430–436
104. Schwartz TH, Resor SRJ, De La Paz R, Goodman RR (1998) Functional magnetic resonance imaging localization of ictal onset to a dysplastic cleft with simulatneous sensorimotor mapping: intraoperative electrophysiological confirmation and postoperative follow-up: technical note. Neurosurgery 43: 639–645
105. Segebarth C, Belle V, Delon C, Massarelli R, Decety J, Le Bas JF, Decorps M, Benabid AL (1994) Functional MRI of the human brain: predominance of signals from extracerebral veins. Neuroreport 5: 813–816
106. Seifert V, Zimmermann M, Trantakis C, Vitzthum HE, Kuhnel K, Raabe

A, Bootz F, Schneider JP, Schmidt F, Dietrich J (1999) Open MRI-guided neurosurgery. Acta Neurochir (Wien) 141: 455–464
107. Stehling MK, Turner R, Mansfield P (1991) Echo-planar imaging: magnetic resonance imaging in a fraction of a second. Science 254: 43–50
108. Steinmeier R, Fahlbusch R, Ganslandt O, Nimsky C, Buchfelder M, Kaus M, Heigl T, Lenz G, Kuth R, Huk W (1998) Intraoperative magnetic resonance imaging with the magnetom open scanner: concepts, neurosurgical indications, and procedures. A preliminary report. Neurosurgery 43: 739–748
109. Stippich C, Freitag P, Kassubek J, Sörös P, Kamada K, Kober H, Scheffler K, Hopfengärtner R, Bilecen D, Radü E-W, Vieth J (1998) Motor, somatosensory and auditory cortex localization by fMRI and MEG. Neuroreport 9: 1953–1957
110. Sumanaweera TS, Adler JR, Napel S, Glover GH (1994) Characterization of spatial distortion in magnetic resonance imaging and its implications for stereotactic surgery. Neurosurgery 35: 696–704
111. Sutherland GR, Kaibara T, Louw D, Hoult DI, Tomanek B, Saunders J (1999) A mobile high-field magnetic resonance system for neurosurgery. J Neurosurg 91: 804–813
112. Sutherland GR, Kaibara T, Wallace C, Tomanek B, Richter M (2002) Intraoperative assessment of aneurysm clipping using magnetic resonance angiography and diffusion-weighted imaging: technical case report. Neurosurgery 50: 893–898
113. Tronnier VM, Wirtz CR, Knauth M, Lenz G, Pastyr O, Bonsanto MM, Albert FK, Kuth R, Staubert A, Schlegel W, Sartor K, Kunze S (1997) Intraoperative diagnostic and interventional magnetic resonance imaging in neurosurgery. Neurosurgery 40: 891–902
114. Tzika AA, Cheng LL, Goumnerova L, Madsen JR, Zurakowski D, Astrakas LG, Zarifi MK, Scott RM, Anthony DC, Gonzalez RG, Black PM (2002) Biochemical characterization of pediatric brain tumors by using in vivo and ex vivo magnetic resonance spectroscopy. J Neurosurg 96: 1023–1031
115. Tzika AA, Zarifi MK, Goumnerova L, Astrakas LG, Zurakowski D, Young-Poussaint T, Anthony DC, Scott RM, Black PM (2002) Neuroimaging in pediatric brain tumors: Gd-DTPA-enhanced, hemodynamic, and diffusion MR imaging compared with MR spectroscopic imaging. AJNR Am J Neuroradiol 23: 322–333
116. Wirtz CR, Bonsanto MM, Knauth M, Tronnier VM, Albert FK, Staubert A, Kunze S (1997) Intraoperative magnetic resonance imaging to update interactive navigation in neurosurgery: method and preliminary experience. Comput Aided Surg 2: 172–179
117. Wirtz CR, Knauth M, Staubert A, Bonsanto MM, Sartor K, Kunze S, Tronnier VM (2000) Clinical evaluation and follow-up results for intraoperative magnetic resonance imaging in neurosurgery. Neurosurgery 46: 1112–1122
118. Wolf M, Vogel T, Weierich P, Niemann H, Nimsky C (2001) Automatic transfer of preoperative fMRI markers into intraoperative MR-images for updating functional neuronavigation. IEICE T Inf Syst E84-D: 1698–1704

119. Woydt M, Krone A, Becker G, Schmidt K, Roggendorf W, Roosen K (1996) Correlation of intra-operative ultrasound with histopathologic findings after tumour resection in supratentorial gliomas. A method to improve gross total tumour resection. Acta Neurochir 138: 1391–1398
120. Xiong J, Gao J, Lancaster J, Fox P (1996) Assessment and optimization of functional MRI analyses. Human Brain Mapping 4: 153–167
121. Yousry TA, Schmid UD, Jassoy AG, Schmidt D, Eisner WE, Reulen HJ, Reiser MF, Lissner J (1995) Topography of the cortical motor hand area: prospective study with functional MR imaging and direct motor mapping at surgery. Radiology 195: 23–29
122. Zimmermann M, Seifert V, Trantakis C, Kuhnel K, Raabe A, Schneider JP, Dietrich J, Schmidt F (2000) Open MRI-guided microsurgery of intracranial tumours. Preliminary experience using a vertical open MRI-scanner. Acta Neurochir (Wien) 142: 177–186

Surgical Anatomy of the Insula

M. GUENOT[1], J. ISNARD[2], and M. SINDOU[1]

[1] Department of Functional Neurosurgery, Hôpital Neurologique P. Wertheimer, University of Lyon, France
[2] Department of Functional Neurology and Epileptology, Hôpital Neurologique P. Wertheimer, University of Lyon, France

With 20 Figures

Contents

Summary

The insula is the only cortical part of the brain which is not visible on the surface of the hemisphere. This is due to the fact that it is totally covered by the fronto-parietal and temporal opercula. The insula has a triangular shape, and is separated from the opercula by the anterior, superior, and inferior periinsular sulci. The limen insulae is the antero-inferiorly located insular cortical surface which conjoins the inferior insular point, the anterior perforated surface, and the temporo-mesial surface.

The insula is morphologically divided into two parts by the central insular sulcus. The anterior part of the insula bears 3 gyri: the anterior, middle, and posterior short insular gyri, separated by the anterior and precentral insular sulcus. The posterior part of the insula contains the anterior and posterior long insular gyri, separated by the postcentral insular sulcus. The vascular supply of the insula is mainly provided by the M2 segment of the middle cerebral artery, which constitutes a substantial obstacle to any open or stereotactic procedure aiming at the insular region. Histologically, the insula is a part of the paralimbic cortex, as it bears in its antero-inferior part an allo and mesocortical area. The insula is functionally involved in cardiac rythm and arterial blood pressure control, as well as in viscero-motor control and in viscero-sensitive functions. There is considerable evidence for the involvement of the insula as a somesthetic area, including a major role in the processing of nociceptive inputs. Its possible role in some epilepsies may explain some failures of temporal lobe resection. Surgery of the insular lobe is a technical challenge, whose risks can be minimized by the use of intra-operative direct cerebral stimulation.

Keywords: Insula; surgical anatomy; pain; epilepsy surgery.

Introduction

Among all the cerebral lobes, the insular lobe is the only one which cannot be seen on the surface of the brain. This is due to the presence of the perisylvian opercula, which cover its whole cortex, and this constitutes one of the main feature of the insula. A second characterisitc is the close relationships between insular cortex and the middle cerebral artery vascular arborization. These two points: thickness of the opercula and sylvian arterial wall, readily explain why the insula has always been considered as very difficult to reach by any surgical or stereotactic approach, as well as for vascular, tumorous, or epileptogenic lesions, and why many of its functional roles have remained unexplored until recently.

The insula owes its name to the german anatomist J. C. Reil, who, in 1819, called this hidden part of the cerebral cortex «insel», which means island in german. Since then, it has been termed «insula of Reil», and then «insula», which means island in latin, as it is presently termed in the nomina anatomica. The first anatomical drawing showing clearly the insula was done in 1641 by the danish anatomist Bartholin (Fig. 1), but without any mention about it in the legend.

– *Phylogenetically*, the appearance of the insular lobe occured rather recently. Indeed, it is not possible to identify any insular lobe as long as the brain is not developed sufficiently to have opercula, which means to

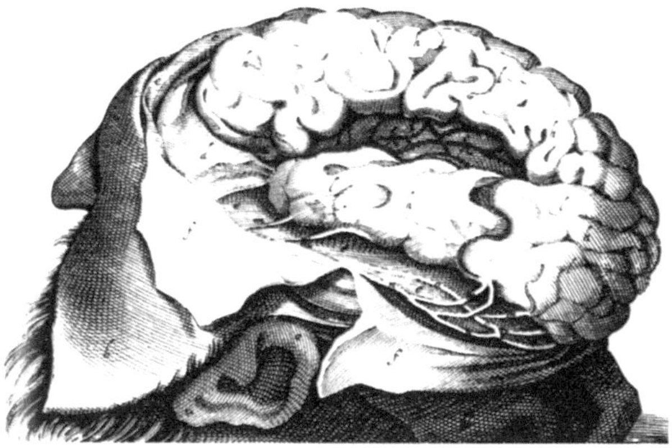

Fig. 1. Anatomical drawing showing the lateral aspect of a left cerebral hemisphere. This is the first drawing in which the insula is clearly shown (letters kk). It was drawn in 1641 by the danish anatomist Bartholin (from Caspar Bartholin (1585–1629): *Institutiones Anatomicae*, 1641)

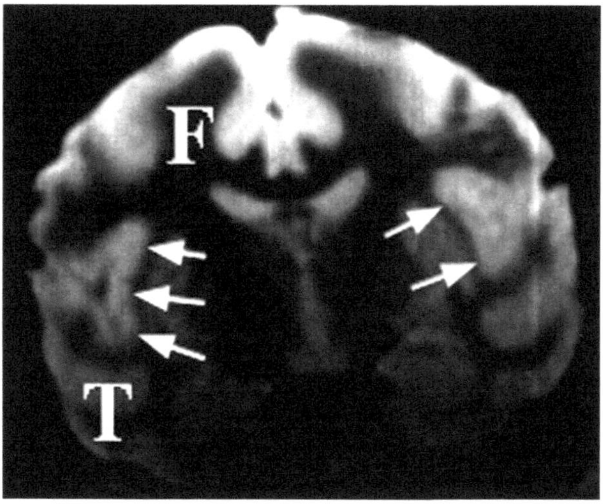

Fig. 2. MRI of a *Macacus Rhesus*'s brain, antero-posterior view. The insula (arrows) is clearly visible. *T* Temporal lobe, *F* frontal lobe

have a temporal lobe. Consequently, the only animal species having an insular lobe belong to the primates (Fig. 2) and the cetacean families. No evidence of the presence of an insular lobe can be found in any other animal family.

– *Ontogenetically*, the appearance of the insula is the result of three different processes (Fig. 3), namely: *rotation* of the hemisphere around a virtual axis which will become the insula, frontal and temporal *operculisation*,

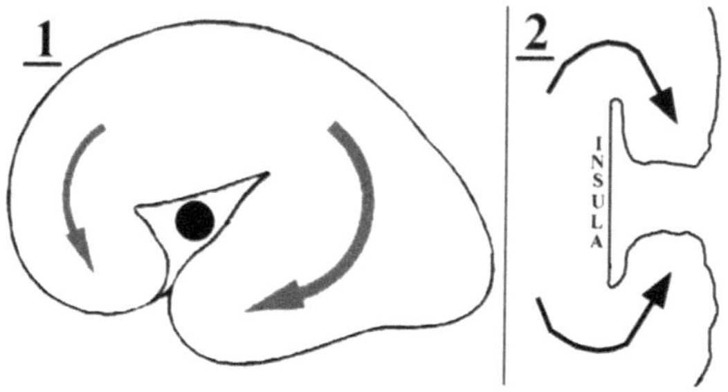

Fig. 3. Schematic drawing of the ontogenetical processes giving birth to the insular lobe. *1* lateral view, *rotation* of the cerebral hemisphere around a virtual axis which will become the insula (black dot). *2* coronal section, temporal (lower arrow) and fronto-parietal (upper arrow) *operculisation*. The constitution of the insula ends with the completion of the cerebral *gyration*

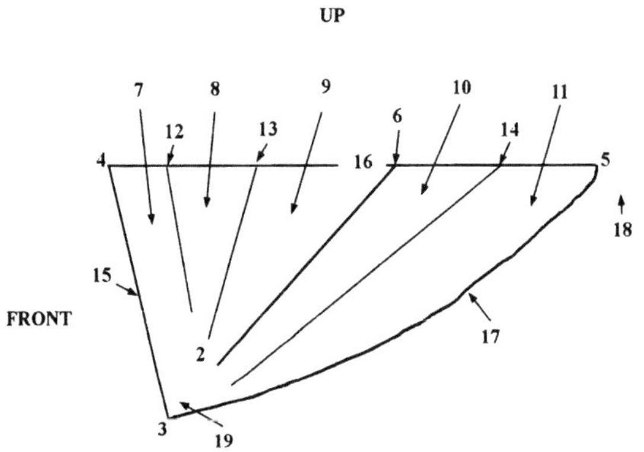

Fig. 4. Schematic drawing displaying the main anatomical landmarks of the insula. *2* Apex insulae, *3* inferior insular point, *4* anterior insular point, *5* posterior insular point, *6* central insular sulcus, *7* anterior short insular gyrus, *8* middle short insular gyrus, *9* posterior short insular gyrus, *10* anterior long insular gyrus, *11* posterior long insular gyrus, *12* anterior insular sulcus, *13* precentral insular sulcus, *14* postcentral insular sulcus, *15* anterior periinsular sulcus, *16* superior periinsular sulcus, *17* inferior periinsular sulcus, *18* sulcus lateralis cerebri (sylvian fissure), *19* limen insulae

and cerebral *gyration*. These processes provide a progressive coverage of the insular cortex by the temporal and fronto-parietal lobes. They begin before the 10th week of the embryonal life, and finalize only at time of birth.

– *Morphologically* (Fig. 4), tridimensionally, the insula has a pyramidal shape, whose summit, called the apex, is located near the inferior insular point, and is the most superficial part of the insula (located about 1 cm under the sylvian surface). The insular surface has a triangular shape. It is

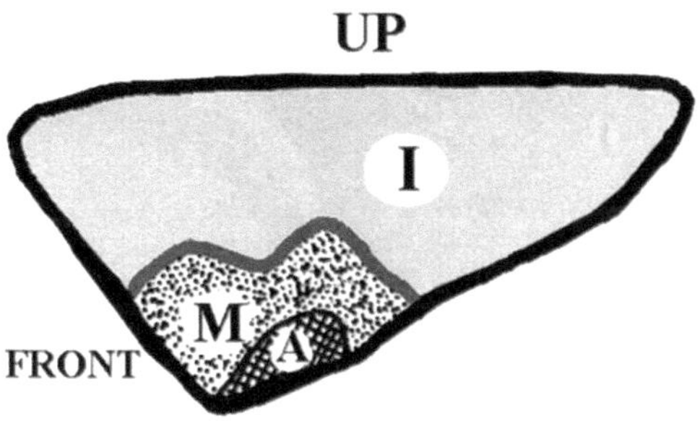

Fig. 5. Schematic drawing showing the topography of the three different cortical areas of the insular lobe. *A* Allocortex, *M* mesocortex, *I* isocortex

divided in two parts by the insular central sulcus. The anterior part bears three main gyri: the anterior, intermediaris, and posterior short insular gyri. The posterior part of the insular cortex bears the anterior and posterior long insular gyri. The insula is separated from the rest of the cerebral cortex by the anterior, superior, and inferior peri-insular sulci.

– *Histologically* (Fig. 5), the insular lobe constitutes several areas which differ in their cortical organization. The allocortex, which has three layers, is located in the antero-inferior part of the insular surface. The mesocortex, which has six layers, surrounds the allocortex. The isocortex, which also has six layers, is located in the most superior and posterior parts of the insula. There is a gradual transition from the primitive allocortex, in the apex, to the recent isocortex, in the periphery of the insular cortex.

– *Functionally*, the insula is probably involved in several neurological processes, particularly in relation to the autonomic system and the somesthetic functions, but some of its functions remain partially unknown.

Descriptive Anatomy [33, 39]

The Borders of the Insula

The insula is clearly demarcated from the surroundings cortical areas by the anterior, superior, and inferior perinsular sulci (Fig. 4).

1) *The anterior periinsular sulcus* has a vertical orientation. It extends from the anterior insular point down to the region of the limen insulae, thus demarcating the anterior part of the insula from the frontoorbital operculum. It is a short (3 cm long) straight sulcus.
2) *The superior periinsular sulcus* is twice as long (6 cm) as the anterior periinsular sulcus. It extends horizontally from the anterior insular point to the posterior insular point, demarcating the insula from the frontal and parietal opercula.

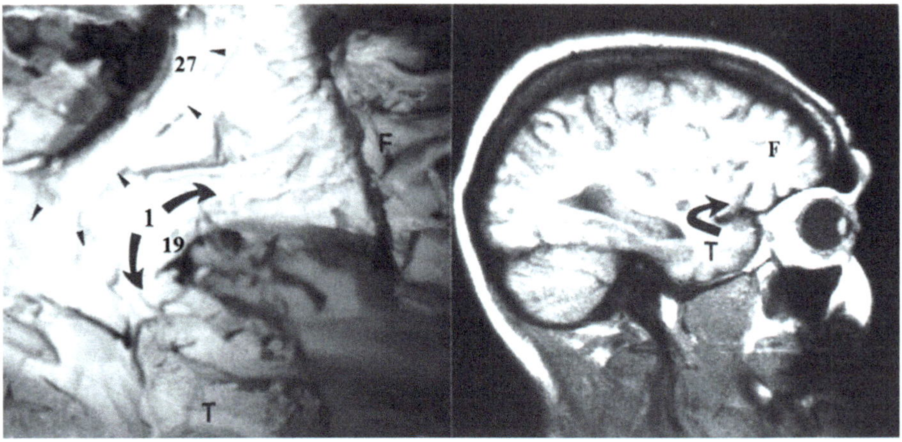

Fig. 6. Content of the limen insulae white matter: *Left*: White matter fibers tracts dissection using Klingler's technique of a left cerebral hemisphere seen from its inner side. *T* Temporal lobe, *F* frontal lobe, *1* uncinate fascicle (joining the temporal pole to the frontal pole), *19* limen insulae, 27 and arrow-heads: anterior commissure. *Right*: T1-weighted sagittal MRI slice showing the uncinate fascicle (curved arrow). *T* Temporal lobe, *F* frontal lobe

3) *The inferior periinsular sulcus* is somewhat curved, extending from the posterior insular point to the region of the limen insulae. It demarcates the insula from the temporal operculum.
4) *The limen insulae* is the antero-inferiorly located insular cortical surface which conjoins the inferior insular point, the anterior perforated surface, and the temporo-mesial surface. This plicated grey matter surface contains, in its white matter, the uncinate fascicle [19] (Fig. 6).

The Insular Cortex (Fig. 7)

1) *The central insular sulcus* is the most visible insular sulcus. It is clearly visible in more than 90% of the cases. It extends obliquely from the superior periinsular sulcus to the limen insulae. In about 2/3 of the cases, the central insular sulcus constitutes a prolongation of the central sulcus. It measures about 5 cm long, and divides the insular surface in an anterior and a posterior part.
2) *The anterior insula* is constituted of 3 main gyri, and often of 2 additional smaller gyri [20, 21].
   a. *The anterior short insular gyrus* is easy to identify most of time. It runs vertically from the junction between anterior and superior periinsular sulci (termed: «anterior insular point») to the limen insulae. It is separated from the fronto-orbital operculum by the anterior

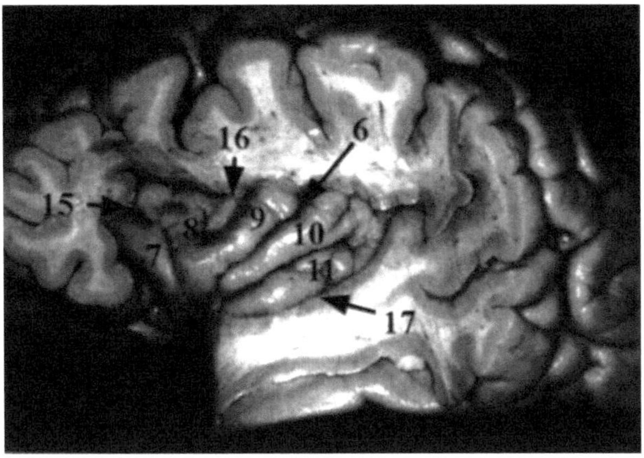

Fig. 7. Lateral view of the left insular cortex (*modified from Yasargil MG. Microneurosurgery IV A (CNS tumors) (1994) Georg Thieme Verlag, Stuttgart*). *6* Central insular sulcus, *7* anterior short insular gyrus, *8* middle short insular gyrus, *9* posterior short insular gyrus, *10* anterior long insular gyrus, *11* posterior long insular gyrus, *15* anterior periinsular sulcus, *16* superior periinsular sulcus, *17* inferior periinsular sulcus

periinsular sulcus, and from the middle short insular gyrus by the *anterior insular sulcus.*

b. *The middle short insular gyrus* is often underdevelopped. It is separated from the posterior short insular gyrus by *the precentral insular sucus.*
c. *The posterior short insular gyrus* is delimited by the precentral and central insular sulci. It is often well developped.
d. *The transverse and accessory insular gyri* are two minor gyri which constitute the insular pole, located on the inferior insular point, near the limen insulae. The former is the lowest one, and constitutes a junction between the most inferior part of the anterior insula and the posterior frontotorbital region. The later is often less developped, and conjoins the anterior short insular gyrus with the fronto-orbital region across the anterior periinsular sulcus.

3) *The posterior insula* is constituted of 2 long gyri
   a. *The anterior long insular gyrus* is always very well developed and is easy to identify. It is separated from the posterior long insular gyrus by the *post-central insular gyrus*; In some cases, this sulcus is interrupted and the anterior and posterior long insular gyri merge inferiorly, toward the limen insulae.
   b. *The posterior long insular gyrus* is demarcated from the temporal operculum by the inferior periinsular sulcus.

Relationships (Figs. 8–12)

1) *The opercula* (frontoorbital, frontoparietal, and temporal) are in very close relationships with the cortical surface of the insula. The *fronto-orbital operculum* is constituted by the posterior orbital gyrus, the posterior portion of the lateral orbital gyrus, and the pars orbitalis of F3. It covers the anterior part of the insular cortex. The *fronto-parietal operculum* is constituted by the pars triangularis and opercularis of F3, the precentral gyrus, the postcentral gyrus, and the superior portion of the supramarginal gyrus. In most cases, there is a gyrus, termed subcentral gyrus, located at the medial aspect of both the inferior precentral and postcentral gyri. The subcentral gyrus covers the central insular sulcus. The *temporal operculum* is constitued from anteriorly to posteriorly by the polar planum, T1, the anterior and posterior Heschl's gyri, and the temporal planum.
2) *White matter and claustrum*: the subcortical white matter of the insular cortex consists of the extreme capsule. The extreme capsule covers the claustrum, whose functional significance remains largely unknown. The external capsule and the putamen are deeper structures, with a more distant relationship to the insula.

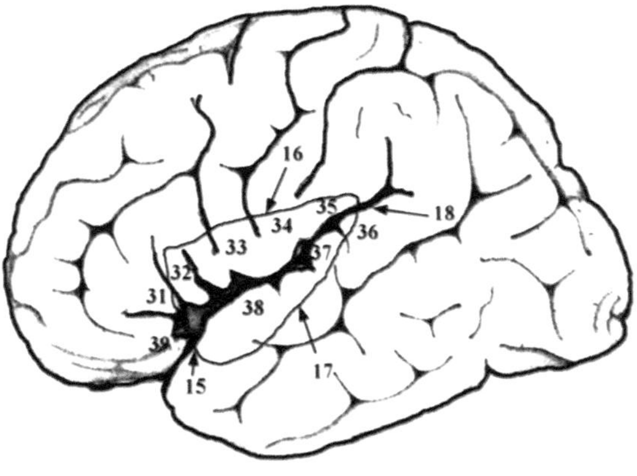

Fig. 8. Relationships between insula and opercula: schematic diagramm of a left hemisphere. The insula is visible by transparency. *15* anterior periinsular sulcus, *16* superior periinsular sulcus, *17* inferior periinsular sulcus, *18* sulcus lateralis cerebri (sylvian fissure), *31* pars triangularis of F3, *32* pars opercularis of F3, *33* precentral operculum, *34* postcentral operculum, *35* superior portion of the supramarginal gyrus, *36* inferior portion of the supramarginal gyrus, *37* transverse temporal gyri, *38* T1 operculum, *39* orbitofrontalis gyrus and pars orbitalis of F3

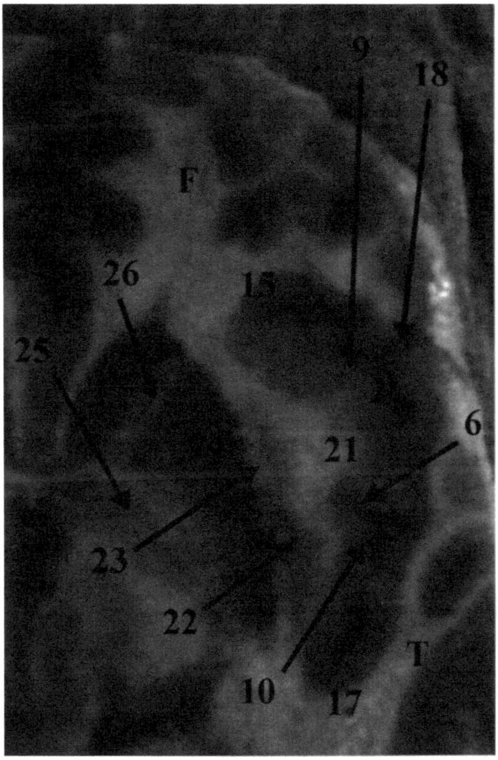

Fig. 9. Horizontal slice of the anterior part of a left hemisphere seen from below. *2* Apex insulae, *6* central insular sulcus, *9* posterior short insular gyrus, *10* anterior long insular gyrus, *15* anterior periinsular sulcus, *17* inferior periinsular sulcus, *18* sulcus lateralis cerebri (sylvian fissure), *21* extreme capsule, *22* claustrum, *23* external capsule, *24* putamen, *25* globus pallidus, *26* internal capsule, *T* temporal lobe, *F* frontal lobe

The Association Fibers

Short Association Fibers

These are U-shaped, superficial fibers; they connect parts of a single gyrus or link adjacent gyri together.

Long Ipsihemispheric Association Fascicles

They are long, deep fibers, gathered into tracts; they interconnect (the) different lobes. They are in close relation with the insular lobe as some of them course in the depth of the white matter beneath the insula and the claustrum:

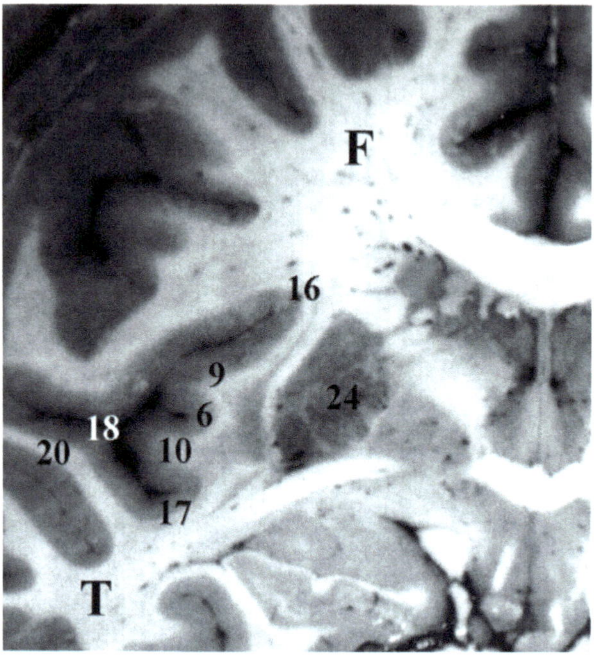

Fig. 10. Coronal slice of the brain: gross anatomy of the main relationships of the insula. *6* Central insular sulcus, *9* posterior short insular gyrus, *10* anterior long insular gyrus, *16* superior periinsular sulcus, *17* inferior periinsular sulcus, *18* sulcus lateralis cerebri (sylvian fissure), *20* gyrus transversus temporalis (Heschl's gyrus), *24* putamen, *T* temporal lobe, *F* frontal lobe

1) *The uncinate fascicle* strongly connects the fronto-orbital region to the temporal pole, the parahippocampus and the anterior part of the insula, through the limen insulae (Fig. 6).
2) *The arcuate fascicle* interconnects middle F2–F3 and middle T2–T3 through external capsule. It is particularly important in the dominant hemisphere for language function.
3) *The superior and inferior longitudinal fascicles* (and also the superior and inferior fronto-occipital fascicles) link the frontal and insular cortical areas to the temporal and parieto-occipital lobes.

Vascular Anatomy (Figs. 13–15)

The vascular arborization made by the divisions, on the insular surface, of the middle cerebral artery trunk, constitutes a real «arterial wall», which is partly responsible for the many dificulties encountered by the neurosurgeon when of performing surgery in the insular region. It poses a sub-

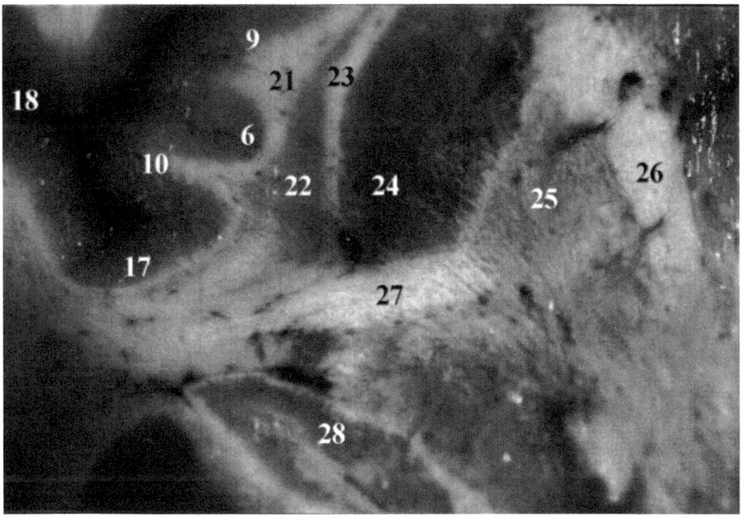

Fig. 11. Coronal slice of the brain: details of some anatomical structures surrounding the insula. *6* Central insular sulcus, *9* posterior short insular gyrus, *10* anterior long insular gyrus, *17* inferior periinsular sulcus, *18* sulcus lateralis cerebri (sylvian fissure), *21* extreme capsule, *22* claustrum, *23* external capsule, *24* putamen, *25* globus pallidus, *26* internal capsule, *27* anterior commissure (in the depth of the limen insulae), *28* hippocampus

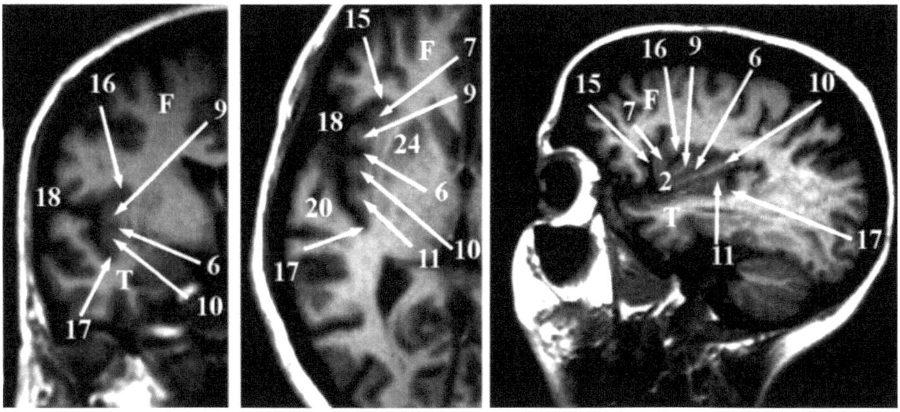

Fig. 12. MRI T1 slices of the cerbrum showing the main relatioships of the insular region. (*Left*) coronal slice, *Middle*: axial slice, *Right*: sagittal slice. *2* Apex insulae, *6* central insular sulcus, *7* anterior short insular gyrus, *9* posterior short insular gyrus, *10* anterior long insular gyrus, *11* posterior long insular gyrus, *15* anterior periinsular sulcus, *16* superior periinsular sulcus, *17* inferior periinsular sulcus, *18* sulcus lateralis cerebri (sylvian fissure), *20* gyrus transversus temporalis (Heschl's gyrus), *24* putamen, *T* temporal lobe, *F* frontal lobe

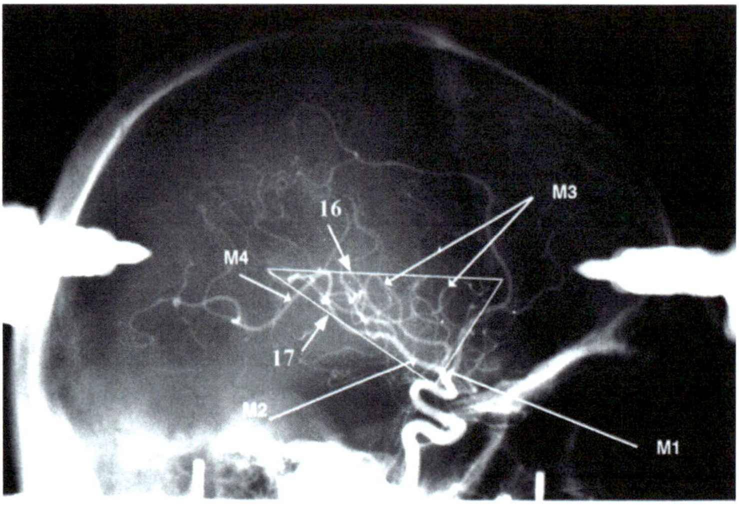

Fig. 13. Carotid artery angiogram, lateral view. *16* Superior periinsular sulcus, *17* inferior periinsular sulcus, *M1* sphenoidal (M1) segment of the middle cerebral artery, *M2* insular (M2) segment of the middle cerebral artery, *M3* opercular (M3) segment of the middle cerebral artery. The superior line of reflexion of the M3 segment provides a good visibility of the location of the superior periinsular sulcus

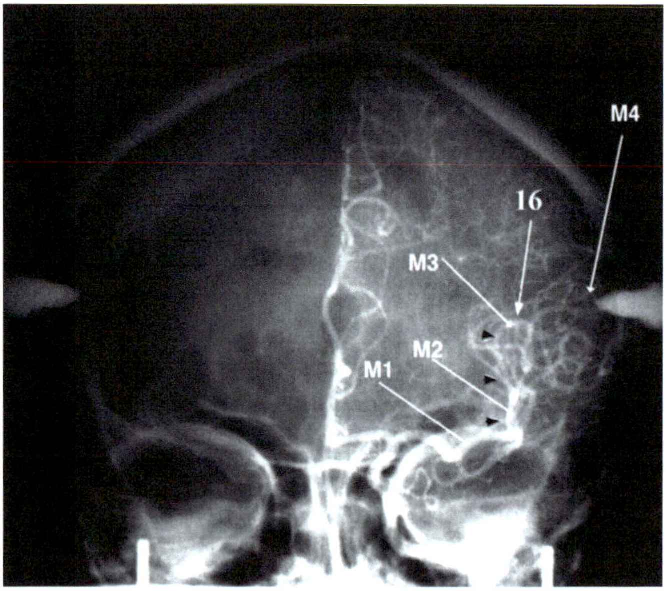

Fig. 14. Carotid artery angiogram, antero-posterior view. *16* Superior periinsular sulcus, *M1* sphenoidal (M1) segment of the middle cerebral artery, *M2* insular (M2) segment of the middle cerebral artery, *M3* opercular (M3) segment of the middle cerebral artery. Arrowheads: insular plane

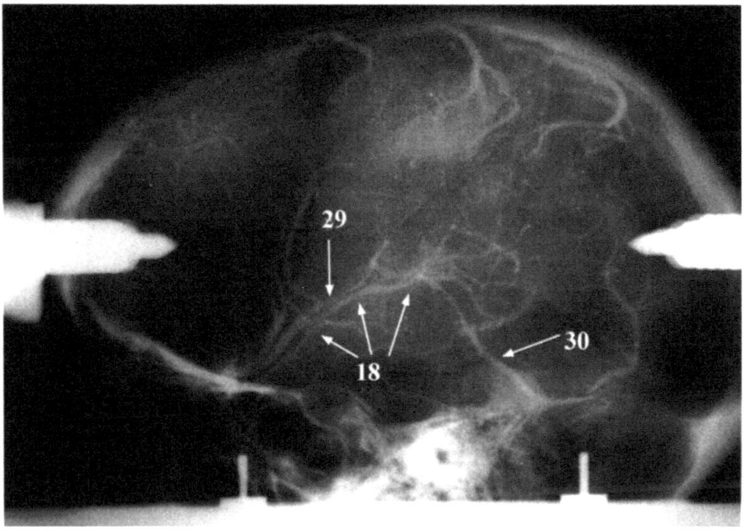

Fig. 15. The sylvian veinous system, lateral view. *18* Sulcus lateralis cerebri (sylvian fissure), *29* superficial sylvian veins, *30* Labbe's vein

stantial obstacle to any stereotactic procedure in this region, due to the risk of intra-operative Sylvian hemorrhage. Nevertheless, good knowledge of the vascularization of the insular region, and of its visualization by angiography before surgery facilitate therapeutic procedures in this location [30, 34].

Arterial Relationships

The *middle cerebral artery* (MCA) is the only arterial supply of the insular lobe.

– The M1 (*sphenoidal*) segment runs from the carotid bifurcation to the limen insulae. It is in relation superiorly with the anterior perforated space, posteriorly the semilunar gyrus and amygdala, and inferiorly the entorhinal area of the uncus. This segment may give one or several (up to 6) branches to the insula itself, and particularly to the limen insulae.

– The M2 (*insular*) segment corresponds to the superior and inferior trunks of the MCA after division of the M1 trunk at the level of the limen insulae. These two trunks are in very close relation with the insular cortex, and provide most of its arterial supply. Some additional trunks can sometimes be encountered at this level. A total of 75 to 104 insular arteries originating from this M2 segment of the MCA was found to provide the arterial supply of the insula by Türe *et al.* [34]. They consist of thin arteries, penetrating perpendicularly to the cortical and M2 plane. The

longest of them may reach the lateral aspect of the putamen and the corona radiata. There are no anastomoses between these insular arteries and the lateral lenticulostriate arteries.
– The M3 (*opercular*) segment runs on the inner side of the opercula. It gives birth to many arteries (orbito-frontal, prefrontal, precentral, central, anterior parietal, posterior parietal, angular, temporo-occipital, posterior temporal, middle temporal, anterior temporal, and temporo-polar arteries) which will constitute, once over the superficial side of the sylvian fissure, the M4 segment. It is noteworthy that the superior reflexion line of the opercular M3 segment provides a precise idea of the topography of the insula on a single lateral angiographic view. The M3 segment provides only a few branches to the insular cortex.
– The M4 segments are the cortical branches after exiting from the sylvian fissure. They usually do not provide any vascular suppply to the insula.

Veins

The insula is almost exclusively drained by the deep and superficial sylvian veinous systems. Most of time, the deep middle cerebral vein is predominant, but the separation between deep and superficial middle cerebral veins territories proves to be very variable [37]. During sylvian or temporal lobe surgery, great care must be taken not to interrupt too largely, avulse or even stretch these veins: This might result in severe hemorrhagic infarcts [29].

Functional Anatomy

Until recently, little was known about the actual functional role of the insula. Proposals and theories about its function were based on animal experimental studies and on some intraoperative cortical recordings. Some information could also be provided by some pathological states, such as epileptic seizures. More recently, some functional features have been explored by means of depth electrode recordings (Fig. 16) [11, 28].

Connections (Fig. 17) [1, 17, 18]

The insula is classified, at least in its allo – and mesocortical parts, as a paralimbic structure. Paralimbic structures of the brain comprise the parahippocampal gyrus, the temporal pole, the cingulate gyrus, the posterior orbitofrontal gyrus and the insula. This paralimbic part of the insula has strong connections with mesio-temporal (particularly amygdala) and orbito-frontal cortex. In its isocortical part, the insula is involved in nonlimbic functions. This latter part has particular connections with the

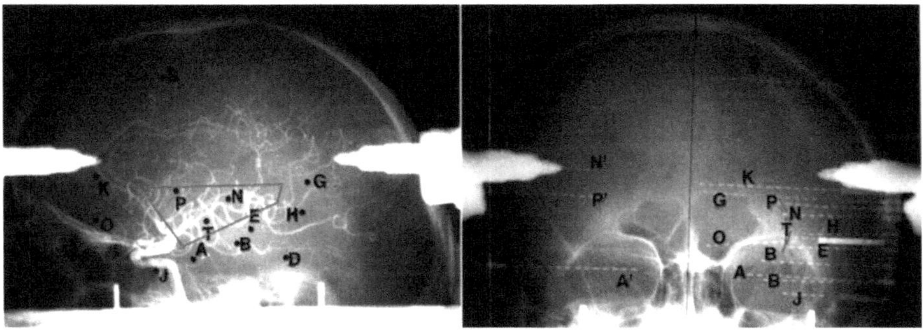

Fig. 16. Epilepsy surgery, pre-surgical investigations: example of depth electrode recordings of the insular lobe. The electrodes are orthogonally implanted using the Talairach's methodology of StereoElectroEncephalography (*SEEG*), thus avoiding the insular arteries. As shown on the lateral, and the antero-posterior views of the carotid angiogramm, the electrodes N, P, and T reach the insular cortex (schematically drawn in gray lines). This allows chronic recordings, as well as stimulations, of the insula

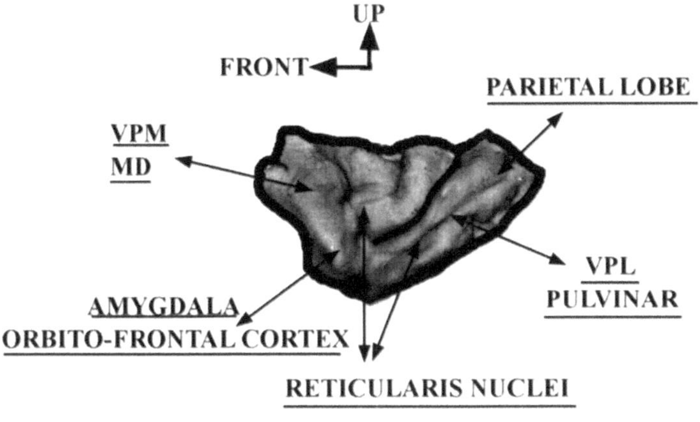

Fig. 17. Connections of the insula: schematic drawing. *VPM* Medial ventro-posterior nucleus of the thalamus, *MD* medio-dorsal nucleus of the thalamus, *VPL* lateral ventro-posterior nucleus of the thalamus. Reticularis nuclei: reticularis nuclei of the thalamus

opercula and with SI and SII areas. All these connections are reciprocal. Moreover, the anterior part of the insula has some connections with the medial ventro-posterior and medio-dorsal nuclei of the thalamus, whereas the posterior part of the insula connects preferentially with the lateral ventro-posterior nucleus of the thalamus and the pulvinar. Both anterior and posterior parts of the insula are connected to the reticularis nuclei of the thalamus.

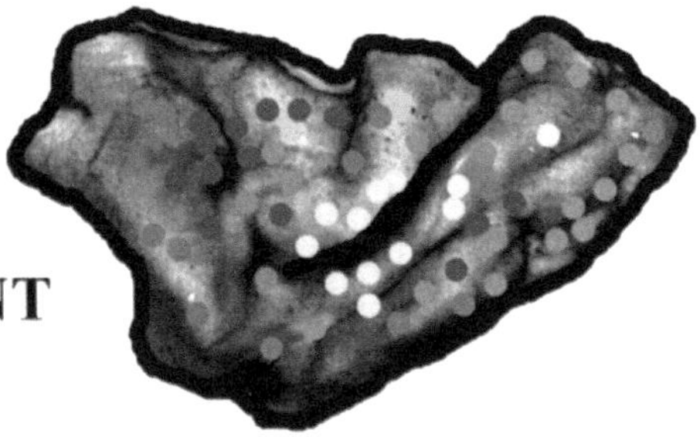

Fig. 18. Summary of the data obtained by our group from cortical stimulations of the insular cortex by means of depth electrodes (*with permisiion of Isnard, Ostrowsky, et al*). Studies performed during SEEG, in awaken patients. Orange plots: non painful somatosensory responses. Green plots: respiratory and laryngeal responses. Blue plots: viscero-sensory responses. Purple plots: miscellaneous responses (dysphasia, autonomic responses, etc ...)

Physiological Roles of the Insula [1–3, 13, 14, 22, 23, 27]

Animal experimental studies, as well as some human studies with insular cortex stimulation, suggest some hypotheses about the physiological functional role of the insula (Fig. 18):

a. The insula is involved in cardiac rythm and arterial blood pressure control. This has been assessed by various studies reporting some modifications of the cardiac rythm, either during seizures of insular origin, or during intra-operative insular stimulations. These data are coherent with the high occurence of cardiac rythm and blood pressure modifications in cases of insular stroke.
b. The insula is involved in the viscero-motor control and in viscero-sensitive functions. Thus, ictal vomiting has been reported in cases with insular epileptic seizures. Modifications of oesophageal peristalsis have been reported after intra-operative insular electrical stimulation. These data are consistent with ictal symptoms associated with insular discharges, as reported in our study [12], and detailed in Section IV.
c. The anterior insular cortex probably bears the primary area for taste. This has been experimentaly demonstrated by cytoanatomical studies (labeling of the cytochrome-oxidase activity) in the monkey. This is coherent with PET studies performed in humans, demonstrating an increased blood flow in the right parietal and left insular cortices after tasty stimuli.
d. The insular cortex may act as an associative motor cortex. Indeed, the

ipsilateral insular cortex has been shown to be activated (PET studies) during shoulder or arm movements, and the insula is involved in the motor cortical reorganization occuring after vascular strokes.

e. The insula may be involved in some automatic language functions, as well as in eye movements and vestibular integration.

Insula and Pain

There is considerable evidence for the involvement of the insula as a secondary, as well as associative, somesthetic area. This includes a role in non-nociceptive perception, and a major role in the processing of nociceptive inputs. The former point is assessed by the reports of some blood flow increase (PET studies) in the posterior insular cortex after tactile and vibratory stimuli. The latter point is assessed by several lines of arguments, resulting from studies made by our group [7–10, 23–26, 35]:

a) *H2O15 PET studies* demonstrated a constant and strong bilateral increase of cerebral blood flow in the insular region following a noxious stimulus applied on the hand. This insular response to a noxious stimulus is very specific to the pain sensation itself, contrary to the orbito-frontal and cingular responses, which are dependant to the attention drawn to the pain.

b) *Depth electrode recordings* of the insula, performed with Talairach's SEEG (StereoElectroEncephaloGraphy) methodology [28, 31, 32], could record evoked insular activities following noxious cutaneous stimuli (made by means of CO2 laser beam). These recordings provided many observations to support the role of the insula as the primary cortical area for pain perception:

– cortical evoked responses can be elicited as soon as 140 ms after stimulus in the operculum, and 180 ms in the posterior insula, whereas the amygdala, hippocampal, cingular, and parietal responses are constantly longer to be elicited. The opercular and insular responses are bilateral, but are preferentially elicited on the controlateral side to the stimulus.

– These operculo-insular responses increase in amplitude as the intensity of the painful stimulus increases, even after the pain sensation threshold has been reached.

– Lastly, it has been shown with PET studies that some allodynic sensations resulting from a lesion of the spino-thalamic tract at the brainstem level are due to an abnormal activation of the insular regions usually involved in pain perception.

c) *Stimulation of the posterior insular cortex*, made by means of chronically implanted depth electrodes, are able to elicit painful sensations. This has been shown in 43 patients who underwent square pulse high frequency (50 Hz) stimulations of the insular cortex. 14 of them experienced mild to

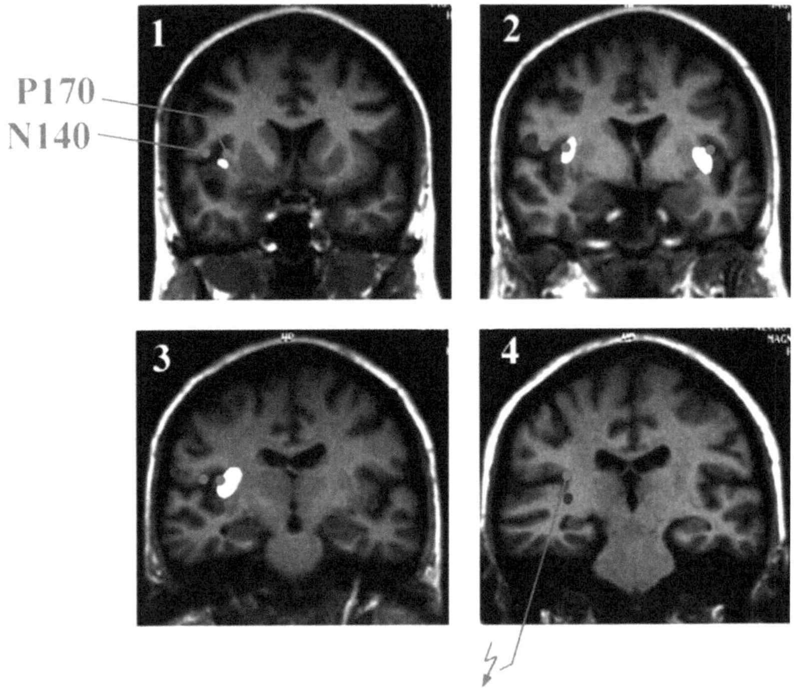

Fig. 19. Summary of the data obtained from different studies related to the role of the insula in pain perception (*with permission of Peyron et al, Frot et al, Ostrowsky et al, Isnard et al*). Yellow areas: H2O15 PET studies: insular responses to left painful stimuli. Blue areas: fMRI studies: insular responses to left painful stimuli. Red plots: CO2 laser evoked potentials: evoked responses (N140 & P170) to left painful stimuli (slices 1, 2 & 3). Slice 4: Painful sensation elicited by direct electrical stimulation of the insular cortex

unbearable painful sensations, mostly following stimulation of the posterior insular cortex of the non-dominant side.

d) *In summary*, these new data, gathered together in Fig. 19, based on metabolic, electrophysiological and cortical stimulation studies in humans, added to more classical studies made in monkeys, provide very strong arguments in favor of the essential role of the insula in pain perception and pain integration. The close relationships between SII area, located at the parietal operculum, and the posterior part of the insula, is consistent with this proposition. The insula can finally be considered as being a part of the median integration system of the somesthetic inputs. Its close connections with the amygdala could be responsible for a fast behavioral adaptation to the painful stimulus.

The Insula and Epilepsy Surgery

Despite the probable major importance of the insular ictal activity in the prognosis of some epilepsies, the precise role of the insula in epilepsy has only been suspected historically by a few authors. The role of the insular cortex in the genesis of some temporal lobe epilepsies have recently been further investigated by our group in patients having chronically implanted depth electrodes and video-EEG recordings [12–14, 22]. This has brought some supplementary light to the functional features of the insula and helped to explain some of the failures after temporal lobe epilepsy surgery:

a) *Ictal symptoms* associated with insular discharges were of various kinds: Respiratory manifestations, manifested by an increase of both amplitude and frequency of the respiratory movements, were the earliest symptoms. The majority of the patients were unaware of these respiratory changes.

The viscero-sensitive manifestations consisted of chest or abdominal constriction, nausea, throat constriction, or even severe abdominal pain.

Oro-alimentary manifestations consisted of chewing and swallowing.

Autonomic manifestations, such as flushing of the face were also reported.

Non painful somato-sensory manifestations were always opposite to the discharging hemisphere. It was mostly a heat sensation in the trunk, limb, or chest.

These ictal manifestations are in concordance with the supposed physiological roles of the insula, as described in the previous chapters.

b) *SEEG recordings* showed that more than 10% of the patients presenting with a clinically typical temporal lobe epilepsy had many of their seizures originating solely from the insula. These insular ictal activities should be differentiated from those originating from the temporal lobe, and propagating rapidly but secondarily to the insula, as a sole temporal lobectomy without insular extension would lead to (at least partly) failure. On the contrary, seizures propagating secondarily to the insular cortex were found to be fully controlled by surgery.

Tumoral Surgery in the Insular Region

Besides epilepsy surgery, surgical procedures may have to be undertaken in or around the insular lobe for oncological (or vascular) reasons [2, 4, 27]. Surgical treatment of glial tumours arising in the insula is especially challenging due to the proximity of the internal capsule, and to the previously described vascular and anatomical features of the insula. Moreover, the surgical approach to the insula implies, at least, the opening of the sylvian fissure, and most often the removal of some of the opercula. Sylvian superficial, insular, and lenticulo-striate perforating branches are at risk

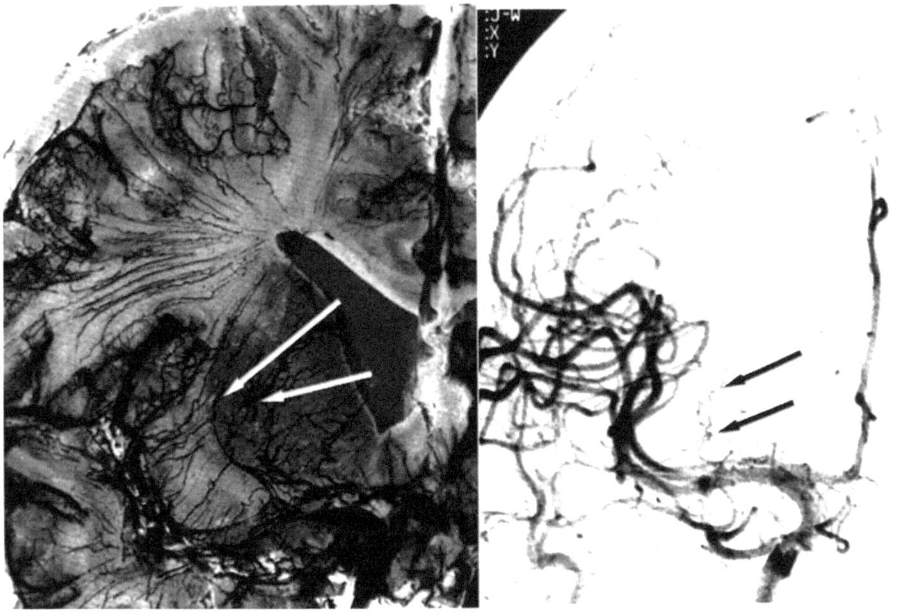

Fig. 20. Relationships between the lenticulo-striate arteries (arrows) and the insular cortex. *Left*: frontal anatomical section, *right*: antero-posterior carotid angiogramm. The lenticulo-striate arteries mostly arise from the M1 segment of the middle cerebral artery. They constitute the main vascular supply of the internal capsule, the putamen, the globus pallidus, and the head of the caudatus nucleus. They are usually not involved in the vascular supply of the insula, staying at some distance from its cortex

during this procedure, as well as the opercular parenchyma itself. Once the insular cortex is reached, great care must be taken when working in the depth of this structure, not to compromise the vasculature to the internal capsule and the lenticulostriate arteries. The internal capsule can be located by means of intraoperative stimulation (see below); the lenticulostriate arteries locations (Fig. 20) can often be visualized on the preoperative arterial angiogram.

These difficulties explain why intrinsic insular tumors have tended to be frequently excluded from surgical treatment. They also explain why, although small insular gliomas have been removed safely by a transylvian approach, only biopsy has been recommended for large dominant insular tumors to avoid postoperative deficits. However, this cannot be considered as the optimal form of treatment, as low grade gliomas (unlike grade IV glioblastomas, which constitute a separate concern) should be removed radically to increase the recurrence-free-interval and, as it is established than partial removal associated with radiotherapy leads to a much higher incidence of recurrence than radical removal without radiotherapy [15, 16, 36].

As it is usually impossible to identify macroscopically a clear cut margin of the tumor during surgery, any attempt to achieve a radical removal of an insular low-grade glioma may take profit from neuronavigation, which can be associated with the use of an ultrasound system, or with multiple biopsies of the peritumoral tissue. Nevertheless, the use of intra-operative direct cerebral stimulation, particularly of the white matter fibers, is consistently emphasized in many studies dealing with insular tumor removal [2–6, 38]. On the dominant side, direct cerebral stimulation can be used, under local anesthesia, to detect and locate the language cortical areas. Whatever the side and the type of anesthesia, direct cerebral stimulation can also be used to locate the white fibers of the internal capsule. Finally, intra-operative direct cerebral stimulation allows more radical tumor resection in the insula with minimization of post-operative neurological deficit risks. A rate of more than 80% of complete or subtotal resection are reported, with a rate of transient post-operative (motor and/or speech) deficit of less than 20% [16, 36, 40]. Such surgery, however, may remain limited when being performed in the dominant hemisphere, with regard to the high risk (more than 10%) of post-operative dysphasia reported by some authors [3, 6]. However, thanks to these technical refinements, a total removal of insular tumours, especially the low-grade astrocytomas, may sometimes be achieved, with an acceptably low risk of permanent post-operative neurological morbidity.

Conclusion

1) The depth of the insular cortex, which is totally covered by the opercula, as well as the thickness of the middle cerebral artery vascular arborization lying on it, explains why any open surgical or stereotactic procedure has always been considered as dangerous to perform in the insular region.
2) Recent studies using chronic and acute electrophysiological recordings, direct cortical stimulation and brain imaging, has improved our knowledge of some of the functional roles of the insula. It is now known to be involved in visceromotor, viscerosensitive and autonomic functions, as well as in non-painful and painful somesthetic integration. Moreover, its possible role in some form of epilepsy has recently be suggested.
3) New refinements in brain imaging, neuronavigation and direct cortical stimulation, which can be performed under awake anesthesia, facilitate safer surgical procedures in the insular region.

References

1. Augustine JR (1996) Circuitry and functional aspects of the insular lobe in primates including humans. Brain Res Brain Res Rev 22(3):229–244

2. Cukiert A, Forster C, Andrioli MS, Frayman L (1998) Insular epilepsy. Similarities to temporal lobe epilepsy. Case report. Arq Neuropsiquiatr 56(1):126–128
3. Duffau H, Bauchet L, Lehericy S, Capelle L (2001) Functional compensation of the left dominant insula for language. Neuroreport 12(10):2159–2163
4. Duffau H, Capelle L, Lopes M, Bitar A, Sichez JP, Van Effenterre R (2002) Medically intractable epilepsy from insular low-grade gliomas: improvement after an extended lesionectomy. Acta Neurochir (Wien) 144:563–573
5. Duffau H, Capelle L, Lopes M, Faillot T, Sichez JP, Fohanno D (2000) The insular lobe: physiopathological and surgical considerations. Neurosurgery 47(4):801–810
6. Ebeling U, Kothbauer K (1995) Circumscribed low grade astrocytomas in the dominant opercular and insular region; a pilot study. Acta Neurochir (Wien) 132(1–3):66–74
7. Frot M, Garcia-Larrea L, Guenot M, Mauguiere F (2001) Responses of the supra-sylvian (SII) cortex in humans to painful and innocuous stimuli. A study using intra-cerebral recordings. Pain 94:65–73
8. Frot M, Rambaud L, Guenot M, Mauguiere F (1999) Intracortical recordings of early pain-related CO2 laser potentails in the human SII-insular cortex. Clin Neurophysiol 110:133–145
9. Frot M, Mauguiere F (1999) Timing and spatial distribution of somatosensory responses recorded in the upper bank of the sylvian fissure (SII area) in humans. Cerebral cortex 9:854–863
10. Frot M, Isnard J, Guenot M, Mauguiere F (2000) Effects of noxious stimulus intensity on signals from operculo-insular cortex: an intra-cerebral recordings study in humans. Clin Neurophysiol 111(1):131
11. Guenot M, Isnard J, Ryvlin P, Fischer C, Ostrowsky K, Mauguiere M, Sindou M (2002) Neurophysiological monitoring for epilepsy surgery: the Talairach SEEG method. Stereotact Funct Neurosurg 73:84–87
12. Isnard J, Guenot M, Ostrowsky K, Sindou M, Mauguiere F (2000) The role of the insular cortex in temporal lobe epilepsy. Ann Neurol 48(4):614–623
13. Isnard J, Guenot M, Ostrowsky K, Sindou M, Mauguiere F (2000) Stereoelectroencephalographic (SEEG) study of electrophysiological relations between insular lobe and temporal lobe in temporal lobe epilepsy (TLE). Clin Neurophysiol 111(1):100
14. Isnard J, Ryvlin P, Guenot M, Ostrowsky K, Fischer C, Sindou M, Mauguiere F (1998) Role of the insular cortex in temporolimbic seizures: a stereoelectroencephalographic (SEEG) study. Epilepsia 39 [Suppl] 6:65
15. Kumabe T, Nakasato N, Suzuki K, Sato K, Sonoda Y, Kawagishi J, Yoshimoto T (1998) Two-staged resection of a left frontal astrocytoma involving the operculum and insula using intraoperative neurophysiological monitoring-case report. Neurol Med Chir (Tokyo) 38:503–507
16. Lang FF, Olansen NE, DeMonte F, Gokaslan ZL, Holland EC, Kalhorn C, Sawaya R (2001) Surgicla resection of intrinsic insular tumors: complication avoidance. J Neurosurg 95:638–650
17. Mesulam MM, Mufson EJ (1982) Insula of the old world monkey. III: efferent cortical output and comments on function. J Comp Neurol 212(1):38–52

18. Mufson EJ, Mesulam MM (1982) Insula of the old world monkey. II: afferent cortical intput and comments on the claustrum. J Comp Neurol 212(1):23–37
19. Nagara S, Rhoton AL, Barry M (1988) Microsurgical anatomy of the choroidal fissure. Surg Neurol 30:3–59
20. Nieuwenhuys R, Voogd J, Van Huijzen C (1988) The human cerebral nervous system: a synopsis and atlas, 3rd edn. Springer, Berlin
21. Ono M, Kubik S, Abernathey CD (1990) Atlas of the cerebral sulci. Georg Thieme Verlag, Stuttgart
22. Ostrowsky K, Isnard J, Ryvlin P, Guenot M, Fischer C, Mauguiere F (2000) Functional mapping of the insular cortex: clinical implication in temporal lobe epilepsy. Epilepsia 41:681–686
23. Ostrowsky K, Magnin M, Ryvlin P, Isnard J, Guenot M, Maugiuere F (2002) Representation of pain and somatic sensation in the human insula: a study of responses to direct electrical stimulation. Cerebral cortex (in press)
24. Peyron R, Garcia-Larrea L, Laurent B (2000) Functional imaging of brain responses to pain. A review. Neurophysiol Clin 30:263–288
25. Peyron R, Garcia-Larrea L, Gregoire MC, Costes N, Convers P, Lavenne F, Mauguiere F, Michel D, Laurent B (1999) Hemodynamic brain responses to acute pain in humans: sensory and attentional networks. Brain 122:1765–1780
26. Peyron R, Garcia-Larrea L, Gregoire MC, Convers P, Richard A, Lavenne F, Barral FG, Mauguiere F, Michel D, Laurent B (2000) Parietal and cingulate processes in central pain. A combined positron emission tomography (PET) anf functional magnetic resonance imaging (fMRI) study of an unusual case. Pain 84:77–87
27. Roper SN, Levesque MF, Sutherling WW, Engel J Jr (1993) Surgical treatment of partial epilepsy arising from the insular cortex. Report of two cases. J Neurosurg 79(2):266–269
28. Rougier A, Saint-Hilaire JC, Loiseau P, Bouvier G (1992) Investigations et traitement chirurgical des épilepsies. Neurochirurgie 38:2–112
29. Sindou M, Auque J (2000) The intracranial venous system as a neurosurgeon's perspective. In Cohadon F *et al* (eds) Advances and technical standards in Neurosurgery, vol 26. Springer Wien, New York, pp 131–216
30. Szikla G, Bouvier G, Hori T, Petrov V (1977) Angiography of the human brain cortex. Springer-Verlag, Wien
31. Talairach J, Bancaud J (1973) Stereotaxic approach to epilepsy. Methodology of anatomo-functional stereotaxic investigations. Progr Neurol Surg 5:297–354
32. Talairach J, Tournoux P (1988) Co-planar stereotaxic atlas of the human brain. Georg Thieme Verlag, Stuttgart, p 122
33. Türe U, Yasargil DCH, Al-Mefty O, Yasargil MG (1999) Topographic anatomy of the insular region. J Neurosurg 90:720–733
34. Türe U, Yasargil MG, Al-Mefty O, Yasargil DCH (2000) Arteries of the insula. J Neurosurg 92:676–687
35. Valeriani M, Restuccia D, Barba C, Le Pera D, Tonali P, Mauguiere F (2000) Sources of early cortical responses to painful CO2 laser skin stimulation of the hand and foot in human brain. Clin Neurophysiol 111:1103–1112

36. Vanaclocha V, Saiz-Sapena N, Garcia-Casacola C (1997) Surgical treatment of insular gliomas. Acta Neurochir (Wien) 139:1126–1134
37. Varnavas GG, Grand W (1999) The insular cortex: morphological and vascular anatomic characteristics. Neurosurgery 44:127–136
38. Walsh AR, Schmidt RH, Marsh HT (1990) Cortical mapping and resection under local anesthetic as an aid to surgery of low and intermediate grade gliomas. Br J Neurosurg 4:485–491
39. Yasargil MG (1994) Microneurosurgery IV A (CNS tumors). Georg Thieme Verlag, Stuttgart, p 396
40. Zentner J, Meyer B, Stangl A, Schramm J (1997) Intrinsic tumors of the insula: a prospective study of 30 patients. J Neurosurg 85:263–271

Subject Index Vol. 1–29

Advances and Technical Standards in Neurosurgery

Author Index Vol. 1–29

Advances and Technical Standards in Neurosurgery

SpringerNeurosurgery

Advances and Technical Standards in Neurosurgery

Volume 24

1998. XIII, 310 pages. 57 figures, partly in colour.
Hardcover **EUR 133,–**
ISBN 3-211-83064-2

Advances: The Septal Region and Memory (D. Y. von Cramon, U. Müller) • The in vivo Metabolic Investigation of Brain Gliomas with Positron Emission Tomography (J. M. Derlon) • Use of Surgical Wands in Neurosurgery (L. Zamorano, F. C. Vinas, Z. Jiang, F. G. Diaz)
Technical Standards: The Endovascular Treatment of Brain Arteriovenous Malformations (A. Valavanis, M. G. Yasargil) • The Interventional Neuroradiological Treatment of Intracranial Aneurysms (G. Guglielmi) • Benign Intracranial Hypertension (J. D. Sussman, N. Sarkies, J. D. Pickard)

Volume 23

1997. XV, 278 pages. 89 figures, partly in colour.
Hardcover **EUR 133,–**
ISBN 3-211-82827-3

Advances: A Critical Review of the Current Status and Possible Developments in Brain Transplantation (S. Rehncrona) • The Normal and Pathological Physiology of Brain Water (K. G. Go)
Technical Standards: Transfacial Approaches to the Skull Base (D. Uttley) • Presigmoid Approaches to Skull Base Lesions (M. T. Lawton, C. P. Daspit, R. F. Spetzler) • Anterior Approaches to Non-Traumatic Lesions of the Thoracic Spine (A. Monteiro Trindade, J. Lobo Antunes) • The Far Lateral Approach to Lumbar Disc Herniations (F. Porchet, H. Fankhauser, N. de Tribolet)

All prices are recommended retail prices.
Net-prices subject to local VAT.

SpringerWienNewYork

Sachsenplatz 4–6, 1201 Vienna, Austria, Fax +43.1.330 24 26, e-mail: books@springer.at, **www.springer.at**
Haberstraße 7, 69126 Heidelberg, Germany, Fax +49.6221.345-4229, e-mail: orders@springer.de
P.O. Box 2485, Secaucus, NJ 07096-2485, USA, Fax +1.201.348-4505, e-mail: orders@springer-ny.com
EBS, 3-13, Hongo 3-chome, Bunkyo-ku, Tokyo 113, Japan, Fax +81.3.38 18 08 64, e-mail: orders@svt-ebs.co.jp

Advances and Technical Standards in Neurosurgery

Volume 20

1993. XIII, 308 pages. 97 figures.
Hardcover **EUR 156,–**
ISBN 3-211-82383-2

Advances: Post-traumatic Brain Swelling (R.D. Lobato) • Transcranial Doppler in Neurosurgery (K.-F. Lindegaard, W. Sorteberg, and H. Nornes) • Clinical and Molecular Neurogenetics in Neurosurgery (A. E. Harding)
Technical Standards: Surgery for Hindbrain Related Syringomyelia (B. Williams) • Medulloblastoma (J.-F. Hirsch E. Hoppe-Hirsch) • Haemangioblastoma, Haemangioblastomatosis, and von Hippel-Lindau Disease (F. Resche, J. P. Moisan, J. Mantoura, A. de Kersaint-Gilly, M. J. Andre, I. Perrin-Resche, D. Menegalli-Boggelli, Y. Lajat, S. Richard)

Volume 19

1992. XIV, 224 pages. 44 figures, partly in colour.
Hardcover **EUR 120,–**
ISBN 3-211-82287-9

Advances: Proton Spectroscopic Imaging in Cerebral Ischaemia: Where we Stand and what Can Be Expected (J. W. Berkelbach van der Sprenkel, N. M. J. Knufman, P. C. van Rijen, P. R. Luyten, J. A. den Hollander, C. A. F. Tulleken) • Radiosurgery (L. Steiner, C. Lindquist, M. Steiner) • Functional Imaging of Blood Brain Barrier Permeability by Single Photon Emission Computerised Tomography and Positron Emission Tomography (F. Iannotti)
Technical Standards: Economic Aspects of Neurosurgery (B. Jennett and J. Pickard) • Current Concepts of Measurement of Cerebrospinal Fluid Absorption and Biomechanics of Hydrocephalus (F. Gjerris, S. E. Børgesen) • A Multidisciplinary Approach to the Treatment of Brain Vascular Malformations (F. Gentili, M. Schwartz, K. TerBrugge, M. C. Wallace, R. Willinsky, C. Young)

GPSR Compliance

The European Union's (EU) General Product Safety Regulation (GPSR) is a set of rules that requires consumer products to be safe and our obligations to ensure this.

If you have any concerns about our products, you can contact us on ProductSafety@springernature.com

In case Publisher is established outside the EU, the EU authorized representative is:

Springer Nature Customer Service Center GmbH
Europaplatz 3
69115 Heidelberg, Germany

Batch number: 10726519

Printed by Printforce, the Netherlands